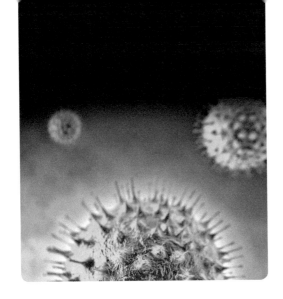

WJEC
Biology
for A2

Marianne Izen

Illuminate Publishing

Published in 2016 by Illuminate Publishing Limited, an imprint of Hodder Education,
an Hachette UK Company, Carmelite House, 50 Victoria Embankment, London EC4Y 0DZ

Orders: Please visit www.illuminatepublishing.com
or email sales@illuminatepublishing.com

British Library Cataloguing in Publication Data

A catalogue record for this book is available from the British Library

ISBN 978-1-908682-51-2

Printed by: Severn, Gloucester

08.21

The publisher's policy is to use papers that are natural, renewable and recyclable products made from wood grown in sustainable forests. The logging and manufacturing processes are expected to conform to the environmental regulations of the country of origin.

Every effort has been made to contact copyright holders of material produced in this book. If notified, the publisher will be pleased to rectify any errors or omissions at the earliest opportunity.

This material has been endorsed by WJEC and offers high quality support for the delivery of WJEC qualifications. While this material has been through a WJEC quality assurance process, all responsibility for the content remains with the publisher.

Editor: Geoff Tuttle
Design: Nigel Harriss
Layout: Neil Sutton, Cambridge Design Consultants

Cover image: Shutterstock: Aqua Images

Image Credits:

© **Alamy:** p179: blickwinkel; p219: Nigel Cattlin; p287, 308, 309: BSIP SA.

© **iStock:** p100(l): AndyRoland; p106: pilipenkoD; p115(t): thp73; p137: PaulCowan; p185: ElementalImaging; p186: Whiteway (Horse chestnut), kiorio (Burdock); p227(t): Elena Elisseeva; p262: stockdevil; p269: Scharvik; p292: muratseyit; p293: JFalcetti; p312: Mihai Andritoiu; p313: BanksPhotos; p322: Liz Leyden; p323: JohnCarnemolla.

© **Fotolia:** p101(l): ijacky; p177: Roman Samokhin; p186(Thrush):Erni; p186(pea pod): Denis Tabler; p188: Bogdan Wankowicz; p192(l): emer; p193: Christian Pedant; p212: zuzanaa.

© **Shutterstock:** p32(r): Gingo_o; p32(l): Morten Normann Almeland; p34: Morten Normann Almeland; p39: Jill pk; p55: dreamerb; p70(l): EMJAY SMITH; p76: Villiers Steyn; p82: dreamerb; p84: Christopher Kolaczan; p91: Ihervas; p97(r): Ozja; p97(l): Panaiotidi; p98: TCreativeMedia; p100(l): Chris G. Walker; p100(br): Sylvie Lebchek; p101(br): Christopher Meder; p101(tr): Jolanta Wojcicka; p102(l): Kletr; p102(r): oticki; p104(tr): Taina Sohlman; p104(bl): EMJAY SMITH; p105: Barbara Tripp; p107(l): Jaochainoi; p107(r): Mikhail Olykainen; p108: Hector Ruiz Villar; p111: Toa55; p112(tl): adisak soifa; p112(tr): Rudmer Zwerver; p112(bl): Incredible Arctic; p113(t): shaferaphoto; p113(b): Khoo Eng Yow; p115(b): Milosz Maslanka; p116: irabel8; p117: golddc; p121: kocakayaali; p122: Anna Jurkovska; p131(b): Blamb; p135: Bartosz Budrewicz; p154: Sally Willis; p160: Jose Luis Calvo; p161: Jose Luis Calvo; p163(t): Designua; p164: Alexilusmedical; p172: Leptospira; p173(tl,tr,br): Sebastian Kaulitzki; p173(bl): Timof; p178: Georgy Markov; p180: kitzcorner, p181: Jubal Harshaw; p184: Allen McDavid Stoddard; p186 (Dandelion): Gajus; p186 (Poppy): wolfman57; p186(Coconut): Sergieiev; p186(Alstromerias): Nadezda Verbenko; p192(r): Nadezda Verbenko; p195: Susan Schmitz; p198(t): yevgeniy11; p198(b): oksana2010; p206(t): dioch; p206(b): Aleksey Stemmer; p211: Sebastian Kaulitzki; p216(t): bikeriderlondon; p216(b): Tinawut Krapanpongsakul; p221(b): visceralimage; p221(b): FotoRequest; p224: Jiri Balek; p225: AndreJakubik; p227(b): lkpro; p229(tl): tony mills; p229(tr): Anatoly Vlasov; p229(br): Alexander Varbenov; p231: Ryan M. Bolton; p232: Boris15; p233: Linda Bucklin, p235. science photo; p238: toeytoey; p240: Vit Kovalcik; p241: extender_01; p249: oticki; p266(tl): Kletr; p266(tr): toeytoey; p271: Designua; p279: Jubal Harshaw; p280: Anna Jurkovska; p281: Jose Luis Calvo; p285: Jose Luis Calvo; p286: sciencepics; p291: Hein Nouwens; p292(tl): Praisaeng; p292(tr): fred Goldstein; p293: stihii; p294(tr): RomanYa; p294(tl): doctorkan; p294(bl): Samuel Cohen; p294(bc): Puwadol Jaturawutthichai; p295: frescomovie; p296(tr): Sebastian Kaulitzki; p296(l): Alila Medical Media; p296(br): Alila Medical Media; p297: Alila Medical Media; p300(t): joshya; p300(b): La Gorda; p304: joshya; p307(t): Matthew Cole; p307(b): Designua; p308: Puwadol Jaturawutthichai; p311(t): Alila Medical Media; p311(bl): KPG_Payless; p311(br): SvedOliver; p312: xpixel; p317: Belozorova Elena; p319(t): Emi; p319(b): jack perks; p320: JSseng; p321: Pim Leijen; p324: BlueRingMedia; p325(t): Gil.K; p325(bl): Carlos Romero; p325(br): Larry Bruce; p326: Olivier Le Moal; p327: Igor Stramyk; p329: Mark Yarchoan.

© **Science Photo Library:** p25: Dr. Kenneth R. Miller; p35: Nigel Cattlin; p61: Martin Shields; p89: Biophoto Associates; p123(l): Thomas Deerinck, NCMIR; p123(r): Dr. Donald Fawcett, Visuals Unlimited; p141: Science Pictures Limited; p142: Wim Van Egmond, Visuals Unlimited; p156: Steve Gschmeissner; p163(b): Biophoto Associates; p165: D. Phillips; p206: Dept. of Clinical Cytogenetics, Addenbrooks Hospital; p263(t): Eye of Science; p263(b): Dr Linda Stannard, UCT; p281: Medical Images, Universal Images Group; p286: Molly Borman; p294: Zephyr; p295: Ciepro; p309: Peter Gardiner; p312: Pr Michel Zanca, ISM; p318: Omikron.

© p56: Ytambe(Wikipedia Commons)

© p70(l): Dr. M. Cheek, RBG Kew

© p131: WJEC A Level 1074/02 Human Biology – HB4, January 2014

© p218(r+l): Olaf Leillinger

© p266: www.celtnet.org.uk/medicine/malaria.ph

© p293(l,c): www.studyblue.com

© p213(all): Dr M S Izen

Acknowledgements

The author wishes to thank Dr Colin Blake, Dr Meic Morgan and Andrew Clarke for their advice in the preparation of this book, Dr Martin Cheek for the photograph of *Salacia arenicola*.

Contents

What this book contains

The contents of this book match the specifications for WJEC A2 Biology. It provides you with information and practice examination questions that will help you to prepare for the examinations at the end of the year. This book addresses:

- The three Assessment Objectives required for the WJEC Biology course. They are described further below.

- The mathematics of biology, which will represent a minimum of 10% of your assessment, and gives you explanations and worked examples.

- Practical work: the assessment of your practical skills and understanding of experimental biology represents a minimum of 15% of your qualification, and will be developed by your use of this book. Some practical tasks are integrated into the text and major experiments are discussed in detail at the end of the relevant chapter.

The content of this book is clearly divided into the units of the course that are tested in the A2 examination, i.e. Unit 3 – Energy, Homeostasis and the Environment and Unit 4 – Variation, Inheritance and Options. The practical work that comprises Unit 5 is described on p9 and throughout the text, where relevant. The specification map on pp10–13 allows you to identify, for each topic, the relevant statements in the specification.

Each chapter covers one topic, divided into a number of sub-topics. These are given at the start of each chapter, as a list of learning objectives. At the end of each unit are questions designed to help you to reinforce what you have learned and to practise for the examinations. Answers to these questions are given at the end of the book. The questions presented here have not been written by the teams who prepare examination papers, nor been subject to the review that examination papers undergo, but they will, none-the-less, contribute usefully to your examination preparation.

Marginal features

The margins of each page hold a variety of features to support your learning:

 Key Terms

These are terms that you need know how to define. They are highlighted in blue in the body of the text and appear in the Glossary at the back of this book. You will also find other terms in the text in bold type, which are explained in the text, but have not been defined in the margin. The use of key terms is an important feature since examination papers may contain a number of terms that need to be defined.

 Knowledge check

These are short questions to check your understanding of the subject, allowing you to apply the knowledge that you have acquired. These questions are of two types: filling in blanks in a passage and matching terms with phrases specific to the topic under study. Answers are given at the back of the book.

 Study point

As you progress through your studies, advice is provided to help you understand and use the knowledge content. In this feature, factual information may be emphasised, or restated to enhance your understanding.

Going further

This may provide extra information not in the main text, but relevant to it. It may to provide more examples but does not contain information that will be tested in an examination.

Exam tip

This feature provides general or specific advice to help you prepare you for the examination. Read these very carefully.

Links to other sections of the course are highlighted in the margin, near the relevant text. They are accompanied by a reference to areas where sections relate to one another. It may be useful for you use these Links to recap a topic before beginning to study the current topic.

YOU SHOULD KNOW ›››

Learning objectives are provided. They are more specific to the area being studied than the more general learning objectives listed at the start of the topic.

WORKING SCIENTIFICALLY

This feature helps you to understand something about science itself, how scientific knowledge has been obtained, how reliable it therefore is and what its limitations are. It may also help you to have a deeper awareness of how science is used to improve our quality of life. It is important to understand the scientific process, to know how evidence has been gathered and how to evaluate it. These features will help you to develop the habit of approaching evidence with a questioning mind. Working Scientifically is further discussed in more detail below.

Working Scientifically

The value of science lies in its consistency. Given suitable conditions, plants photosynthesise; ingesting certain bacteria produces particular predictable symptoms; burning more fossil fuel puts more carbon dioxide into the atmosphere. Such observations always hold, which is why scientific knowledge is so valuable. It is not a matter of opinion and it is demonstrably true, within the limits of experiment. However, as methodology and technology improve, understanding improves, so concepts and explanations are modified. Science is, thus, updated in a continuing process. The scientific community has procedures for testing and checking the findings and conclusions of individual scientists. The serious way that scientific fraud is treated builds confidence in the procedures.

It is important that you are able to demonstrate your understanding of the scientific process and the impact that scientific knowledge has on individuals and on society. These are demonstrated throughout this course and are highlighted throughout this book. Examples are given below:

- Data from observations and measurements are of critical importance, e.g. 3.5(m) The effects of human activities on the carbon cycle, including the change in species distribution and extinction as a result of climate change.

- A good explanation allows a prediction of what may happen in other situations, e.g. 3.5(a) An explanation of energy flow and nutrient cycling through ecosystems and how this leads to changes in population size and composition.

- There may be a correlation, a cause or a chance link between a factor and an outcome, e.g. 4.3(d) The use of a chi squared test to test for Mendelian inheritance.

- A new theory may provide an explanation for the available data, e.g. 3.1(a) The synthesis of ATP involving a flow of protons down a proton gradient through ATP synthetase, in chemiosmosis. The converse is also true, i.e. new data may support a theory or require its amendment.

- Devising and testing a scientific explanation is not a simple and straightforward process. There is always uncertainty in data. An observation may be untrustworthy because of the limitations of either the measuring equipment or the person using it, e.g. 4.4(i) The use of stem cells for replacing damaged tissues and organs.

- Devising an experiment and deducing an explanation for the results are creative steps. There are different ways to demonstrate the same effect and different people may provide different explanations for the same data, e.g. 3.2(k) The role of leaf structure in allowing plants to photosynthesise efficiently: the diameter and distribution of stomata in leaves of a given species may differ in different habitats. There are likely to be conflicting pressures determining their values.

- The application of scientific knowledge in new technologies, materials and devices greatly enhances our lives, but may have unintended and undesirable side effects, e.g. 4.5(d) The knowledge derived from the inventions of the polymerase chain reaction and genetic fingerprinting have significant forensic use, but skill is required for the proper interpretation of gene profiles.

- The application of science has social, economic, political and ethical implications, e.g. 3.6(f) The concept of planetary boundaries shows how human behaviour has an impact on all life on Earth through its effects on planet-wide systems.

Mathematical requirements

As assessment of your mathematical skills is very important, some common uses of mathematics in biology are included throughout this book. There is nothing difficult here. You are preparing for a biology examination, not a mathematics exam, but it is still important to apply numerical analysis, and these examples will help you to do so. Mathematical requirements are given in Appendix C, on p70 of your specification. With the exception of statistics, which is equivalent to Level 3 or A Level, the level of understanding is equivalent to Level 2 or GCSE Mathematics.

Assessment

The contents of the WJEC AS and A Level Biology course are shown in the table below:

Component	Topic
AS/A2 Year 1	
Unit 1 – Basic biochemistry and cell organisation	Chemical elements are joined together to form biological compounds
	Cell structure and organisation
	Cell membranes and transport
	Enzymes
	Nucleic acids
	Cell division
Unit 2 – Biodiversity and physiology of body systems	Evolutionary history
	Gas exchange
	Transport
	Nutrition
A2 Year 2	
Unit 3 – Energy, homeostasis and the environment	Importance of ATP
	Photosynthesis
	Respiration
	Microbiology
	Population size and ecosystems
	Human impact on the environment
	Homeostasis and the kidney
	Nervous system
Unit 4 – Variation, inheritance and Options	Sexual reproduction in humans
	Sexual reproduction in plants
	Inheritance
	Variation and evolution
	Application of reproduction and genetics
	Option A – Immunology and disease
	Option B – Human musculoskeletal anatomy
	Option C – Neurobiology and behaviour

Assessment objectives

Examinations test not only your subject knowledge, but also skills associated with how you use that knowledge. These skills are described in Assessment Objectives, and examination questions are written to reflect them, with marks in the proportions shown:

Unit		% of total marks			
		Unit weightings	AO1	AO2	AO3
AS	1	20	17.5	22.5	10
	2	20	17.5	22.5	10
A2	3	25	6.7	10.8	7.5
	4	25	6.7	10.8	7.5
	5	10	2.6	5.4	2.0
A Level total		100	30	45	25

You must meet these Assessment Objectives in the context of the subject content, which is given in detail in the specification.

Assessment objective AO1

Demonstrate knowledge and understanding of scientific ideas, processes, techniques and procedures.

Assessment objective AO2

Apply knowledge and understanding of scientific ideas, processes, techniques and procedures:

- in a theoretical context
- in a practical context
- when handling qualitative data
- when handling quantitative data.

Assessment objective AO3

Analyse, interpret and evaluate scientific information, ideas and evidence to:

- make judgements and reach conclusions
- develop and refine practical design and procedures.

Mathematical skills assessed will contribute 10% of the marks, at a minimum.

Practical skills assessed will contribute 15% of the marks, at a minimum.

Examination questions will present information in novel situations. You are not expected to be familiar with these new scenarios, but you will be tested on how well you apply your own knowledge to them. It is essential to select and communicate information and ideas concisely and accurately, using appropriate scientific terminology. This skill will be tested within all three assessment objectives.

The examinations

The AS qualification has two written examinations and the A Level qualification has these two and three others, the third of which has both practical and written components. Their structure is described in the table below.

No more than 10% of marks in a single paper will test simple recall (AO1). Most of the marks are awarded for the skills developed in application of scientific ideas (AO2) and analysis, interpretation and evaluation (AO3), as explained above. You are expected to be able to link different parts of the specification together: there will be a small number of marks in each paper that require you to refer to information or concepts from other parts of the specification.

Unit	Unit title	Structure		Exam Length	Marks	% of total
1	Basic biochemistry and cell organisation	Short and longer questions, some in a practical context; one extended response.		1 h 30 mins	80	20
2	Biodiversity and physiology of body systems	Short and longer questions, some in a practical context; one extended response.		1 h 30 mins	80	20
3	Energy homeostasis and the environment	Short and longer questions, some in a practical context; one extended response.		2 hours	90	25
4	Variation, inheritance and options	Section A	Short and longer questions, some in a practical context; one extended response.	2 hours	70	25
		Section B	Choice of one option out of three; short and longer questions.		20	
5	Practical examination	Experimental task		2 hours	20	10
		Practical analysis task		1 hour	30	

Examination questions

As well as being able to recall biological facts, name structures and describe their functions, you need to appreciate the underlying principles of the subject and understand associated concepts and ideas. In other words, you need to develop skills so that you can apply what you have learned, perhaps to situations not previously encountered. You may be asked to inter-convert numerical data and graph form, analyse and evaluate numerical data or written biological information, interpret data and explain experimental results.

The mark value at the end of each part of each question and the number of lines provided for you to write on are useful guides as to the amount of information required in the answer.

You will be expected to answer different styles of question, for example:

- Short answer questions: these require a simple calculation or a brief answer, e.g. the name of a structure and its function, for one mark.

- Structured questions: these have several parts, usually about a common theme, which become more difficult as you work your way through. Structured questions can be short, requiring a brief response, or may include the opportunity for extended writing. The number of lines provided indicates the length of answer expected. The mark allocations at the end of each part of the question are there to help you: if three marks are allocated, you must give three separate points.

- Extended prose questions: each examination paper will contain one question, worth nine marks, which requires extended prose for its answer. This is not an essay and so does not require the structure of introduction – body – conclusion. Often candidates rush into such questions. You should take time to read carefully, to discover exactly what the examiner requires in the answer, and then construct a plan. This will not only help you organise your thoughts logically but will also give you a checklist to which you can refer when writing your answer. In this way you will be less likely to repeat yourself, wander off the subject or omit important points. You may wish to use diagrams to clarify your answer, but if you do, make sure they are well drawn and fully annotated.

For 7–9 marks, you should provide most of the relevant factual information with clear scientific reasoning; 4–6 marks will be awarded if there are significant omissions and 1–3 marks if there is little factual recall and few valid points. In the highest band of marks, a piece of writing that answers the question directly, using well-constructed sentences and suitable biological terminology, addressing all three Assessment Objectives will be awarded 9 marks. But the same information with poor spelling, grammar or waffle, will only merit 7 marks.

Examination questions are worded very carefully. It is essential not to penalise yourself by reading them too quickly or too superficially. Take time to think about the precise meaning of each word in the question so that you can construct a concise, relevant and unambiguous response. To access all the available marks it is essential that you follow the instructions accurately. Here are some command words that are commonly used in examinations:

- *Annotate*

  To annotate a diagram is to give a short description with each label.

  Example: Annotate the diagram of the flower to show the functions of its parts.

- *Compare*

  If you are asked to make a comparison, make an explicit comparison in each sentence, rather than writing separate paragraphs about what you are comparing.

  Example: if you are asked to compare the nervous systems of a vertebrate and a Cnidarian, produce a sentence that refers to both, e.g. 'A Cnidarian has a nerve net but a vertebrate has a central nervous system'.

- *Describe*

  This term may be used where you need to give a step-by-step account of what is taking place. In a graph question, for example, if you are required to recognise a simple trend or pattern then you should also use the data supplied to support your answer. At this level it is insufficient to state that 'the graph' or 'the line goes up and then flattens'. You are expected to describe what goes up, in terms of the dependent variable, i.e. the factor plotted in the vertical axis, and illustrate your answer by using figures and a description of the gradient of the graph.

  Example: Describe the concentration of progesterone throughout the menstrual cycle.

- *Evaluate*

  State the evidence for and against a proposal and conclude whether or not the proposal is likely to be valid.

  Example: Evaluate the statement that increased use of nitrogenous fertilisers has contributed to eutrophication.

- *Explain*

  A question may ask you to describe and also explain. You will not be given a mark for merely describing what happens – a biological explanation is also needed.

  Example: Use the graph to explain the biological significance of changes in the concentration of progesterone in the menstrual cycle.

- *Justify*

  You will be given a statement for which you should use your biological knowledge as evidence in support. You should cite evidence to the contrary and draw a conclusion as to whether the initial statement can be accepted.

  Example: Justify the statement that human activity has been responsible for significant increase in the rates of plant and animal extinction.

- *Name*

  You must give a simple answer. You do not have to waste time by repeating the question or putting your answer into a sentence.

  Example: Name the organism that causes malaria.

- *State*

  Give a brief, concise answer with no explanation.

  Example: State the name given to the graph showing the rate of photosynthesis as a function of wavelength of light.

- *Suggest*

  This command word may be at the end of a question. There may not be a definite answer to the question, but you are expected to put forward a sensible idea based on your biological knowledge.

  Example: Suggest why offshore islands, such as Madagascar, have a large number of plant and animal species not seen on the mainland.

Practical work

Practical work is given prominence in WJEC Biology, as it provides a way for you to consolidate what you have learned and a way for you to develop skills that improve your ability to plan and design a course of action, to undertake the physical manipulation of apparatus and to analyse and evaluate results. The experiments identified in your specification will contribute to you developing competencies, which will be assessed. You should be able to:

- Follow written instructions.
- Apply an investigative approach when using instruments and equipment.
- Safely use a range of equipment and materials.
- Make and record observations.
- Undertake research, cite the references you have used and write a report of your findings.

Experiments are designed for many reasons. They may, for example, test a hypothesis, measure a parameter or investigate a response. As in the first year of this course, particular aspects must be carefully considered:

- Rationale.
- Design, including defining variables.
- Method, including your choice of how many readings to make for each value of the independent variable.
- Results, including the sensible design of a results table.
- Numerical analysis including calculations, graphical representation and statistical analysis, where appropriate.
- A biological explanation for your findings. You should recognise if an apparent link between two variables is due to chance, is merely a correlation or if it has a biological basis.
- Consistency.
- Accuracy.
- Further work.
- The moral and ethical implications of your work in relation to social, environmental and economic concerns, where relevant.

Practical work is far more than just doing experiments, although these are important both in the laboratory and in the field. During this course, you may undertake practical work that encompasses:

- Microscopy, which allows you to understand better an actual structure, as opposed to the structure portrayed in a diagram. Indeed, in some experiments, microscopy is essential for making readings.
- The study and interpretation of photographic images.
- The dissection of tissues, organs and whole organisms.

In A2 Biology, practical work represents a minimum of 15% of all available marks. It is assessed in:

- All the theory papers: at least 5% of all the available marks are questions in written papers that directly address practical work. It is important, therefore, that you maintain excellent lab notes throughout the course, as they will provide you with essential revision material.

- A practical examination: this represents 10% of all the marks available. It is tested in two tasks:
  - The experimental task will last 2 hours. During the Spring Term of the second year of the course, you will be given apparatus and an experimental method to do an individual experiment. Your teacher will observe you to assess your practical skills.
  - The analysis of an experiment will last an hour. A practical scenario, with data, will be given to you and you will undertake a written paper about the experiment. In addition, your microscopy skills may be tested by presenting you with a light or electron micrograph. You may be required to make a biological drawing of part of it or to make measurements and calculations from it.

Specification map

Chapter	Topic	Spec section	Page
1 – Importance of ATP	Chemiosmosis	a	15
	The mitochondrial and chloroplast membranes	b	15
	The proton gradient	c	16
	The electron transport chain	d	17
	Investigation of dehydrogenase activity in different strains of yeast	Practical	18
2 – Photosynthesis	The distribution of chloroplasts in relation to light trapping	a	21
	Chloroplasts as transducers	b	23
	Light harvesting	c	24
	The photosystems	d	25
	The light dependent stage	e, f, g	26
	The light-independent stage	h, i	29
	Product synthesis	j	30
	Limiting factors in photosynthesis	k	31
	Mineral nutrition	l	35
	Investigation into the separation of chloroplast pigments by chromatography	Practical	36
	Investigation into the role of nitrogen and magnesium in plant growth	Practical	39
	Investigation into the effect of carbon dioxide concentration the rate of photosynthesis	Practical	40
3 – Respiration	An overview of respiration	a	43
	Glycolysis	b	44
	Krebs' cycle	c	46
	The electron transport chain	d	46
	Anaerobic respiration	d	48
	Energy budget of glucose breakdown	e	49
	Alternative respiratory pathways	f	50
	Investigation into factors affecting respiration in yeast	Practical	52
4 – Microbiology	Classification of bacteria	a	55
	Gram staining	a	56
	Conditions necessary for culturing bacteria	b, c	58
	Methods of measuring growth	d	60
	Investigation into the numbers of bacteria in milk	Practical	62
5 – Population size and ecosystems	Population growth	a, b	65
	Factors that regulate population increase	b,c	68
	Abundance and distribution of organisms in a habitat	d	69
	Ecosystems	e, f	72
	Biotic components of an ecosystem	g	73
	Energy flow through an ecosystem and ecological pyramids	h	73
	Succession	i, j	79
	Recycling nutrients	k	83
	The carbon cycle and human impact	l, m	83
	The nitrogen cycle and human impact	n, o, p	88
	Assessing the distribution of organisms in a rocky shore transect	Practical	92

Unit 3 – Energy, homeostasis and the environment

CH 1

The importance of ATP

Green plants capture light energy and convert it to chemical energy in the products of photosynthesis, such as glucose. If respiration broke down these molecules directly, energy would be released in an uncontrollable way. Instead, energy is released gradually and is held as potential energy in a proton gradient across a membrane. It is transferred as chemical energy to the intermediate molecule, ATP, which allows it to be released in small, manageable quantities, as needed. ATP is a reservoir of potential chemical energy and acts as a common intermediate in metabolism, linking energy-requiring and energy-yielding reactions. The early evolution of ATP synthetase probably gave ATP its prominent role in all living systems. ATP is described as the 'universal energy currency'.

Topic contents

By the end of this topic you will be able to:

- Explain that dehydrogenase enzymes produce reduced cofactors, such as reduced NAD, in respiration.
- Understand how the inner membranes of chloroplasts and mitochondria establish proton gradients.
- Know that electrons pass through a series of carriers to oxygen, the final electron acceptor.
- Explain how the action of ATP synthetase leads to the production of ATP.
- Understand the meaning of the term chemiosmosis.
- Describe how dehydrogenase activity may be monitored with artificial electron acceptors.

The importance of ATP

ATP is described as the universal energy currency. This means it is used in all cells to drive their reactions. It is made when energy becomes available, for example in respiration and in the light-dependent reactions of photosynthesis; it is broken down when the cell needs energy, such as in biosynthesis, muscle contraction and powering the membrane Na^+/K^+ pumps. ATP is one of the molecules that is characteristic of all living systems and was probably present in LUCA, the last universal common ancestor of all cells.

ATP is ideally suited to its role because it:

- is inert
- can pass out of mitochondria into the cytoplasm
- releases energy efficiently
- releases energy in useable quantities, so little is wasted as heat
- is easily hydrolysed to release energy
- is readily reformed by phosphorylation.

Chemiosmosis

In the synthesis of ATP, electrons and protons derived from hydrogen atoms have different pathways:

- Electrons from hydrogen atoms are transferred from a donor molecule to a recipient. Then, a sequence of reactions transfers the electrons from one molecule to the next along a chain. Each transfer is a redox reaction, in which one molecule is oxidised, i.e. loses electrons, and the next in the sequence is reduced, i.e. gains electrons. Oxidation reactions make energy available, and this energy is eventually used to synthesise ATP.

- The energy released by oxidation pumps the protons from the hydrogen atoms across a membrane so that they are more concentrated on one side of the membrane than the other. The difference in the concentration of protons and the charge on either side of the membrane constitutes an electrochemical gradient and is a source of potential energy. Protons flow back down this gradient, in a process called **chemiosmosis**, through the enzyme ATP synthetase, sometimes called ATP synthase. The energy they release as they do so is converted into chemical energy in ATP.

The mitochondria and chloroplast membranes

ATP synthetase makes ATP from the energy associated with proton gradients across membranes. Bacteria do not have internal membranes and they use the cell membrane to establish a proton gradient, by pumping protons out into the cell wall. Respiration uses the inner membranes of the mitochondria; photosynthesis uses the thylakoid membranes of the chloroplasts. This common function of the inner membranes of the mitochondria and chloroplasts and the bacterial cell membrane support the theory of endosymbiosis. These membranes must only let protons through, and in a highly controlled fashion. Protons are very small and easily pass through water molecules, so the membranes must also be watertight. This is why they are described as '**sealed membranes**'.

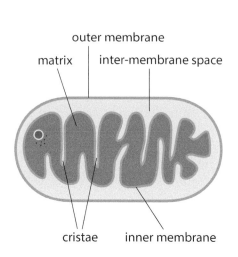

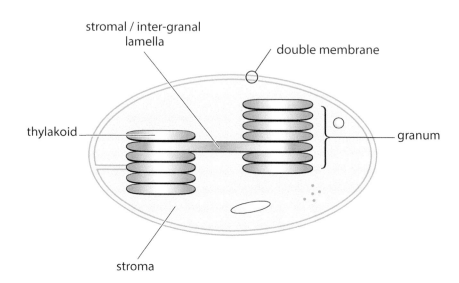

Basic structure of a mitochondrion

Basic structure of a chloroplast

The proton gradient

Proton gradients occur in non-living systems, such as oceanic alkaline hydrothermal vents. It may be that naturally occurring proton gradients such as these had a significant role in the origin of life, because proton gradients are a fundamental characteristic of all living things.

In the light-dependent stage of photosynthesis, electrons are excited by energy from light. These electrons move through a series of carriers in the thylakoid membranes and their energy pumps protons from the stroma into the spaces between the thylakoid membranes. The energy is released in chemiosmosis, in which protons flow back down an electrochemical gradient into the stroma, through ATP synthetase. The energy is incorporated into ATP. This ATP drives the light-independent reactions of photosynthesis and energy is incorporated into macromolecules made by the cell.

In respiration, electrons are excited by energy derived from food molecules. Their energy is made available as they move through a series of carriers on the inner mitochondrial membrane. The energy pumps protons across the membrane, from the matrix into the inter-membrane space, setting up a proton gradient. Energy is released in chemiosmosis, as the protons flow back into the matrix through ATP synthetase, and is incorporated into ATP. Energy that is not incorporated into ATP is lost as heat.

Disrupting proton gradients causes death:

Link The light-dependent reactions of photosynthesis are described on p26. The breakdown of food in respiration is described on p43.

- Apoptosis is programmed cell death and occurs, for example, during embryonic development. It operates by preventing proton gradients across cell membranes from forming.
- DNP is a mitochondrial poison that allows electron transport but does not allow ATP synthesis, i.e. they are 'uncoupled'. Some people have used DNP to try to lose weight. The body oxidises fats and carbohydrates, so weight is lost, but all the energy released from those molecules is converted to heat, as no ATP can be made. The body overheats, sometimes fatally.

The electron transport chain

The electron transport chain is a series of protein carriers on the inner membranes of mitochondria and chloroplasts. It releases energy from electrons and incorporates it into ATP.

In respiration, hydrogen atoms derived from the respiratory breakdown of glucose are transferred by dehydrogenase enzymes to the coenzymes NAD and FAD and carried to the inner membrane of the mitochondrion. The electrons and protons of the hydrogen atoms have different pathways but because they both move through the electron transport chain, the system is often described as carrying hydrogen atoms. For every two protons delivered by reduced NAD, enough energy is released to synthesise three molecules of ATP. When reduced FAD delivers two protons, enough energy is released for only two molecules of ATP.

The energy for the proton pump and the electron transport chain is derived from oxidation reactions, i.e. electron loss. Phosphorylation is the addition of a phosphate group. So synthesising ATP by adding a phosphate ion to ADP using energy derived from oxidation reactions is called oxidative phosphorylation.

In photosynthesis, groups of pigments and proteins called photosystems transfer excited electrons to electron acceptors and, from there, to a series of protein carriers, all on the thylakoid membranes. Protons from water and the electrons are transferred to the coenzyme NADP and subsequently, to glycerate phosphate, in the pathway that synthesises carbohydrates. The energy that powers the proton pump and electron transport chain in the chloroplast comes from light, so chloroplasts synthesise ATP by photophosphorylation.

WORKING SCIENTIFICALLY

Different fields of science support each other. Early electron micrographs of mitochondria showed 'stalked' or 'elementary particles' on the cristae, with more in active cells , e.g. muscle fibres than inactive cells. Biochemists identified the stalked particles as ATP synthetase.

▼ **Study point**

ATP is produced when protons flow across the internal membranes of mitochondria and chloroplasts, down their concentration gradient through ATP synthetase, in a process called chemiosmosis.

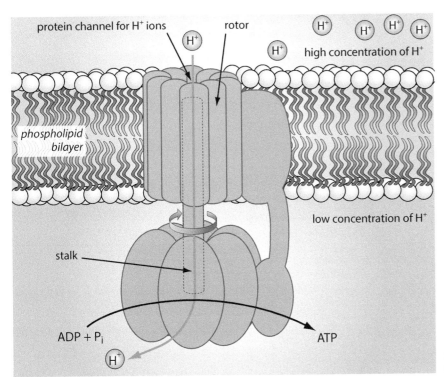

ATP synthetase complex

ATP synthetase occurs in all Bacteria, Archaea and Eukaryotes so it presumably evolved very early in the history of life. The diagram shows the ATP synthetase complex in a mitochondrion. It is a nano-machine. As the protons diffuse down their electrochemical gradient through the ATP synthetase, the energy released causes the rotor and stalk to rotate. The mechanical energy from this rotation is converted into chemical energy as a phosphate ion, P_i, is added to ADP to form ATP in the catalytic head. It takes three protons to move the rotor through 120°, releasing 1 ATP molecule.

Knowledge check

Match the terms A–D with statements 1–4

A. ATP

B. ATP synthetase

C. Proton gradient

D. Chemiosmosis

1. Intermediate molecule in cell metabolism associated with energy use or generation.

2. The passage of protons down their electrochemical gradient through ATP synthetase, with the synthesis of ATP.

3. The difference in proton concentration and charge on either side of a membrane provides the energy for ATP synthesis.

4. Protein complex spanning the inner mitochondrial membrane and thylakoid membrane of a chloroplast which catalyses the synthesis of ATP.

The table shows the similarities and differences between the electron transport systems and ATP synthesis in mitochondria and chloroplasts:

		Mitochondria	Chloroplasts
Similarities		Use of ATP synthetase	
		2 protons provide energy to synthesise 3 ATP	
		Proton pump across inner membrane of organelle	
		Electron transport chain on inner membrane of organelle	
Differences	Phosphorylation type	Oxidative phosphorylation	Photophosphorylation
	Energy source	Chemical energy in redox reactions	Light
	Site of electron transport chain	Cristae	Thylakoid membranes
	Coenzyme	NAD, FAD	NADP
	Proton gradient	Inter-membrane space $\rightarrow$ matrix	Thylakoid space $\rightarrow$ stroma
	Number of proton pumps	3 with NAD; 2 with FAD	1
	Final electron acceptor	Oxygen and H$^+$	Cyclic photophosphorylation: chlorophyll$^+$; Non-cyclic photophosphorylation: NADP + H$^+$

Specified practical exercise

Investigation of dehydrogenase activity in different strains of yeast

Rationale

Dehydrogenases are enzymes that carry hydrogen atoms from various substrates and transfer them to coenzymes, such as NAD and FAD in the mitochondria. The coenzymes are reduced by the hydrogen. They donate the hydrogen atoms to carriers in the electron transport chain and by doing so, are reoxidised. Reduction by dehydrogenases may be demonstrated with an artificial hydrogen acceptor, such as methylene blue.

The reduction potential of a compound indicates how likely it is to become reduced. Methylene blue has a higher reduction potential than NAD and FAD and so, when they are all present, as in this experiment, methylene blue is reduced in preference to NAD or FAD.

Methylene blue is dark blue when oxidised and its reduction may be monitored as it loses its colour on reduction. The time taken for the methylene blue to lose its colour can be used as a measure of the rate of dehydrogenase activity.

Design

	Name of variable	Value of variable
Independent variable	Yeast strain	Different strains of brewer's yeast and baker's yeast
Dependent variable	Time for methylene blue to decolourise	Seconds
Controlled variable	Yeast and glucose concentration	100 g dried yeast + 54 g glucose in 1 dm^3 water. (The glucose concentration is 0.3 mol dm^{-3})
	Methylene blue concentration	0.1 g 100 cm^{-3}
	Temperature	30°C
Reliability	Perform test with each strain three times and calculate the mean time to decolourise	
Main hazard	**Risk**	**Control measures**
Methylene blue	Harmful if swallowed or if dust inhaled	Weigh powder and make solution in a fume cupboard; wear a mask.
	Skin irritation and staining	Avoid skin contact.
	Eye irritation	Use eye protection.

Apparatus

- Redox indicator: methylene blue 0.1 g 100 cm^{-3}
- Yeast suspension 100 g dm^{-3}
- 30°C water bath
- Test tubes
- Corks for test tubes
- 10 cm^3 syringe
- 1 cm^3 syringe
- Stop clock

Outline method

1. Syringe 10 cm^3 of stirred yeast suspension into a test tube.
2. Place test tube in water bath for 5 minutes to equilibrate to 30°C.
3. Add 1 cm^3 methylene blue.
4. Invert the test tube once to mix and start the stop clock.
5. Replace the test tube in the water bath.
6. Time how long the mixture takes to lose its colour, bearing in mind that the mixture at the top of the test tube will remain blue as it is exposed to oxygen in the air.

Further work

- To investigate the effect of pH on the rate of dehydrogenase activity, yeast suspensions may be made in buffer at different pH values, e.g. pH 1, 3, 5, 7, 9, 11.
- To investigate the effect of glucose concentration on the rate of dehydrogenase activity, the yeast suspension may be made with different glucose concentrations, e.g. 0.1, 0.3, 0.5, 0.7 and 0.9 mol dm^{-3}.

CH 2

Photosynthesis uses light energy to synthesise organic molecules

Photosynthesis takes place in chloroplasts, where pigments such as chlorophyll trap light energy. The biochemical processes of photosynthesis may be divided into the light-dependent and the light-independent stages. The light-dependent stage, on the thylakoid membranes of the chloroplasts, involves electron transport and the photolysis of water. It produces ATP by chemiosmosis, reduced NADP, and the waste product, oxygen. The light-independent stage takes place in the stroma: carbon dioxide combines with a 5-carbon acceptor and, with reducing power and energy from the light-dependent stage, makes glucose.

Topic contents

By the end of this topic you will be able to:

- Describe the distribution of chloroplasts and their function as transducers.
- Describe the absorption spectrum and the action spectrum.
- State that the light-dependent stage takes place on and across the chloroplast's thylakoid membranes.
- Understand that chloroplast pigments are grouped together to form photosystems I and II, in which antenna complexes bring energy from photons of light to the two types of reaction centre.
- Explain that both cyclic and non-cyclic photophosphorylation involve chemiosmosis and ATP production.
- Describe the light-dependent stage as the photoactivation of chlorophyll and energy transfer, producing ATP, reduced NADP and the by-product oxygen.
- Explain the use of reduced NADP and ATP in the light-independent stage, in which carbon dioxide is fixed and reduced to glucose.
- Describe how cellular metabolites are derived from the products of photosynthesis.
- Explain how light intensity, carbon dioxide concentration and temperature may be limiting factors in photosynthesis.
- Describe the roles of nitrogen and magnesium in flowering plants.

An overview of photosynthesis

The overall equation for photosynthesis is $6CO_2 + 6H_2O \longrightarrow C_6H_{12}O_6 + 6O_2$

The diagram shows that photosynthesis involves two stages: the light-dependent stage and the light-independent stage.

The light-dependent stage converts light energy to chemical energy. The photolysis of water releases electrons and protons. Energy carried by electrons establishes a proton gradient across the thylakoid membrane. The energy is used to phosphorylate ADP, which generates ATP, in **photophosphorylation**. The protons and electrons reduce NADP.

In the light-independent stage, the ATP and reduced NADP reduce carbon dioxide and produce energy-containing glucose.

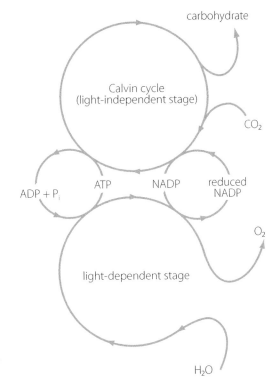

carbohydrate

Calvin cycle (light-independent stage)

CO_2

ADP + P_i   ATP   NADP   reduced NADP

O_2

light-dependent stage

H_2O

Photosynthesis summary

Key Term

Photophosphorylation: An endergonic reaction bonding a phosphate ion to a molecule of ADP using energy from light, making ATP.

The distribution of chloroplasts in relation to light trapping

Photosynthesis takes place in chloroplasts. Chloroplasts are surrounded by a double membrane. The inner membrane folds inwards to make thylakoid lamellae. These combine in stacks of up to 100 disc-shaped structures, forming grana, where the photosynthetic pigments are located and where the reactions of the light-dependent stage of photosynthesis take place. The stroma is the fluid-filled interior, bathing the thylakoids and grana, where the light-independent reactions take place. Starch grains in the chloroplast appear white because the stain used for electron micrographs, osmium tetroxide, binds to lipid, but not to carbohydrates.

Electron micrograph of a chloroplast

Going further ▶

The base sequences of the small number of genes in chloroplasts indicate that they are indeed descended from cyanobacteria, as suggested by several biologists in the early 20th century and formulated in the concept of endosymbiosis by Lynn Margulis, in 1967.

Chloroplasts need light and so they are found only in those parts of the plant that are exposed to light – the leaves and the stem. The leaf is the main organ of photosynthesis and the chloroplasts are found largely in the palisade mesophyll. They also occur in the spongy mesophyll, but the only epidermal cells containing chloroplasts are the guard cells.

The structure of leaves makes them well suited to their role in photosynthesis:

	Structural feature	Significance for photosynthesis
Leaf	Large surface area	Capture as much light as possible
	Thin	Light penetrates right through the leaf
	Stomatal pores	Allow carbon dioxide to diffuse into the leaf
	Air spaces in the spongy mesophyll	Allow carbon dioxide to diffuse to the photosynthesising cells
	Spaces between palisade cells	Allow carbon dioxide to diffuse to the photosynthesising cells
Cells	Cuticle and epidermis are transparent; cellulose cell walls are thin	Light penetrates through to the mesophyll
	Palisade cells have a large vacuole	Chloroplasts form a single layer at the periphery of each cell so do not shade each other
	Palisade cells are cylindrical, elongated at right angles to the surface of the leaf	Leaves can accommodate a large number of palisade cells; light only passes through two epidermal cell walls and one palisade cell wall before reaching chloroplasts. If the cells were stacked horizontally, light would be absorbed by passing through many cell walls, preventing it reaching chloroplasts.
Chloroplasts	Chloroplasts have a large surface area	Maximum absorption of light
	Chloroplasts move within palisade cells	Chloroplasts move towards the top of the cell on dull days, for maximum absorption of light. If the light intensity is very high, they move to the bottom of the cell, protecting pigments from bleaching.
	Chloroplasts rotate within palisade cells	Thylakoids maximise the absorption of light
	Pigments in the thylakoids are in a single layer at the surface of the thylakoid membrane	Pigments maximise their absorption of light
	About five times as many chloroplasts in palisade cells than spongy mesophyll cells	Palisade cells are at the top of the leaf and they are more exposed to light than the spongy mesophyll cells, so chloroplasts capture as much light as possible

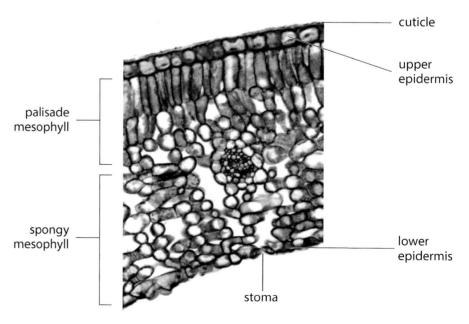

Leaf structure

Chloroplasts as transducers

A transducer changes energy from one form into another. A light bulb, for example, transduces electrical energy into light and heat, and an engine transduces the chemical energy in petrol into kinetic energy, heat and sound. Biological transducers are much more efficient than such artificial devices and waste less energy in the conversions that they make. Chloroplasts are transducers, turning energy in the photons of light into chemical energy, made available through ATP and incorporated into molecules such as glucose.

WORKING SCIENTIFICALLY

Engelmann's experiment

The site of photosynthesis was demonstrated in 1887 by the botanist, Engelmann. His experiments, using the filamentous green alga, *Cladophora*, determined which wavelengths of light were the most effective. Engelmann placed *Cladophora* in a suspension of evenly distributed motile, aerobic bacteria and exposed them to a range of wavelengths. After a short time, he noticed that in blue and red light, the bacteria clustered near the chloroplasts. He deduced that these wavelengths resulted in a high rate of photosynthesis, which produced a lot of oxygen, which attracted the bacteria.

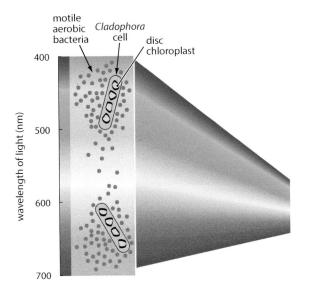

Engelmann's experiment

YOU SHOULD KNOW ›››

››› That chloroplasts are energy transducers

››› The names of the photosynthetic pigments

››› The definitions of absorption and action spectra

››› The structure of the antenna complex

Going further ▶

Biological transduction in animals includes the conversion of heat, light and sound into electrical energy in sense organs and the conversion of chemical energy into kinetic energy and heat in muscles. Some animals, including fireflies and glow worms, and the phosphorescent bacteria can transduce chemical energy into light.

Photosynthetic pigments

A pigment is a molecule that absorbs specific wavelengths of light. In the chloroplasts, light energy is trapped by photosynthetic pigments, with different pigments trapping different wavelengths. This allows a large range of wavelengths to be absorbed and is consequently more useful than if there were just one pigment, absorbing a small range of wavelengths. In flowering plants there are two main classes of pigments that act as transducers, the chlorophylls and the carotenoids.

Pigment class	Pigment	Peak wavelengths absorbed / nm	Pigment colour	Occurrence in plants
Chlorophylls	chlorophyll a	435, 670–680	yellow-green	All (mosses, ferns, conifers, flowering plants)
	chlorophyll b	480, 650	blue-green	Higher plants (conifers, flowering plants)
Carotenoids	β-carotene	425–480	orange	All
	xanthophylls	400–500	yellow	Most

Link Pigment separation and identification is explained on p36.

Light harvesting

Absorption and action spectra

Key Terms

Absorption spectrum: A graph showing how much light is absorbed at different wavelengths.

Action spectrum: A graph showing the rate of photosynthesis at different wavelengths.

The different pigments can be shown to absorb different wavelengths of light by making separate solutions of each and shining light through them. An **absorption spectrum** is a graph that indicates how much light a particular pigment absorbs at different wavelengths. Chlorophylls a and b absorb light energy mainly in the red and blue-violet regions of the spectrum and reflect green, giving leaves their colour. The carotenoids, ß-carotene and the xanthophylls, absorb the light energy from the blue-green region of the spectrum, so they appear yellow-orange.

Absorption at a particular wavelength does not indicate, however, whether that wavelength is actually used in photosynthesis. An **action spectrum** is a graph that shows the rate of photosynthesis at different wavelengths of light, as measured by the mass of carbohydrate synthesised by plants exposed to different wavelengths.

If the action spectrum is superimposed on the absorption spectrum, a close correlation between the two can be seen. This suggests that the pigments responsible for absorbing the light are used in photosynthesis.

Graph showing the relationship between the absorption spectrum and the action spectrum

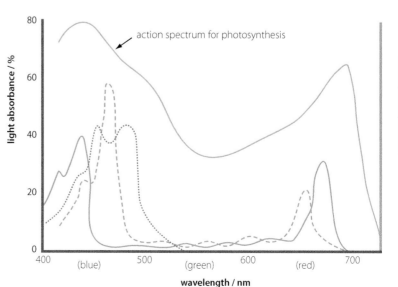

The photosystems

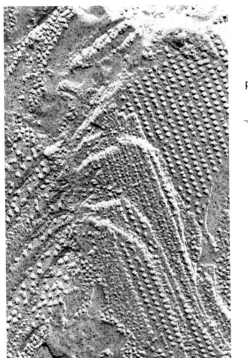

Thylakoid membrane of a chloroplast showing arrays of photosystems

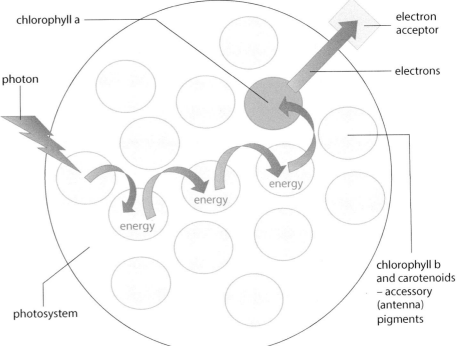

A photosystem

Photosystems lie in the plane of the thylakoid membrane. Each comprises:

- An antenna complex, containing the photosynthetic pigments. Chlorophylls and carotenoids are anchored into the phospholipids of the thylakoid membrane, held together by protein molecules in clusters of up to 400 molecules. Each cluster is called an **antenna complex**. The combination of pigments allows light at a range of wavelengths to be absorbed.

- A reaction centre, within the antenna complex. It contains two molecules of the primary pigment, chlorophyll a. When the chlorophyll a molecules absorb light, their excitation allows each one to emit an electron. There are two types of reaction centre:

  – Photosystem I (PSI) is arranged around a chlorophyll a molecule with an absorption peak of 700 nm. It is also called P700.

  – Photosystem II (PSII), which was discovered after PSI, hence the numbering, is arranged around a chlorophyll a molecule with an absorption peak of 680 nm. It is also called P680.

Some photons are absorbed by chlorophyll a directly but many are absorbed first by chlorophyll b and the carotenoids, which are the accessory (or antenna) pigments. The photons excite the accessory pigments and energy is passed through them to the reaction centre, where electrons of chlorophyll a are excited and raised to a higher energy level. Chlorophyll a is the most significant molecule of the reaction centre because it passes energy to the subsequent reactions of photosynthesis. It is referred to as the primary, or core, pigment.

WORKING SCIENTIFICALLY

Electron microscopists have examined chloroplast membranes using the freeze-fracture technique. Frozen chloroplasts fracture along their membranes, which can be covered with a thin layer of carbon and viewed in the electron microscope. Images produced in this way show an array of particles, which are the photosystems.

Key Term

Antenna complex: An array of protein and pigment molecules in the thylakoid membranes of the grana that transfer energy from light of a range of wavelengths to chlorophyll a, at the reaction centre.

▼ Study point

Each photosystem is a collection of accessory pigments, which absorbs light at various wavelengths and transmits the energy to a reaction centre.

The two stages of photosynthesis

Photosynthesis includes a sequence of reactions that take place on the thylakoid membranes, using light as an energy source and using water. This sequence is the **light-dependent stage** and it produces:

- ATP, which provides the chemical energy transduced from light energy, to synthesise energy-rich molecules such as glucose.
- Reduced NADP, which provides the reducing power to synthesise molecules such as glucose from carbon dioxide.
- Oxygen, a by-product, derived from water. Oxygen diffuses out of the chloroplast, out of the photosynthetic cells and out of the leaf through the stomata.

The reactions using ATP and reduced NADP, making molecules such as glucose, occur in solution in the stroma. They can happen in the light but do not require it. These reactions constitute the **light-independent stage**, and include a cycle of reactions called the Calvin cycle, named after one of their discoverers.

The light-dependent stage of photosynthesis

Photophosphorylation

Phosphorylation is the addition of a phosphate ion to ADP. The term **photophosphorylation** implies that the energy for this reaction comes from light. Photosynthesis has two pathways for photophosphorylation. Non-cyclic photophosphorylation involves both PSI and PSII and the pathway of electrons is linear. Cyclic photophosphorylation uses PSI only and electrons go through a cycle.

The passage of electrons

Cyclic photophosphorylation

- PSI absorbs photons, which excites electrons in the chlorophyll a molecules in its reaction centre.
- These are emitted and picked up by an electron acceptor, which passes them down a chain of electron carriers back to PSI. The energy released as electrons pass through the electron transport chain phosphorylates ADP to ATP.
- Electrons have flowed from PSI to the electron acceptor, back to PSI so this phosphorylation is described as cyclic photophosphorylation.

Non-cyclic photophosphorylation

In an alternative pathway, electrons are transferred from the electron acceptor to oxidised NADP in the stroma, which, with protons from the photolysis of water, is reduced:

- The electrons have not been returned to PSI so its chlorophyll is left with a positive charge.
- The positive charge is neutralised by electrons from PSII. They have been excited to a higher energy level by light absorption, picked up by an electron acceptor and passed down the electron transport chain to PSI.

- Electron passage down the electron transport chain makes energy available for phosphorylation of ADP. As the electrons from PSII move is one direction only (PSII → electron acceptor → electron transport chain → PSI), this is non-cyclic photophosphorylation.

- The chlorophyll in PSII is left with a positive charge and that is neutralised by the electrons released in the photolysis of water.

▼ **Study point**

NADP is reduced in non-cyclic photophosphorylation only.

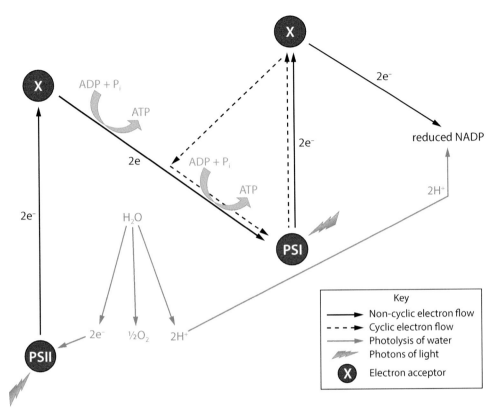

Key Term

The Z scheme: The pathway taken by electrons in non-cyclic photophosphorylation.

Exam tip

Learn the Z scheme diagram in stages:
- Remember the electron flow for non-cyclic photophosphorylation first.
- Then the photolysis of water.
- Then the electron flow for cyclic photophosphorylation.
- Last, the formation of ATP and reduced NADP.

Summary of light-dependent stage

If you turn the page through 90°, you can see that the passage of electrons is drawn like the letter Z, so this passage of electrons is called the **Z scheme**.

Photolysis of water

In the thylakoid spaces, water molecules absorb light, which indirectly causes them to dissociate into hydrogen, oxygen and electrons: $H_2O \rightarrow 2H^+ + 2e^- + \frac{1}{2}O_2$ The splitting of water by light is called **photolysis**. It is enhanced by a protein complex in PSII, which is the only known enzyme that causes water to be oxidised.

Key Term

Photolysis: The splitting of water molecules by light, producing hydrogen ions, electrons and oxygen.

- The electrons produced replace those lost from PSII.

- The protons from water and electrons from PSI reduce NADP.

- Oxygen diffuses out of the chloroplast and cell, out through the stomata as a waste product.

The passage of protons and phosphorylation

- As electrons pass through a proton pump in the thylakoid membrane, they provide energy to pump protons from the stroma into the thylakoid space. The protons join H⁺ ions from the photolysis of water and accumulate. They generate an electrochemical gradient, since there are more inside the thylakoid space than there are outside, in the stroma. This gradient is a source of potential energy.

⟨Link⟩ Nitrifying bacteria convert ammonium ions to nitrites and nitrites to nitrates, releasing protons and electrons. Energy is released as electrons pass along an electron transport chain and chemiosmosis generates ATP from the proton gradient formed.

- Chemiosmosis occurs. The H^+ ions diffuse down their electrochemical gradient through ATP synthetase in the thylakoid membrane, into the stroma. This makes available the energy derived from light and carried by the electrons. As they pass through ATP synthetase, ADP is phosphorylated to ATP.
- Once in the stroma, H^+ ions are passed to oxidised NADP, reducing it: $NADP + 2H^+ + e^- \longrightarrow$ reduced NADP. This removal of H^+ ions, in conjunction with the proton pump, contributes to maintaining the proton gradient across the thylakoid membranes.

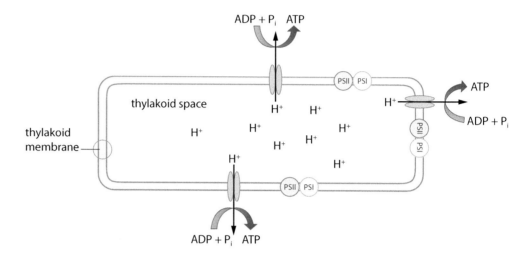

ATP production in the chloroplast by chemiosmosis

In summary, three factors maintain the proton gradient between the thylakoid space and the stroma:

- The proton pump associated with the electron transport chain pushing protons into the thylakoid space.
- The photolysis of water in the thylakoid space.
- The removal of protons from the stroma reducing NADP.

The light-dependent stage of photosynthesis can be summarised as:

$$NADP + H_2O + 2ADP + 2P_i \longrightarrow \text{reduced NADP} + \tfrac{1}{2}O_2 + 2H_2O + 2ATP$$

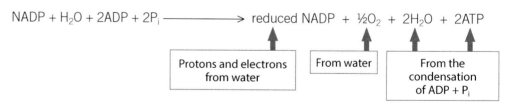

		Type of photophosphorylation	
		Cyclic	Non-cyclic
Photosystem		PSI only	PSI and PSII
Chlorophyll at reaction centre		P700	P680
Electron flow		Cyclic	Linear
ATP production		✓	✓
Photolysis of water		✗	✓
Oxygen production		✗	✓
NADP reduction		✗	✓
Occurrence		All photosynthetic organisms	Plants, algae, cyanobacteria

In 1939 Robert Hill showed that isolated chloroplasts produce oxygen from water in the presence of an oxidising agent. This is the Hill Reaction. In the cell, NADP is the oxidising agent that removes hydrogen from water. In the laboratory, the blue dye DCPIP acts as a substitute for NADP, and, in the presence of light loses its colour when it is reduced:

$$\text{oxidised DCPIP} \xrightarrow{\text{light + chloroplasts}} \text{reduced DCPIP}$$

dark blue   colourless

The light-independent stage of photosynthesis

The light-independent stage occurs in solution in the stroma of the chloroplast and it involves many reactions, each catalysed by a different enzyme. The reactions use the products of the light-dependent stage:

- ATP is a source of energy.
- Reduced NADP is the source of the reducing power, reducing carbon dioxide.

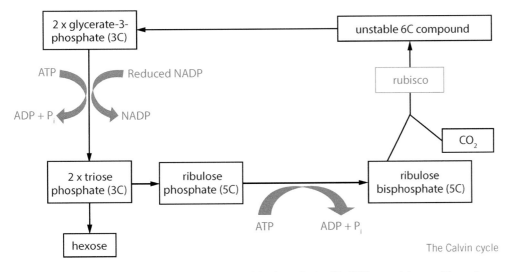

The Calvin cycle

> **YOU SHOULD KNOW ›››**
>
> ››› How carbon dioxide is incorporated into organic molecules
>
> ››› The roles of ATP and reduced NADP in the light-independent reactions
>
> ››› The importance of the regeneration of the carbon dioxide acceptor
>
> ››› How carbohydrates, fats and proteins are made from the products of the Calvin cycle

- A five-carbon acceptor molecule, ribulose bisphosphate (RuBP), combines with carbon dioxide, catalysed by the enzyme ribulose bisphosphate carboxylase, abbreviated to rubisco. It is the most abundant protein in the biosphere, and the high concentration reflects its importance.
- An unstable six-carbon compound is formed.
- The six-carbon compound immediately splits into two molecules of a three-carbon compound, glycerate-3-phosphate (GP).
- GP is reduced to triose phosphate by reduced NADP. Reducing a molecule requires energy and in this case, the energy is provided by the ATP made in the light-dependent stage. Triose phosphate is the first carbohydrate made in photosynthesis.
- NADP is reformed.
- Some of the triose phosphate is converted to glucose phosphate, and then into starch by condensation.
- Most of the triose phosphate goes through a series of reactions which regenerates RuBP so the cycle can continue. ATP made in the light-dependent stage provides the energy for this to happen.

▼ **Study point**

For every six molecules of triose phosphate formed, five are used to regenerate ribulose bisphosphate and only one molecule is converted to glucose.

Knowledge check

Identify the missing word or words.

In the light-independent reaction, carbon dioxide combines with the 5-carbon acceptor,................ ATP and reduced .... from the .... stage are used to produce ....

WORKING SCIENTIFICALLY

The sequence of reactions in the light-independent stage of photosynthesis is known as the Calvin cycle because it was worked out by Calvin and his associates. They used ^{14}C, a radioisotope of carbon, incorporated into hydrogen carbonate ions, $H^{14}CO_3^-$, as a source of carbon dioxide for photosynthesis by the protoctistan *Chlorella*. $H^{14}CO_3^-$ was added to a flat lollipop vessel at the top. The *Chlorella* photosynthesised and every 5 seconds, samples were dropped into hot methanol, to stop any further chemical reactions. The compounds produced were separated by chromatography and identified.

The lollipop apparatus

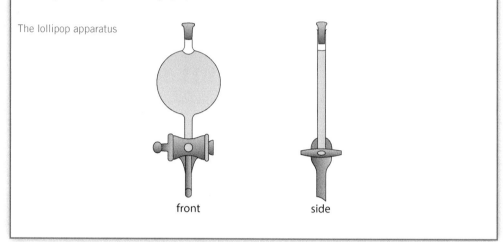

front   side

◀Link▶ Acetyl coenzyme A is an important molecule in the respiratory pathway. Its role in respiration is described on p45.

Product synthesis

Whereas animals eat to obtain the raw materials for energy and growth, photosynthetic organisms can make all the molecules that they need. Like animals, they require fats, proteins and carbohydrates, but these can all be made, more or less directly, from the 3C compounds produced in the Calvin cycle.

Carbohydrates: the first hexose made is fructose phosphate. This can be converted to glucose and combined with the glucose to make sucrose, for transport around the plant. The α-glucose molecules may be converted to starch, for storage, or to β-glucose, which is polymerised into cellulose for cell walls.

Fats: acetyl coenzyme A (AcCoA) can be synthesised from glycerate-3-phosphate, made in the Calvin cycle, and converted to fatty acids. Triose phosphate can be converted directly to glycerol, another 3C compound. Fatty acids and glycerol undergo condensation reactions to form triglycerides.

Proteins: glycerate-3-phosphate can also be converted into amino acids for protein synthesis. The amino group is derived from NH_4^+ ions, made from nitrate ions (NO_3^-) taken in at the roots and transported throughout the plant.

Limiting factors in photosynthesis

Plants need a suitable environment to be efficient at photosynthesis. They need:

- The reactants carbon dioxide and water
- Light at a high enough intensity and of suitable wavelengths
- A suitable temperature.

If any of these factors is lacking, photosynthesis cannot take place. Each factor has an optimum value, at which the rate of photosynthesis is highest. If any is at a sub-optimal level, the rate of photosynthesis is reduced. In that case, the value of that factor is controlling the rate of photosynthesis. If it is higher, photosynthesis will happen faster. The factor is a **limiting factor**, because it is limiting, or controlling, the rate of photosynthesis.

Carbon dioxide concentration

As the carbon dioxide concentration increases from zero, the rate of the light-independent reactions increases, and so the rate of photosynthesis increases, showing that carbon dioxide concentration is a limiting factor. If the concentration is increased above about 0.5%, the rate of photosynthesis remains constant, implying that carbon dioxide concentration is not affecting the rate of photosynthesis and is therefore not a limiting factor at these concentrations. The rate decreases above about 1% as the stomata close, preventing carbon dioxide uptake.

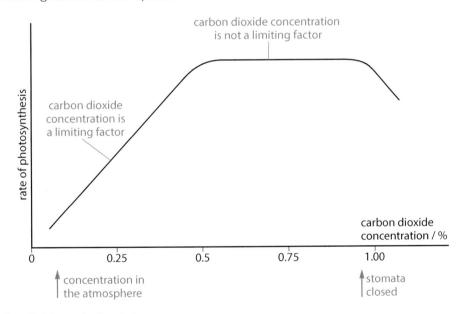

Carbon dioxide as a limiting factor

Carbon dioxide concentration is usually a limiting factor for photosynthesis in terrestrial plants, because its concentration in the air is about 0.04%. The photosynthesis of crop plants is most efficient at 0.1% carbon dioxide. Experiments with tomatoes show even faster photosynthesis at 0.5%, but only in the short term. Aquatic plants and algae use carbon dioxide from the hydrogen carbonate ion, HCO_3^-. Some algae can increase the concentration of carbon dioxide in their cells with carbonic anhydrase, so that it is not a limiting factor in their photosynthesis. The optimum concentration for algae is approximately 0.1 mol dm^{-3} i.e. about 0.1%.

The rate of the slowest reaction in a sequence determines the overall rate of the process. It is called the rate-limiting step. In the light-independent reactions of photosynthesis, the reaction catalysed by rubisco is the rate-limiting step.

YOU SHOULD KNOW ›››

››› The definition of a limiting factor

››› That carbon dioxide concentration, light intensity and water availability can act as limiting factors in photosynthesis

Key Term

Limiting factor: A factor that limits the rate of a physical process by being is short supply. An increase in a limiting factor increases the rate of the process.

Going further ▶

Carbon dioxide diffuses into leaves and dissolves, making carbonic acid. H$^+$ ions formed from its dissociation decrease the pH, denaturing proteins. Stomata close at 1% carbon dioxide and prevent this.

 ‹Link› In the first year of this course you learned that carbonic anhydrase has a role in red blood cells, in the transport of carbon dioxide.

Going further ▶

Plant breeders are trying to improve the efficiency of rubisco, to increase the rate of the light-independent stage and, consequently, the yield of crop plants. It would be useful to have varieties of rubisco that continue to function efficiently as global temperatures increase.

Light intensity

Light intensity is a significant factor in controlling the rate of photosynthesis.

- If the plant is in darkness, the light-independent reactions of photosynthesis are still possible but the light-dependent reactions are not, and so no oxygen is evolved.
- As the light intensity increases, the light-dependent reactions occur with increasing efficiency, and so overall, the rate of photosynthesis increases. Light intensity is controlling the rate of photosynthesis and so it is a limiting factor.
- At a certain intensity, around 10 000 lux, the reactions of the light-dependent stage are at their maximum rate. Higher light intensity does not produce faster reactions and so the rate of photosynthesis remains constant. Light intensity is not a limiting factor.
- If the light intensity is even higher, the rate of photosynthesis will decrease because the photosynthetic pigments are damaged. They will not absorb light so efficiently and so the light-dependent stage fails.

▼ Study point

The light intensity on a bright summer's day in Wales is about 100 000 lux, so for exposed plants, the light intensity is rarely a limiting factor in growth.

Going further ▶

Some plants, e.g. bramble, have sun leaves and shade leaves, depending on the conditions in which the leaves develop. Sun leaves have fewer grana and less chlorophyll in their chloroplasts, although their rate of photosynthesis can be five times that of shade leaves. Sun leaves often have extra layers of palisade cells, making them thicker than shade leaves, although their area may be smaller.

(graph: y-axis labelled "rate of photosynthesis"; x-axis labelled "light intensity"; annotations "light intensity is not a limiting factor", "light intensity is a limiting factor", "light saturation at about 10 000 lux", "chloroplast pigments bleached")

Light intensity as a limiting factor

Sun and shade plants

Different species have adapted so that their photosynthesis is most efficient at different light intensities. Sun plants, such as *Salvia*, are most efficient at photosynthesis in high light intensity and shade plants such as lily of the valley are most efficient at low light intensity.

Salvia

Lily of the valley

Photosynthesis can be assessed in terms of carbon dioxide uptake into photosynthetic cells. As the light intensity decreases, the rate of the light-dependent reactions decreases. The rate of the light-independent reactions also decreases and so the rate of carbon dioxide uptake decreases. At a particular light intensity, so little carbon dioxide is needed that respiration provides all that is required and none is absorbed. Similarly, all the oxygen needed for respiration is provided by photosynthesis. There is, therefore, no gas exchange. The light intensity at which this happens is called the **light compensation point**. It occurs at a lower light intensity for shade plants than for sun plants, so they can grow in more shaded habitats than sun plants.

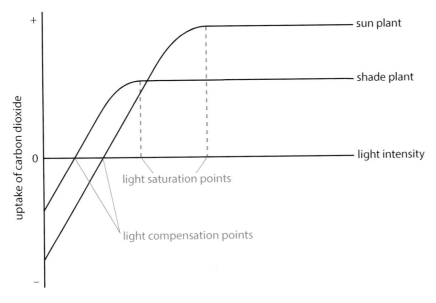

Light compensation point

Temperature

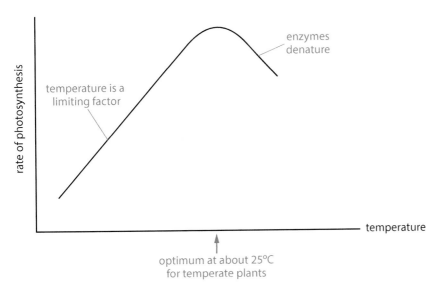

Temperature as a limiting factor

Increased temperature increases the rate of photosynthesis because the kinetic energy of the molecules involved increases. Above a particular temperature, which is different in different species, the enzymes progressively denature and the rate of photosynthesis decreases. Thus, temperature does control the rate of photosynthesis and it is therefore a limiting factor.

Key Term

Light compensation point, sometimes called just compensation point: The light intensity at which a plant has no net gas exchange as the volume of gases used and produced in respiration and photosynthesis are equal.

Going further ▶

Plants only grow well within a limited temperature range, e.g. warm-season vegetables, such as tomatoes, grow best between 16°C and 27°C and cool-season vegetables, such as lettuce and spinach, grow best between 10°C and 21°C.

⟨Link⟩ You learned about the effect of temperature on enzyme-controlled reactions in the first year of this course.

Going further ▶

The implications for crop production are significant as water availability in crop growing areas decreases in response to climate change.

Water

When water is scarce, a plant's cells plasmolyse, stomata close, wilting occurs and many physical functions are affected. Experiments with water-deficient plants show that even slight water deprivation can reduce the carbohydrate made, so water availability is a limiting factor in photosynthesis. But as so many systems are affected, it is not straightforward to see its effect on photosynthesis alone.

Limiting factors can combine

Sometimes, a plant has optimal values for all these environmental factors and then its photosynthesis will occur efficiently and it will grow well. But if a plant has several of these factors limiting, for example a wood anemone in an oak forest on a cool spring morning, when temperature and light intensity are low, its ability to perform photosynthesis is reduced and it will grow slowly. The actual rate of photosynthesis is controlled by the factor that is nearest to its minimum value.

Wood anemones

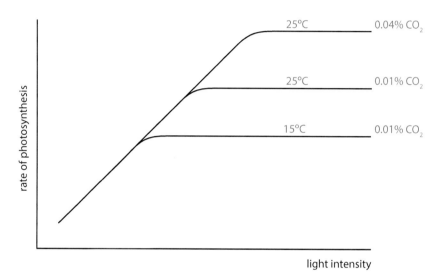

Limiting factors can combine

Going further ▶

Limiting factors apply to all processes and living organisms, not just to photosynthesis in higher plants. Light intensity is a major limiting factor in aquatic communities. The photic zone of a body of water represents the depth to which light penetrates, rarely more than 150 m, and so the organisms that can photosynthesise are limited to these surface waters. Very little light penetrates through the 'dysphotic' zone and in the 'aphotic' zone no light penetrates at all. Without photosynthesis, the only oxygen below the photic zone is that which has diffused down from the surface, so organisms there have a low metabolic rate. Only the photic zone can support a grazing food web. Below it, organisms have to employ unusual techniques to obtain food, such as chemosynthesis, described on p72, and feeding on detritus, such as in whale-fall. The whale carcasses may be as deep as 3000 m and, in complete darkness, they support characteristic animal communities including the snot-worm, *Osedax*, that bores into the bones for their food, Crustacea such as giant isopods, squat lobsters, crabs and shrimp, fish such as hagfish and sleeper sharks, and the sea cucumbers.

Mineral nutrition

Inorganic nutrients are needed by plants and may be limiting factors for metabolism if in short supply. They have various roles:

- Structural, e.g. calcium in the middle lamella of cell walls.
- Synthesis of compounds needed for the growth of the plant, e.g. as enzyme activators, such as the magnesium requirement by ATPase and DNA polymerase.
- They may form an integral part of a molecule, e.g. magnesium in chlorophyll, iron in the carriers of the electron transport chain and manganese in photosystem II.

Macronutrients include nitrogen, potassium, sodium, magnesium, calcium, nitrate and phosphate. They are required in greater quantities than micronutrients, for example manganese and copper.

YOU SHOULD KNOW ›››

››› The roles of nitrogen and magnesium in plant metabolism

Exam tip

Nitrogen and magnesium are the only minerals you are required to know about.

Nitrogen

Most of the nitrogen in the soil is in the humus, in organic molecules of decaying organisms. Inorganic nitrogen occurs as ammonium ions, NH_4^+ or nitrate ions, NO_3^-. Nitrogen is largely taken up by roots as nitrates, although *Rhizobium* in root nodules delivers ammonium ions to the plant. The ions are transported in the xylem and delivered to the cells. Nitrate is converted into ammonium ions, which become the amino ($-NH_2$) groups of amino acids. Amino acids are transported in the phloem and used for the synthesis of proteins, chlorophylls and nucleotides.

Because of its role in protein and nucleic acid synthesis, symptoms of nitrogen deficiency include reduced growth in the whole plant. Nitrogen is a component of chlorophyll and so its deficiency also causes chlorosis, a yellowing of the leaves due to inadequate chlorophyll production. Chlorosis first appears in the older leaves.

Chlorotic leaves of *Brassica napus*

Magnesium

Magnesium is absorbed as Mg^{2+} and it is transported in the xylem. It is required by all tissues, but especially leaves. Magnesium forms part of the chlorophyll molecule and so the main symptom of magnesium deficiency is chlorosis. This begins between the veins of older leaves as existing magnesium in the plant is mobilised and transported to newly formed leaves. Magnesium ions are also important enzyme activators such as for ATPase.

4

Knowledge check

Match the terms 1–4 with the descriptions A–D.

1. Limiting factor.
2. Shade plant.
3. Chlorosis.
4. Compensation point.

A. Yellowing of leaves related to insufficient magnesium.
B. Light intensity at which a plant has no net gas exchange.
C. A plant that performs photosynthesis most efficiently at low light intensity.
D. A factor which, when increased in value, directly increases the rate of photosynthesis.

Specified practical exercises

Investigation into the separation of chloroplast pigments by chromatography

Rationale

Flowering plants have two main groups of chloroplast pigments, the chlorophylls and the carotenoids, located on the membranes of the stromal lamellae and the grana. They can be extracted by dissolving in an organic solvent and can be separated by chromatography.

In the technique of paper chromatography, a mixture in solution is applied to paper with very fine, evenly sized pores, called chromatography paper. The solution moves through the channels by capillary action, and as it does so, the solvent evaporates so that the air surrounding it is saturated with its vapour. Different solutes come out of solution at different positions up the chromatogram. The more soluble the solute is and the less it adsorbs to the chromatography paper, the further it will travel before coming out of solution. If a suitable solvent is used, all the components of the mixture precipitate at different positions.

This means that the appearance of a chromatogram depends on:

- The solvent used for extracting the mixture to be separated.
- The solvent used to separate the mixture.
- The physical and chemical properties of the chromatography paper.

The same principles apply if a silica gel is used instead of chromatography paper. It is only meaningful, however, to compare chromatograms that have been prepared identically.

Apparatus

- Spinach leaves
- Scissors
- Sand: provides abrasion, which enhances the breaking of cells
- Pestle
- Mortar
- Propanone: to extract the pigments and as a component of the separation mixture
- Distilled water
- Petroleum ether (40–60ºC fraction): a component of the separation mixture
- Chromatography paper strips, 20 mm × 300 mm
- Pencil
- Ruler
- Capillary tube
- Hair drier
- Boiling tube
- Stopper
- Dropping pipette
- Vial

Method

To prepare the pigment solution

1. Grind 2 g finely chopped spinach leaves with a pinch of sand in 5 cm³ propanone, to make a slurry.
2. Shake the preparation vigorously with 3 cm³ distilled water.
3. Leave for 8 minutes.
4. Mix with 3 cm³ petroleum ether, by shaking gently.
5. Stand to let the layers separate.
6. Transfer the upper, petroleum ether, layer which contains the chloroplast pigments, to a vial, using a dropping pipette.

To prepare the chromatography paper

1. Draw a pencil line across the chromatography paper 20 mm from one end.
2. Draw chloroplast pigment solution into a capillary tube and put a small spot in the centre of the pencil line. Ensure that the capillary tube does not pierce or tear the chromatography paper.
3. Dry the spot with a hair drier as quickly as possible, preventing its spread.
4. Repeat until there is a small, intense spot of pigment.

To run the chromatogram

1. Place freshly made 2:1 propanone : petroleum ether in the boiling tube to a depth of 5 mm.
2. Slide the chromatography paper into the boiling tube, taking care that it does not touch the sides of the boiling tube, so that its end is just below the surface of the ethanol but the spot is above, and not touching it.
3. Fold the end of the chromatography paper over the rim of the boiling tube and hold it in place with a rubber stopper.
4. Without moving the boiling tube, allow the solvent to climb up the paper until it is 10 mm from the top.
5. Remove the chromatography paper from the boiling tube and immediately, draw a pencil line across the paper to mark the solvent front.
6. Allow the chromatography paper to dry.

Risk assessment

Hazard		Risk	Control measure
Propanone Petroleum ether	May cause eye damage	Macerating leaf material; Pouring solvent for chromatography	Eye protection
	May degrease the skin	Macerating leaf material; Pouring solvent for chromatography	Wear gloves
	Inhalation may exacerbate respiratory problems	Macerating leaf material; Pouring solvent for chromatography	Work in fume cupboard
	Fire hazards	Accidental ignition	Work in fume cupboard

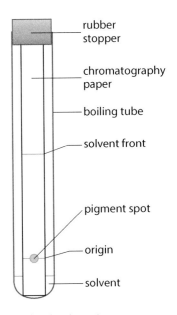

rubber stopper

chromatography paper

boiling tube

solvent front

pigment spot

origin

solvent

Running the chromatogram

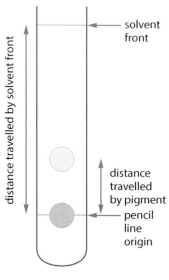

Calculating R_f

To identify the pigments

- Measure the distance from the origin to the solvent front.
- Measure the distance from the origin to the centre of each pigment spot.
- Calculate R_f for each pigment, where $R_f = \dfrac{\text{distance travelled by pigment}}{\text{distance travelled by solvent front}}$
- Published data allows the pigments to be identified as R_f is constant for each pigment in a given solvent. The table here shows data for separation by paper chromatography in 2:1 propanone:petroleum ether.

Spot colour	Pigment	R_f
yellow	β-carotene	0.96
yellow-brown	xanthophyll	0.75
grey	phaeophytin	0.70
blue-green	chlorophyll a	0.58
green	chlorophyll b	0.48

Results

- If the distance from the origin to the solvent front = 185 mm, use the following measurements to calculate the R_f of each pigment
- Use the table above to suggest the identity of each pigment.

Spot number	Distance from origin / mm	$R_f = \dfrac{\text{distance travelled by pigment}}{\text{distance travelled by solvent front}}$	Pigment
1	176		
2	144		
3	124		
4	104		
5	92		

Further work

- Test the effect of different periods of darkness on leaf pigments.
- Compare pigments in deciduous leaves at different times of the year.
- Investigate the effect of magnesium or iron deficiency on leaf pigments by growing plants from seed in magnesium-free or iron-free solution.

Investigation into the role of nitrogen and magnesium in plant growth

Rationale

Plants absorb nitrate ions and magnesium ions at their roots. The nitrate provides the nitrogen needed for amino acid, nucleotide and chlorophyll synthesis. Magnesium ions are needed to make chlorophyll and also to act as enzyme activators. Without these ions, growth and development are compromised.

Duckweed, *Lemna sp.,* is a very small flowering plant, lacking true stem and roots. The plant body comprises fronds with a small number of roots hanging down into the water. It normally reproduces asexually, from a meristem at the base of the frond.

Apparatus

- Sterile liquid culture media:
  - Complete medium, containing cations Ca^{2+}, Mg^{2+}, Na^+, K^+ and Fe(III) and anions SO_4^{2-}, PO_4^{3-}, Cl- and NO3.
  - Medium lacking nitrate ions
  - Medium lacking magnesium ions
- 9 cm sterile Petri dishes
- Duckweed, *Lemna sp.*
- 1% sodium hypochlorite
- Sterile distilled water
- Sterile 250 cm^3 beakers
- Sterile bacterial loop

A duckweed plant

Method

To surface sterilise the duckweed

- Place plants in 100 cm^3 1% sodium hypochlorite in a beaker for 30 seconds, and swirl gently for 1 minute.
- Tip off sodium hypochlorite and flood plants with sterile distilled water. Swirl gently for 1 minute.
- Tip off the distilled water and flood plants again with sterile distilled water. Swirl gently for 1 minute.
- Tip off the distilled water and replace with fresh distilled water.

To culture the plants

Using sterile technique:

- Place 5 cm^3 medium in each Petri dish.
- Transfer 20 *Lemna* individuals to each dish, using a sterile bacterial loop to lift them from below, to prevent damage.
- Place in ambient light at room temperature.
- Monitor over a two-week period.

Risk assessment

Hazard	Risk	Control measure
Culture solution may be a skin or eye irritant	Contact may occur in preparation or pouring	Wear eye protection and cover skin

Results

Medium	Description of plants
Complete	Plants green; new fronds have formed; several roots visible
Lacking nitrate	Leaves yellow-green; no new fronds; roots few and short
Lacking magnesium	Leaves with yellow patches; some have new fronds; several roots visible

Further work

- The dry mass of leaves could be tested following incubation in different media.
- Pigments could be extracted from the leaves, separated by chromatography and identified.

Investigation into the effect of carbon dioxide concentration on the rate of photosynthesis

Rationale

The effect of carbon dioxide concentration on the rate of photosynthesis in pondweed can be assessed by measuring the volume of oxygen produced in a photosynthometer. The carbon dioxide is provided by sodium hydrogen carbonate solution. It is used in the light-independent stage of photosynthesis, in the rate limiting step:

$$\text{ribulose bisphosphate + carbon dioxide} \xrightarrow{\text{rubisco}} \text{unstable 6C intermediate.}$$

Design

Variable	Name of variable	Value of variable		
Independent	Concentration of carbon dioxide	0, 0.01, 0.1, 0.5, 1 mol dm^{-3} sodium hydrogen carbonate		
Dependent	Volume of oxygen	Measured in cm^3		
Controlled	Light intensity	Same light bulb at constant distance throughout the experiment		
	Temperature	Boiling tube in water bath, monitored using a thermometer		
Control	Repeat experiment under identical conditions with boiled pondweed.			
Reliability	Perform 3 readings at each concentration of carbon dioxide and calculate a mean volume of oxygen.			
Risk assessment	**Hazard**	**Risk**	**Control measure**	
	Sodium hydrogen carbonate may cause skin and eye irritation	When making or pouring solutions	Solid to be weighed in fume cupboard	
	Gas accumulation in culture vessel could cause the glass to break	During experimental period	Wear eye protection	
	Excess heat from lamp may cause burns	When handling apparatus	Ensure no contact with skin	

Method

- Place pondweed such as hornwort (*Ceratophyllum demersum*) or red cabomba (*Cabomba furcata*), under lights, with a bubbled air supply for several hours prior to the experiment.
- Place 30 cm^3 sodium hydrogen carbonate solution in the boiling tube.
- Place a piece of pondweed in the boiling tube, with the stem uppermost.
- Cut off the end of the stem with sharp scissors.
- Place the boiling tube in a beaker of water, to maintain its temperature, which can be monitored with a thermometer.
- Fill the syringe barrel with hydrogen carbonate solution and push in the plunger so that all the air is removed from the apparatus and the tube is entirely filled with solution.
- Place the funnel at the end of the tube over the cut end of the pondweed. Use curved forceps ensure the end of the pondweed is under the funnel.
- Place a lamp 200 mm from the beaker holding the boiling tube, using a heat shield if the bulb is incandescent.
- Allow the pondweed to equilibrate for 5 minutes.
- Push the plunger in to force out of the tube any gas that has accumulated during the equilibration.
- Start the timer.

- After 3 minutes, withdraw the plunger of the syringe so that the bubble of gas collected is in the capillary tube adjacent to the graduated scale.
- If the scale is graduated in cm^3, read the volume of the gas bubble directly. If the scale is graduated in cm, the volume may be found:
  - Measure the diameter of the capillary tube to find the radius (radius, $r = \dfrac{diameter}{2}$).
  - Calculate the area of cross section (area, $A = \pi r^2$).
  - Calculate the volume of the bubble (volume, $V = \pi r^2 \times$ length of the bubble).
- Push the syringe barrel in so that the gas bubble is forced into the solution in the boiling tube to take another reading.

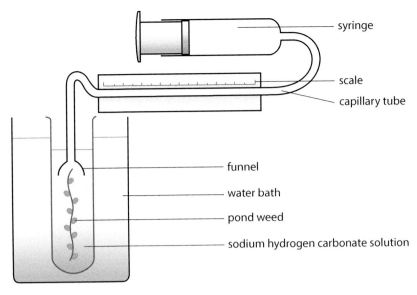

Photosynthometer

Source of error

The volume of oxygen produced is an under-estimate because:

- Some of the oxygen produced in photosynthesis is used in respiration.
- Oxygen is slightly water-soluble.

Thus there is a small, but consistent, error in volume measurement.

Sample results

Concentration of sodium hydrogen carbonate / mol dm⁻³	Volume of oxygen collected in 3 minutes / cm³				Volume of oxygen collected per minute / cm³
	1	2	3	Mean	
0	0	0	0	0	0
0.1	2.0	2.6	2.6		
0.5	4.0	3.9	3.5		
0.75	3.9	3.6	3.6		
1	4.1	3.5	3.8		

- Complete the table and plot the data on a graph.
- Determine the concentration of sodium hydrogen carbonate solution at which the reactions of the light-independent stage of photosynthesis become constant. This can be thought of as the concentration at which rubisco, catalysing the rate-limiting step in photosynthesis, becomes saturated.

Further work

- Test the effect of temperature by placing the boiling tube containing the pondweed in water baths maintained at different temperatures.
- Test the effect of light intensity by moving the lamp to different distances from the pondweed.
- Test the effect of the wavelength of light by covering the beaker with different coloured filters. So that the pondweed receives the same light intensity at each wavelength, readings must be taken with a light meter covered with the filters over the probe, to find distance of the lamp, for each wavelength, which produces the same light intensity.

CH 3

Respiration releases chemical energy in biological processes

Chemical energy incorporated into organic molecules in photosynthesis is released in respiration. If energy-rich molecules were broken down directly, energy release would be uncontrollable. Instead energy is transferred to an intermediate, ATP, in a sequence of oxidation-reduction reactions, which release the energy in small, manageable quantities. ATP is a reservoir of potential chemical energy and acts as a common intermediate in metabolism, linking energy-requiring and energy-yielding reactions. The breakdown of energy-rich compounds in the absence of oxygen releases far less energy than in its presence.

Topic contents

By the end of this topic you will be able to:

- Outline respiration as a series of four distinct but linked stages: glycolysis, the link reaction, the Krebs cycle and the electron transport chain.

- Explain that glycolysis takes place in the cytoplasm and involves the breakdown of glucose to pyruvate with the production of ATP and reduced NAD.

- Describe the diffusion of pyruvate into the matrix of the mitochondrion and its conversion to acetyl coenzyme A by the link reaction.

- Describe the Krebs cycle as a series of reactions resulting in the formation of ATP and carbon dioxide, and reducing the coenzymes NAD and FAD.

- Explain that the Krebs cycle involves decarboxylation and dehydrogenation reactions.

- Explain that oxidative phosphorylation is located on the cristae of the mitochondria and involves electron carriers, proton pumps and ATP synthetase in the production of ATP.

- Describe how anaerobic respiration involves only glycolysis, with the conversion of pyruvate to ethanol and carbon dioxide in fungi and plants under certain conditions, and to lactate in animals.

- Understand that anaerobic respiration yields far less energy than aerobic respiration.

- Describe how fats and amino acids may be used as alternative energy sources.

Overview of respiration

Life can be defined in many different ways but one common feature in all models of life is the presence of metabolism. The term metabolism refers to all the reactions of the organism. Respiration is a metabolic pathway, which means it is a sequence of reactions controlled by enzymes. The reactions of respiration are catabolic. They break down energy-rich macromolecules, such as glucose and fatty acids. In respiration, C–C, C–H and C–OH bonds are broken and lower energy bonds formed. The energy difference allows the phosphorylation of ADP to ATP. ATP does not 'produce' energy, but when it is hydrolysed, it releases energy. This energy is available for use by the cell or is lost as heat.

There are three types of phosphorylation:

1. Oxidative phosphorylation, which occurs on the inner membranes of the mitochondria in aerobic respiration. The energy for making the ATP comes from oxidation-reduction reactions and is released in the transfer of electrons along a chain of electron carrier molecules.

2. Photophosphorylation, which occurs on the thylakoid membranes of the chloroplasts in the light-dependent stage of photosynthesis. The energy for making the ATP comes from light and is released in the transfer of electrons along a chain of electron carrier molecules.

3. Substrate-level phosphorylation, which occurs when phosphate groups are transferred from donor molecules, e.g. glycerate-3-phosphate to ADP to make ATP in glycolysis, or when enough energy is released for a reaction to bind ADP to inorganic phosphate, e.g. in the Krebs cycle.

Three groups of organisms are recognised, depending on their respiration:

- Most living organisms use aerobic respiration, and break down substrates using oxygen, with the release of a relatively large amount of energy. These are obligate aerobes.

- Some micro-organisms, including yeast and many bacteria, respire aerobically, but can also respire without oxygen; these are facultative anaerobes.

- Some species of bacteria and Archaea use anaerobic respiration. They respire without oxygen and cannot grow in its presence. They are obligate anaerobes.

▼ **Study point**

An energy currency molecule, such as ATP, acts as an immediate donor of energy to meet the metabolic needs of the cell. An energy storage molecule, such as starch, is a long-term store of potential chemical energy.

Key Terms

Aerobic respiration: The release of large amounts of energy, made available as ATP, from the breakdown of molecules, with oxygen as the terminal electron acceptor.

Anaerobic respiration: The breakdown of molecules in the absence of oxygen, releasing relatively little energy, making a small amount of ATP by substrate-level phosphorylation.

Going further ▶

Respiration is the release of energy from food. More specifically, it is the generation of ATP. Eukaryote cells have the most efficient biological mechanism for this. In essence, they remove electrons from glucose and pass them, in their mitochondria, to oxygen. But among the Archaea and Eubacteria, there is far more scope. The electron source does not have to be glucose. It does not even have to be organic, and there are cells that use minerals such as hydrogen sulphide, hydrogen gas or Fe(II) ions to obtain electrons. The terminal acceptor does not have to be oxygen. Some prokaryotes use nitrate, nitrite, sulphate or sulphite ions. The same bacterium can use different electron sources and sinks depending on its environment, because it can readily take in the genes that code for the necessary enzymes from other bacteria. A single bacterium may have more potential variety in its respiratory pathways than the entire Eukaryote domain.

▼ **Study point**

An acid in solution (e.g. pyruvic acid) makes an ion (e.g. pyruvate). Acids in the cell are dissolved, so the names of their ions, ending in '-ate', are used here, rather than the names of the acids.

Exam tip

Although glucose is rich in energy it is relatively unreactive unless phosphorylated.

Key Term

Dehydrogenation: The removal of one or more hydrogen atoms from a molecule.

Aerobic respiration

Aerobic respiration can be divided into four distinct but linked stages:

- Glycolysis, which occurs in solution in the cytoplasm and generates pyruvate, ATP and reduced NAD.
- The link reaction, in solution in the matrix of the mitochondrion. Pyruvate is converted into acetyl coenzyme A.
- The Krebs cycle, in solution in the mitochondrial matrix generates carbon dioxide and reduced NAD and FAD.
- The electron transport chain, on the cristae of the inner mitochondrial membrane, in which the energy from protons and electrons generates ATP from ADP and inorganic phosphate, P_i.

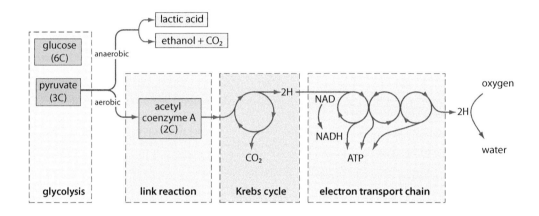

Summary of respiration

Glycolysis

Glycolysis is the initial stage of both aerobic and anaerobic respiration. Glycolysis occurs in the cytoplasm, because glucose cannot pass through the mitochondrial membranes. But even if it could, the enzymes for its breakdown are not present in the mitochondria and so it could not be metabolised there.

- A glucose molecule is phosphorylated by the addition of two phosphate groups, using two molecules of ATP, making a hexose phosphate called glucose diphosphate. As a result, the phosphorylated glucose is:
  - More reactive so less activation energy is required for the enzyme-controlled reactions.
  - Polar and, therefore, less likely to diffuse out of the cell.
- The glucose diphosphate splits into two molecules of a triose phosphate, a 3-carbon sugar, glyceraldehyde-3-phosphate.
- The two triose phosphate molecules are **dehydrogenated**, i.e. hydrogen is removed from each of them, oxidising them to pyruvate, also a 3-carbon molecule. The hydrogen atoms are transferred to NAD, a hydrogen carrier molecule, making reduced NAD. These steps release enough energy to synthesise four ATP molecules. The ATP is formed by substrate-level phosphorylation: the phosphate from the triose phosphate converts ADP to ATP, without the involvement of an electron transport chain, producing pyruvate.

Of the 4 ATPs made by substrate-level phosphorylation, 2 were used to phosphorylate the glucose molecule. Therefore there is a net gain of 2 ATPs from each molecule of glucose.

Two molecules of reduced NAD are also produced. If oxygen is available, each has the potential for the synthesis of an additional three molecules of ATP, making six altogether from the electron transport chain.

Some energy is lost as heat but a considerable amount of chemical potential energy remains in the pyruvate. If oxygen is available, some of this energy can be released via the Krebs cycle, in the mitochondria.

The link reaction

The link reaction links glycolysis to the Krebs cycle:

- Pyruvate diffuses from the cytoplasm into the mitochondrial matrix.
- The pyruvate is dehydrogenated and the hydrogen released is accepted by NAD to form reduced NAD.
- The pyruvate is also **decarboxylated**, i.e. a molecule of carbon dioxide is removed from it. All that remains of the original glucose molecule is a 2-carbon acetate group which combines with coenzyme A (CoA) , making acetyl coenzyme A (AcCoA), which enters the Krebs cycle.

Summary of the link reaction: pyruvate + NAD + CoA $\longrightarrow$ AcCoA + reduced NAD + CO_2

> **Key Term**
>
> **Decarboxylation:** The removal of a carboxyl group from a molecule, releasing carbon dioxide.

> *Exam tip*
>
> You do not need to know the names or formulae of the intermediates of respiration or the names of the proton pumps and electron carriers in the electron transport chain. The only enzyme names in the respiratory pathway you are required to know are dehydrogenase and decarboxylase.

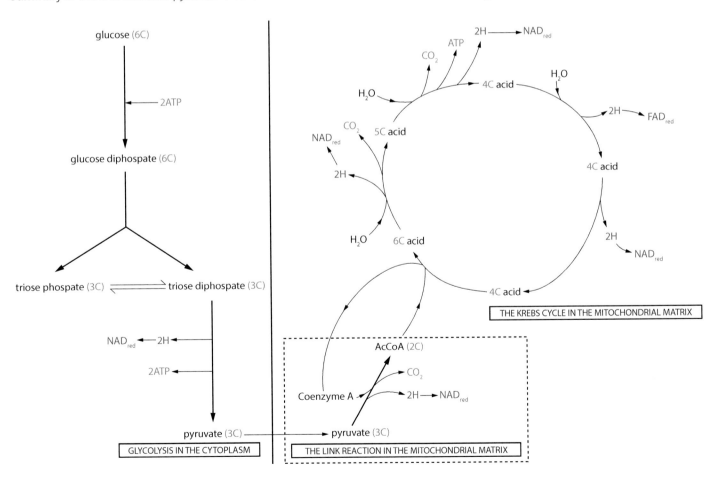

Summary of glycolysis, the link reaction and the Krebs cycle

The Krebs cycle

The Krebs cycle is a means of liberating energy from C–C, C–H and C–OH bonds. It produces ATP, containing the energy which was held in the chemical bonds of the original glucose molecule. It also produces reduced NAD and reduced FAD, which deliver hydrogen atoms to the electron transport chain on the inner mitochondrial membrane. Three molecules of water are used in reactions in the Krebs cycle. Carbon dioxide is released as a waste product.

- Acetyl CoA enters the Krebs cycle by combining with a 4-carbon acid, to form a 6-carbon compound, and the CoA is regenerated.
- The 6-carbon acid is dehydrogenated, making reduced NAD, and decarboxylated to make carbon dioxide and a 5-carbon acid.
- The 5-carbon acid is dehydrogenated, making reduced NAD and FAD, and decarboxylated to make carbon dioxide and to regenerate the 4-carbon acid.
- The 4-carbon acid can combine with more AcCoA and repeat the cycle.

In the Krebs cycle, there are two significant types of reaction:

- Decarboxylation happens twice. Decarboxylases remove carbon dioxide from the –COOH groups of Krebs cycle intermediates, as 6C acid $\rightarrow$ 5C acid $\rightarrow$ 4C acid.
- Dehydrogenation happens four times. Dehydrogenases remove pairs of hydrogen atoms from Krebs cycle intermediates. They are collected by hydrogen carriers giving three molecules of reduced NAD and one molecule of reduced FAD.

The acetate group from the original glucose molecule is now entirely broken down to carbon dioxide and water. The energy in the bonds of the glucose molecule is carried by electrons in the hydrogen atoms in the reduced NAD and FAD.

In summary, each turn of the Krebs cycle produces:

- One ATP produced by substrate-level phosphorylation
- Three molecules of reduced NAD
- One molecule of reduced FAD
- Two molecules of carbon dioxide.

The electron transport chain

The electron transport chain is located on the cristae of the inner mitochondrial membranes. It is a series of protein molecules that are carriers and pumps, which are sometimes called 'respiratory enzymes'. The carrier molecules include cytochromes and so the electron transport chain is sometimes called the cytochrome chain. Cytochromes are proteins conjugated to iron or copper and the metal ions are oxidised and reduced by electron transport.

The reactions they catalyse release energy, which is carried by ATP. Hydrogen atoms are carried into the electron transport chain by the **coenzymes** NAD and FAD. NAD feeds electrons and protons into the electron transport chain at an earlier stage than FAD does. Each pair of hydrogen atoms carried by reduced NAD provides enough energy to synthesise three molecules of ATP, using three proton pumps. Reduced FAD passes the hydrogen atoms directly to the second proton pump so the carrier system involving FAD has two pumps and produces two molecules of ATP for each pair of hydrogen atoms.

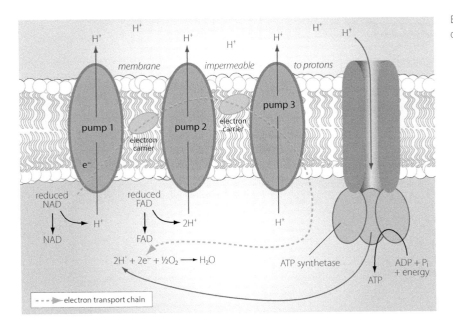

Electron transport
chain

The passage of electrons

- The reduced NAD donates the electrons of the hydrogen atoms to the first of a series of electron carriers in the electron transport chain.

- The electrons from these atoms provide energy for the first proton pump and protons from the hydrogen atoms are pumped into the inter-membrane space.

- The electrons pass along the chain of carrier molecules providing energy for each of three proton pumps in turn.

- At the end of the chain, the electrons combine with protons and oxygen to form water:

$$2H^+ + 2e^- + \tfrac{1}{2} O_2 \rightarrow H_2O$$

The passage of protons

- The inner membrane is impermeable to protons and so the protons accumulate in the inter-membrane space.

- The concentration of protons in the inter-membrane space becomes higher than in the matrix, so a gradient of concentration and charge is set up, and maintained by the proton pumps.

- In the membrane are protein complexes, which are channels through which protons flow back into the mitochondrial matrix. The enzyme ATP synthetase is associated with each channel. Protons diffuse back through these channels, and as they do so their electrical potential energy produces ATP: $ADP + P_i \rightarrow ATP + H_2O$.

- At the end of the chain, the protons combine with electrons and oxygen to form water.

Oxygen is referred to as the 'final electron acceptor' or the 'final hydrogen acceptor' in the electron transport chain. It is essential as it removes protons and electrons: oxygen is reduced by the addition of hydrogen ions and electrons, making water. Cyanide is a non-competitive inhibitor of the final carrier in the electron transport chain. In its presence, electrons and protons cannot be transferred to water. They accumulate, destroying the proton gradient. ATP synthetase cannot operate and the cell dies very quickly.

For each molecule of glucose entering glycolysis the electron transport system receives

- $10NAD_{red}$ which generates $10 \times 3 = 30$ ATP

- $2FAD_{red}$ which generates $2 \times 2 = 4$ ATP

▼ **Study point**

The proton gradient in mitochondria is established by three proton pumps. The gradient in photosynthesis is established by only one.

5

Knowledge check

Identify the missing word or words:

The first stage in the breakdown of glucose is a process called .... This takes place in the .... of the cell and results in the net production of molecules of ATP and .... If oxygen is present, this product diffuses into the .... and is converted to acetyl CoA, which enters the .... cycle. The final stage of respiration takes place on the cristae of the mitochondria and is called the ....

▼ **Study point**

The important point about anaerobic respiration is that it regenerates NAD so that glycolysis can continue. If glycolysis were to stop, no ATP at all could be made.

Anaerobic respiration

If there is no oxygen to remove hydrogen atoms from reduced NAD, and make water, the electron transport chain cannot function. There is no oxidative phosphorylation and no ATP is formed. Without oxygen, reduced NAD cannot be reoxidised and no NAD is regenerated to pick up more hydrogen. Consequently, the link reaction and the Krebs cycle cannot take place and only the first stage of respiration, glycolysis, is possible. For glycolysis to continue, pyruvate and hydrogen must be constantly removed and NAD must be regenerated. This is done by pyruvate accepting the hydrogen from reduced NAD. The only ATP that can be made in the absence of oxygen is by substrate-level phosphorylation.

There are two different anaerobic pathways to remove the hydrogen from reduced NAD. Both take place in the cytoplasm:

- In animals, muscle cells may not get sufficient oxygen during vigorous exercise. When deprived of oxygen, pyruvate is the hydrogen acceptor and is converted to lactate, regenerating NAD. If oxygen subsequently becomes available, the lactate can be respired to carbon dioxide and water, releasing the energy it contains.

- In various micro-organisms, such as yeast, and in plant cells under certain conditions, such as in roots in waterlogged soils, pyruvate is converted to carbon dioxide and to ethanal, a hydrogen acceptor, by decarboxylase. Ethanal is reduced to ethanol and NAD is regenerated, in alcoholic fermentation. This pathway is not reversible, so even if oxygen becomes available again, ethanol is not broken down. It accumulates in the cells and can rise to toxic concentrations.

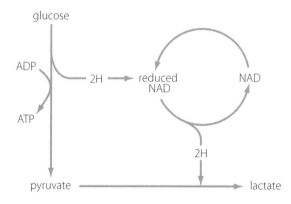

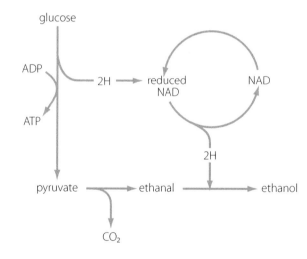

▼ **Study point**

In the absence of oxygen the pyruvate produced during glycolysis is converted to either lactate or ethanol. This is necessary in order to reoxidise reduced NAD so that glycolysis can continue.

WORKING SCIENTIFICALLY

It is important to communicate with precision. A problem may arise as some words have more than one meaning. Some people, for example, maintain that that the word 'respiration' should only be used if an electron transport chain operates. They would, therefore, call the reactions of anaerobic respiration 'fermentation'. However, the word 'fermentation' has different uses:

- For the reactions of anaerobic respiration
- For reactions forming ethanol
- For any reactions of bacteria.

In this book, 'anaerobic respiration' is used to mean respiration without oxygen. The term 'fermentation' is used when ethanol is produced.

The energy budget

Aerobic respiration

The table shows the number of pairs of hydrogen atoms and ATP molecules produced at each stage of respiration from one molecule of glucose, which gives two turns of the Krebs cycle. Use the flow diagrams of glycolysis, the link reaction and Krebs cycle, to understand how the numbers are derived. Remember that a pair of hydrogen atoms carried by reduced NAD generates 3 ATP and by reduced FAD, generates 2 ATP:

Stage	Pairs of H atoms carried by NAD	Pairs of H atoms carried by FAD	ATP from oxidative phosphorylation	ATP from substrate-level phosphorylation	Total number of ATP molecules
Glycolysis	2		6		8
				2	
Link reaction	2		6		6
Krebs cycle	6		18		24
		2	4		
				2	
Total	10	2	34	4	38

This gives a total of 38 molecules of ATP per molecule of glucose respired. It is a theoretical total and the cell is generally not this efficient because:

- ATP is used to move pyruvate, ADP, reduced NAD and reduced FAD across the mitochondrial membrane.
- The proton gradient may be compromised by proton leakage across the inner mitochondrial membrane, rather than passing through ATP synthetase.
- Molecules may also leak through membranes.

On average, cells produce 30–32 molecules of ATP per molecule of glucose respired.

If a mole of glucose is combusted in oxygen, it produces 2880 kJ. The energy required to make ATP = 30.6 kJ/mole. If the theoretical maximum is considered, a mole of glucose makes 38 moles of ATP which is equivalent to $(30.6 \times 38) = 1162.8$ kJ.

$\therefore$ the efficiency of ATP production =

$$\frac{\text{energy made available through ATP}}{\text{energy released in combustion}} \times 100 = \frac{1162.8}{2880} \times 100 = 40.4\% \text{ (1 dp)}.$$

This is a theoretical maximum and is higher than artificial methods of transducing energy. A petrol engine in a car, for example, is about 30% efficient.

YOU SHOULD KNOW ›››

››› The difference between substrate-level and oxidative phosphorylation

››› The yield of ATP from glycolysis, the link reaction and the Krebs cycle

››› The difference in ATP yield between aerobic and anaerobic respiration

››› The reason for the difference in yield of ATP from reduced NAD and reduced FAD

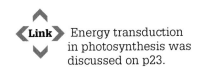

Link Energy transduction in photosynthesis was discussed on p23.

Exam tip

Candidates often state that energy is 'produced' in respiration. Energy cannot be produced but it can be changed from one type to another. It can also be transferred from one molecule to another.

Knowledge check

Link the processes 1–5 with the locations, A, B or both, where they take place.

1. Krebs cycle.
2. Fermentation.
3. Electron transport chain.
4. Glycolysis.
5. Phosphorylation.

A. Cytoplasm.
B. Mitochondrion.

Anaerobic respiration

Without the ATP synthetase associated with the electron transport system, the only ATP formed is in glycolysis, which makes two molecules of ATP per molecule of glucose, by substrate-level phosphorylation. This is a small amount compared with the 38 molecules of ATP produced during aerobic respiration.

In anaerobic respiration, pyruvate is not transferred to the mitochondria but is converted, in the cytoplasm, to ethanol in plants or lactate in mammals. The 2H released in the conversion of glucose to pyruvate reduce NAD and they are given up again in the formation of ethanol in plant cells or lactate in the cells of mammals. A huge number of different metabolic pathways have been identified in bacteria and Archaea, and many organic acids and alcohols are produced by their fermentation.

The efficiency of ATP production =

$$\frac{\text{energy made available through ATP}}{\text{energy released in combustion}} \times 100 = \frac{30.6 \times 2}{2880} \times 100 = 2.1\% \text{ (1 dp)},$$

which is much less than the efficiency of aerobic respiration.

YOU SHOULD KNOW ›››

››› The products of hydrolysis of lipids, fatty acids and glycerol, can enter the respiratory pathway at different points, with the production of ATP

››› Amino acids can provide energy in extreme circumstances

Alternative respiratory pathways

The Krebs cycle is sometimes called a 'metabolic hub' because the metabolic pathways of carbohydrates, lipids and proteins can feed into it and in some situations, fats and proteins can be used as respiratory substrates. Acetyl coenzyme A is a most significant molecule as it links the metabolism of the three types of macromolecule.

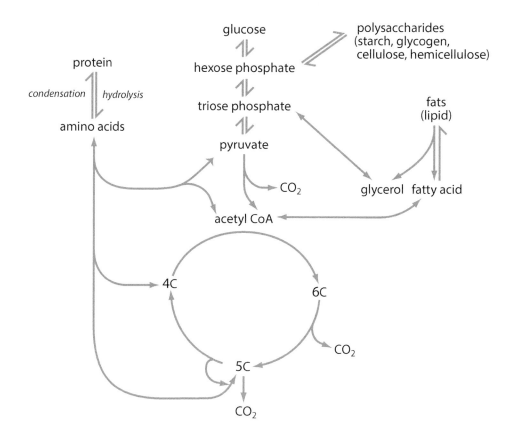

Alternative energy substrates and pathways

▼ Study point

Many cells can only use glucose as their respiratory substrate, but others break down fatty acids, glycerol and amino acids in respiration.

Lipids

Fat provides an energy store and is used as a respiratory substrate when carbohydrate in the body, such as glycogen and blood glucose are low. First, fat is hydrolysed into its constituent molecules, glycerol and fatty acids. Then the glycerol is phosphorylated with ATP, dehydrogenated with NAD and converted into a 3-carbon sugar, triose phosphate, which enters the glycolysis pathway:

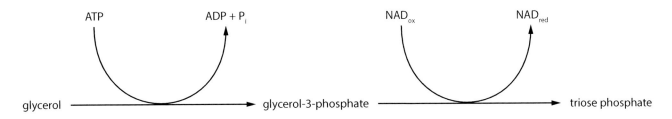

The long chain fatty acid molecules are split into 2-carbon fragments that enter the Krebs cycle as AcCoA. Hydrogen is released, picked up by NAD and fed into the electron transport chain. This produces very large numbers of ATP molecules but the precise number depends on the length of the hydrocarbon chain of the fatty acid. Longer fatty acid chains have:

- More carbon atoms, so more carbon dioxide is produced. Muscles have a limited blood supply and if they respired fat, rather than glucose, they would produce more carbon dioxide than could be removed quickly enough.

- More hydrogen atoms, so more NAD is reduced, so more ATP is produced. This explains why tissues with a rich blood supply, such as the liver, respire fat: the large amount of ATP they produce is readily transported around the body.

- More hydrogen atoms, so more water is produced. This 'metabolic water' is very important for desert animals and explains why they respire fat.

Proteins

Protein can be used as a respiratory substrate whenever dietary energy supplies are inadequate; the protein component of the food is diverted for energy purposes if carbohydrate and fat are lacking in the diet. In prolonged starvation, tissue protein is mobilised to supply energy. Heart muscle and kidney tissue are among the first the body breaks down to release protein. Protein is hydrolysed into its constituent amino acids, which are deaminated in the liver. The amino group is converted into urea and excreted. The residue is converted to acetyl CoA, pyruvate or some other Krebs cycle intermediate, and oxidised:

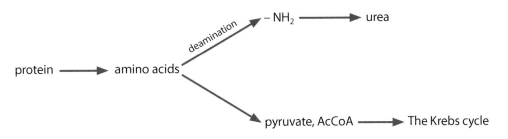

▼ **Study point**

Most metabolic pathways lead to acetyl CoA. It is formed by the oxidation of carbohydrates, fats and proteins and allows the oxidation products to be fed into the Krebs cycle.

Exam tip

Protein is usually only used as a respiratory substrate when all reserves of carbohydrate and fat have been depleted.

Going further ▶

Exercise breaks down muscle protein, increasing blood urea concentration. It also causes dehydration, increasing blood urea further. Marathon runners may show symptoms of high blood urea, e.g. cramp, seizures, itching.

Specified practical exercise

Investigation into factors affecting respiration in yeast

Rationale

Carbon dioxide is produced by yeast whether it respires aerobically or anaerobically. The number of bubbles of carbon dioxide produced in a given time is used as a measure of the rate of respiration. It can be measured in a variety of physical situations. In the experiment described here, the effect of sucrose concentration on the rate of respiration is measured.

Apparatus

– Yeast (100 g dm^{-3})
– Sucrose solutions (0, 0.2, 0.4, 0.6 and 0.8 mol dm^{-3})
– 20 cm^3 syringe
– Weight
– Trough
– Timer
– Electronic water bath at 30°C

Method

- Stir the yeast suspension and draw 5 cm^3 into the 20 cm^3 syringe.
- Wash the outside of the syringe with running water to remove any yeast suspension there.
- Draw into the syringe an additional 10 cm^3 sucrose solution.
- Pull the plunger back until it almost reaches the end of the syringe barrel.
- Invert the syringe gently to mix the contents.
- Place the syringe horizontally in the water bath, ensuring the nozzle is uppermost and place the weight on top of the syringe to hold it in place.
- Allow 2 minutes for the yeast and sucrose to equilibrate to temperature.
- When gas bubbles emerge regularly from the nozzle of the syringe, count the number released in one minute.

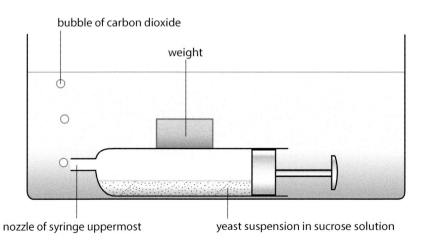

bubble of carbon dioxide

weight

nozzle of syringe uppermost

yeast suspension in sucrose solution

Diagram of apparatus

Risk assessment

Hazard	Risk	Control measure
Allergic response to yeast	Contact with skin or eyes	Use eye protection; cover skin
Electric shock from water bath	Electric shock if unplugged with wet hands	Dry hands before unplugging

Results

Sucrose concentration / mol dm^{-3}	Number of bubbles per minute			
	1	2	3	Mean
0	1	3	2	2
0.2	5	9	10	
0.4	14	13	12	
0.6	12	12	18	
0.8	13	10	13	

- Complete the table of results.
- Plot a graph of the results, with the sucrose concentration on the x-axis.
- Determine the sucrose concentration at which the reactions are the most rapid. This is the sucrose concentration at which the enzyme catalysing the rate-limiting step is saturated.

Further work

- The effect of substrate concentration may be tested by using different temperatures, e.g. 10, 30, 50, 70 and 90°C.
- The effect of pH may be assessed by making all solutions in buffers of different pH values, e.g. pH 1, 3, 5, 7, 9.

CH 4

Microbiology

Micro-organisms include bacteria, fungi, viruses and protoctista. They have significant roles in the biosphere yet are not visible to the naked eye. Bacteria and fungi decay dead organisms, releasing and recycling nutrients. Using similar metabolic pathways, they can cause the deterioration of harvested crops. Some bacteria are pathogens, and, like other micro-organisms, cause disease in humans, crops and domestic animals. Some, however, have no effect and many are beneficial. Bacteria reproduce asexually, by binary fission, and can do so very rapidly. Some can also reproduce sexually.

This section deals mainly with bacteria, describing their classification, growth and how to monitor their population numbers. Methods of culturing bacteria are described, with emphasis on the principles of aseptic technique.

Topic contents

By the end of this topic you will be able to:

- Describe how bacteria may be classified according to their shape and by the Gram stain technique.
- Describe the structure of bacterial cell walls.
- Explain how bacterial growth is affected by temperature, pH, oxygen and nutrients.
- Explain the importance of taking certain safety precautions when working with micro-organisms.
- Describe how the growth of bacteria may be monitored by a number of methods, with particular reference to the technique of counting colonies.
- Show how to estimate the number in a bacterial population by diluting, plating and counting colonies.

Classification of bacteria

There are said to be a million bacteria in 1 cm³ fresh water and 40 million in 1 g soil. The human body has 10^{13} cells but living in the gut our 'microbiome' contains 10^{14} bacteria, of 500–1000 different species. It has been estimated that worldwide there are 5×10^{30} bacteria and that their combined mass is greater than the combined mass of all animals and plants. These huge numbers alone make it seem sensible to study bacteria.

The physical differences between bacteria relate to their size, their cell wall structure and, hence, their staining characteristics and their shape. Their genetic differences produce different metabolic features and different surface molecules, so they have different antigenic properties.

Bacteria vary considerably in size. The smallest are Archaea, such as *Nanoarchaeum equitans*, which, at 0.4 µm diameter, is the smallest known organism that can reproduce independently. The size of the smallest cells is limited by the size of the molecules needed for life. The largest bacterium described is the sulphur bacterium *Thiomargarita namibiensis*. It is 750 µm diameter and visible to the naked eye, although most of the cell is a vacuole. *E. coli* is 1.8 µm diameter and 7 µm long.

The diagram shows a comparison of various bacteria.

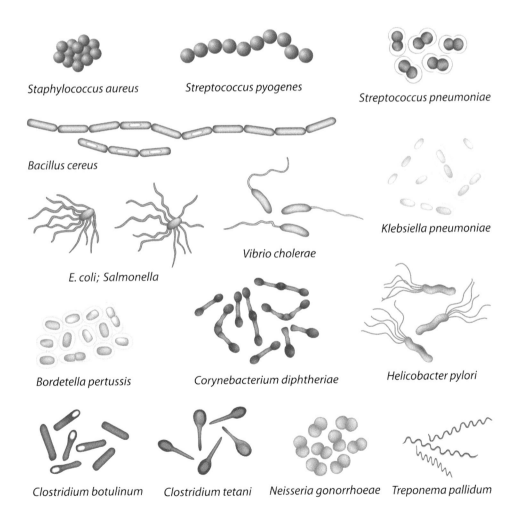

Staphylococcus aureus   Streptococcus pyogenes

Streptococcus pneumoniae

Bacillus cereus

E. coli; Salmonella

Vibrio cholerae

Klebsiella pneumoniae

Bordetella pertussis   Corynebacterium diphtheriae

Helicobacter pylori

Clostridium botulinum   Clostridium tetani   Neisseria gonorrhoeae   Treponema pallidum

Drawings of bacteria

YOU SHOULD KNOW ›››

››› How bacteria are classified

››› The structure of the bacterial cell wall in relation to the Gram stain technique

< **Link** > You will learn more about the antigenic properties of bacteria if you study Option A: Immunology and Disease.

Going further ▶

Haemophilus influenzae is one of the smallest bacteria known. It is important as it is found in the human respiratory tract, and was the first organism to have its entire genome sequenced.

Exam tip

Bacteria, in common with all organisms, are named according to the binomial system, e.g. *Staphylococcus aureus*.

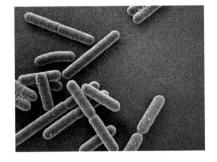

Scanning electron microscope image of *E. coli*

Key Term

Gram stain: A method of staining the cell walls of bacteria as an aid to their identification.

Going further ▶

There were three major advances in microbiology between 1881 and 1887: Fanny Hesse introduced agar jelly into the laboratory, Hans Gram invented his stain and Julian Petri invented his dish.

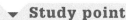

▼ Study point

Gram stain binds to the peptidoglycan of bacterial cell walls. Archaea and Eukaryotes do not contain peptidoglycan and are not stained by Gram stain.

Classification by shape

A genus of bacteria has one of three main shapes, and the shape is sometimes indicated in its name:

- Bacillus or rod-shaped, e.g. *Escherichia; Bacillus*
- Coccus or spherical, e.g. *Staphylococcus; Streptococcus*
- Spiral or corkscrew-shaped, e.g. *Spirillum.*

Further differentiation is often possible according to the way bacteria tend to group. They may be single, e.g. *Helicobacter;* in pairs, e.g. *Diplococcus pneumoniae;* in chains, e.g. *Streptococcus;* or in clusters, e.g. *Staphylococcus.*

Classification by the Gram stain reaction

The **Gram stain** allows microbiologists to distinguish between Gram-positive and Gram-negative bacteria. The different staining properties are due to differences in the chemical composition of their cell walls. Before staining, bacteria are colourless. After staining, Gram-positive bacteria are stained violet and Gram-negative bacteria are red:

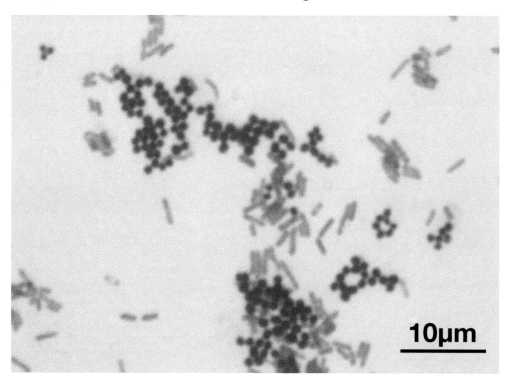

Photomicrograph of bacteria stained with Gram stain

The cell wall of all bacteria is a 3-dimensional network of polysaccharides and polypeptides, known as peptidoglycan or murein. The cross-linking of these molecules provides strength and gives the cell its shape. The wall protects against swelling and bursting or lysis caused by the osmotic uptake of water. Gram-positive bacteria possess this basic cell wall structure and Gram-negative bacteria have an additional outer layer of lipopolysaccharide.

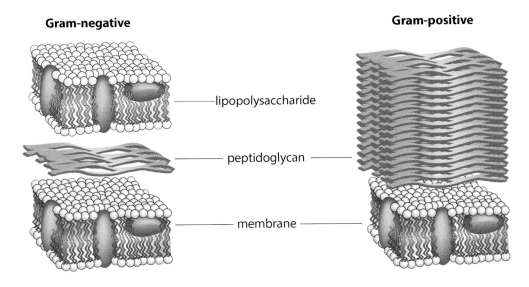

Gram-negative   **Gram-positive**

— lipopolysaccharide

— peptidoglycan —

— membrane —

Cell wall of bacteria

Gram staining has four stages:

Reagent	Function	Effect
Crystal violet	Basic dye	Binds to peptidoglycan so all bacteria stain purple
Lugol's iodine	Mordant	Binds the crystal violet to the peptidoglycan more strongly
Acetone-alcohol	Decoloriser	Removes unbound crystal violet and lipopolysaccharide: • Gram-negative bacteria lose their stain and become colourless • Gram-positive bacteria remain purple
Safranin	Counter-stain	• Gram-negative bacteria stain red • Gram-positive bacteria remain purple.

Gram-positive bacteria

After staining, Gram-positive bacteria are violet or purple under the microscope. They include *Bacillus, Staphylococcus* and *Streptococcus*.

The absence of an outer lipopolysaccharide layer in the cell walls of Gram-positive bacteria allows them to bind stain efficiently and makes them more susceptible to the antibiotic, penicillin, and the enzyme, lysozyme, than Gram-negative bacteria.

Bacteria constantly make and break chemical links in their cell walls. The antibacterial enzyme lysozyme, which occurs in human tears and saliva, hydrolyses the bonds holding the peptidoglycan molecules together.

Penicillin prevents the bonds inter-linking peptidoglycan molecules from forming. This is especially significant when the bacteria make new cell walls when they divide. Penicillin therefore makes the cell walls structurally weak and prone to collapse. Water uptake by osmosis bursts the cell.

Gram-negative bacteria

Gram-negative bacteria have a more chemically complex cell wall than Gram-positive bacteria. Their outer membrane is supplemented with large molecules of lipopolysaccharide which protect the cell and exclude dyes like crystal violet, so they appear red or pink in the microscope.

Gram-negative bacteria include *Salmonella* species and *E. coli*. The outer lipopolysaccharide protects the peptidoglycan below and so they are not affected by lysozyme and are resistant to penicillin. To control them requires a different class of antibiotics, that interfere with the cell's ability to make proteins. Eukaryotic cells also make proteins, but the protein-making cellular machinery is different from that in bacteria, so these antibiotics do not harm them.

▼ **Study point**

Animal cells do not have a cell wall, so human cells are not damaged by penicillin.

Knowledge check

Match the terms 1–3 with the statements A–C.

1. Peptidoglycan.
2. Bacillus.
3. Lipopolysaccharide.

A. Rod-shaped bacteria.
B. Found in all bacterial cell walls.
C. Found only in Gram-negative bacteria.

Going further ▶

Many bacteria cannot be cultured because they may be: mutualistic, needing other species, intracellular parasites, have very particular growth requirements, have a very long generation time or be poisoned by media components.

▼ **Study point**

Two potential problems which must be avoided when working with bacteria are the contamination of cultures from the environment and the contamination of the environment from the cultures.

Key Term

Aseptic technique: Laboratory practice that maintains sterility in apparatus and prevents contamination of the equipment and the environment.

Conditions necessary for culturing bacteria

Micro-organisms can undergo binary fission and reproduce quickly, given a suitable environment. In optimum conditions they divide every twenty minutes. In the laboratory, bacteria can be grown on a wide variety of substrates providing they are supplied with nutrients, water and suitable physical conditions, such as temperature. Micro-organisms vary in their requirements and usually grow over a range of temperatures and pH values, with an optimum within the range.

Micro-organisms require the following conditions for growth:

- Nutrients – in the laboratory, nutrients are supplied in nutrient media. The bacteria may be cultured in a liquid medium, called a nutrient broth, or on medium solidified with agar. The media provide water and they include:
  - A carbon and energy source, usually glucose.
  - Nitrogen for amino acid synthesis, in organic molecules and in inorganic form, such as nitrate ions.
- Growth factors including vitamins, e.g. biotin and mineral salts e.g. Na^+, Mg^{2+}, Cl^-, SO_4^{2-}, PO_4^{3-}.
- Temperature – as bacterial metabolism is regulated by enzymes, the range of 25–45°C is suitable for most bacteria. The optimum for mammalian pathogens is around 37°C, the temperature of the human body.
- pH – most bacteria are favoured by slightly alkaline conditions (pH 7.4), whereas fungi grow better in neutral to slightly acid conditions.
- Oxygen – many micro-organisms require oxygen for metabolism and are obligate aerobes, e.g. *Mycobacterium tuberculosis*. Some grow best in the presence of oxygen, but can survive in its absence; these are facultative anaerobes, e.g. *E. coli*. Others cannot grow in the presence of oxygen and are obligate anaerobes. *Clostridium* bacteria are obligate anaerobes that produce toxins or poisons in a wound. They destroy body tissue in the condition called 'moist gangrene'.

There are different ways to describe culture media:

- A 'defined' medium contains only known ingredients.
- An 'undefined' medium contains components that are not all known, because they include, for example, yeast extract or beef peptone.
- A selective medium only allows certain bacteria to grow, e.g. MacConkey agar only allows Gram-negative bacteria to grow; media containing tetracycline only allow tetracycline-resistant bacteria to grow.

Principles of aseptic technique

Bacteria and fungi are cultured on or in media that are designed to supply the cell with all its nutritional and physical requirements. **Aseptic technique**, also known as sterile technique, in which the apparatus and equipment are kept free of micro-organisms, prevents contamination of bacterial cultures by other microbes and contamination of the environment.

To prevent the contamination of pure cultures and apparatus by bacteria from the environment:

- Sterilise all apparatus and media before use to prevent initial contamination.
- Handle cultures carefully, flaming the necks of culture vessels before opening and closing and use equipment such as sterile loops to prevent subsequent contamination.

To prevent contamination to the environment by the bacteria being used in experiments:

- Sterilise the work surface before and after an experiment using a disinfectant, for example, 3% Lysol.
- Use the correct handling techniques to prevent the contamination of personnel and the immediate environment by the organisms being cultured. When carrying out the process of inoculation:
  - Hold the culture bottle in one hand; remove the cap with the little finger of the other hand. Do not place the cap down on the work surface.
  - Flame the mouth of the bottle for 2 or 3 seconds.
  - Pass the inoculating loop through a flame until it is red hot, and allow it to cool in the air.
  - Lift the lid of the Petri dish just enough to allow entry of the inoculating loop.
  - Secure the Petri dish lid with two pieces of adhesive tape. Do not seal all the way round as this could create anaerobic conditions and, potentially, encourage the growth of **pathogenic** micro-organisms.
  - Incubate at around 25°C. Cultures should not be cultured at 37°C as this is an ideal temperature for the growth of many pathogenic species.
  - Do not open Petri dishes after incubation.
- In a laboratory the preferred method of sterilisation is to use an autoclave. This is a sealed container in which glass and metal equipment is heated at 121°C in steam under pressure for 15 minutes after the required pressure has been reached.
- After use disposable materials, such as plastic Petri dishes, can be sealed inside autoclavable plastic bags, autoclaved and then placed in a dustbin.

Key Term

Pathogen: An organism that causes disease in its host.

▼ **Study point**

Gamma radiation is used commercially to sterilise plastic equipment that would melt in an autoclave.

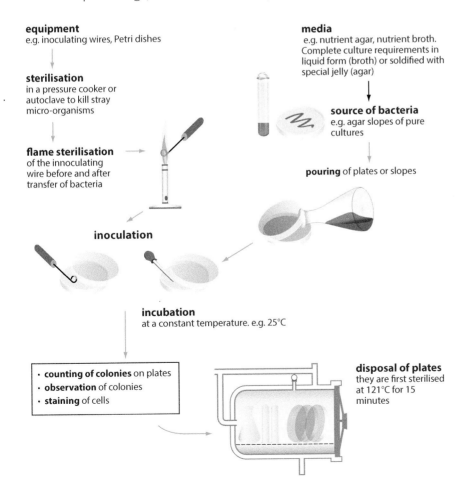

equipment
e.g. inoculating wires, Petri dishes

sterilisation
in a pressure cooker or autoclave to kill stray micro-organisms

flame sterilisation
of the innoculating wire before and after transfer of bacteria

inoculation

incubation
at a constant temperature. e.g. 25°C

- **counting of colonies** on plates
- **observation** of colonies
- **staining** of cells

media
e.g. nutrient agar, nutrient broth. Complete culture requirements in liquid form (broth) or soldified with special jelly (agar)

source of bacteria
e.g. agar slopes of pure cultures

pouring of plates or slopes

disposal of plates
they are first sterilised at 121°C for 15 minutes

Sterile techniques

8

Knowledge check

State which of the following six statements regarding aseptic techniques are false.

1. Flame the mouth or neck of the culture bottle.
2. Leave the lid off the Petri dish for a few minutes after inoculation.
3. Incubate at 37°C.
4. Seal the Petri dish and lid all the way round.
5. Sterilise all equipment in boiling water for 15 minutes.
6. Sterilise the work surface before and after an experiment using a disinfectant.

Methods of measuring growth

Estimating the growth of a population of bacteria is extremely important. Environmental health officers regularly inspect food premises and take samples for analysis. Water authorities check water supplies daily. Food manufacturers must ensure the food they sell is fit to eat. Many items, such as foods and drugs, are produced using bacteria grown in 200 dm³ industrial fermenters. Accurately measuring their population growth is an important part of the process.

The size of a population of micro-organisms in liquid culture may be measured:

- Directly, by counting cells:
  - Viable counts describe living cells only.
  - Total counts describe living and dead cells.
- Indirectly, by measuring the turbidity (cloudiness) of the culture. This method can be used as a field work technique, using light absorption by a sample of river water to indicate the number of bacteria present.

Measuring growth directly

Plating and counting colonies

It is not possible to count a whole population of micro-organisms. Instead, the number of cells in a very small sample of culture is counted. Even then, the population density is likely to be so high that cell counts have to be made using dilutions of the culture. This type of count relies on each live cell forming a **colony**, so it provides a viable count.

Key Term

Colony (of bacteria or fungus): A cluster of cells, or clone, which arises from a single bacterium or fungal spore by asexual reproduction.

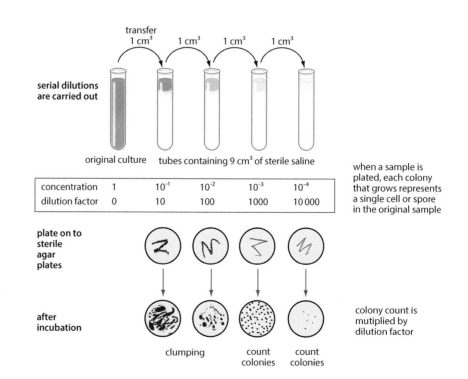

Estimating a bacterial population

The sample is diluted using serial dilution. When 1 cm³ suspension is added to 9 cm³ medium, it has been diluted 10 times and is a 10^{-1} dilution. If this process is repeated, dilutions of 10^{-2}, 10^{-3} 10^{-4} and so on can be made. But if 0.1 cm³ is progressively added to 9.9 cm³, the first dilution is 10^{-2} and subsequent dilutions are 10^{-4}, 10^{-6}, 10^{-8}, etc.

1cm³ of each diluted sample is spread over a sterile agar plate and incubated at 25°C for two days, allowing the bacteria to grow. A dish containing 20–100 colonies that are distinct and separate is chosen and the colonies counted. To find the total viable cell count, the number of colonies is multiplied by the appropriate dilution factor.

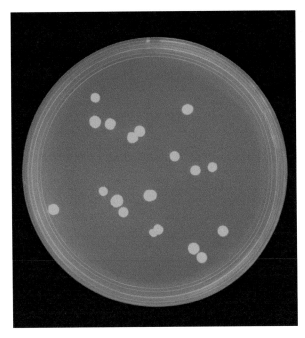

Colonies of bacteria containing the jellyfish gene for a green fluorescent protein

If a 1 cm³ of a 10^{-5} dilution produces a mean of 85 colonies per plate:

the initial concentration = 85×10^5 cm⁻³ = 8.5×10^4 (1 dp) cm⁻³.

If 0.5 cm³ of a 10^{-7} dilution produces a mean of 129 colonies per plate:

the initial concentration = $\dfrac{129}{0.5} \times 10^7 = 258 \times 10^7 = 2.6 \times 10^9$ (1 dp) cm⁻³.

If the dilution is too great, there will be too few colonies on each plate for the count to be statistically sound. If the dilution is insufficient, colonies merge, referred to as 'clumping', and counting may be inaccurate, resulting in an underestimate of numbers.

Using a haemocytometer

This is a more accurate method than colony counting. It uses a specialised microscope slide, called a haemocytometer. It is not possible to distinguish between living and dead cells so the result is a total cell count.

Measuring growth indirectly

Turbidimetry

A colorimeter can be used to measure the cloudiness or turbidity of the culture as cell numbers increase. Measurements of the bacterial population are derived by finding the absorbance of the suspension and then reading from a standard graph of light absorbance plotted against the number of bacterial cells. The result is a total cell count, because the colorimeter cannot distinguish living from dead cells.

Specified practical exercise

Investigation into the numbers of bacteria in milk

Rationale

Commercially available fermented milks include familiar brands such as Yakult and Actimel. They are made from raw milk by lactic acid bacteria, including *Lactobacillus* and *Lactococcus*. These bacteria are facultative anaerobes and they convert lactose in the milk into lactate. The small amount of ATP produced provides energy and the bacteria reproduce. Their number increases as the milk ages. The population can be monitored by plating and counting colonies produced by a suitable dilution of different ages of milk.

The experiment described here requires a large number of plates to be made and could be performed by groups rather than by individuals.

Design

Variable	Name of variable		Value of variable		
Independent	Age of milk		0, 3, 6, 9, and 12 days before 'use-by' date		
Dependent	Number of bacteria per cm^3		Number cm^{-3}		
Controlled	Temperature		Cultures incubated at 25°C		
	Incubation time		2 days		
	Chemical composition of nutrient medium		All samples plated on the same nutrient medium		
Control	Repeat experiment under identical conditions with boiled and cooled bacterial suspensions.				
Reliability	Make 5 plates for each dilution of milk and calculate a mean number of colonies for the dilution at which counts are made.				
Risk assessment		Hazard	Risk	Control measure	
		Exposure to pathogenic bacteria	Pathogens could be incorporated into culture when plates are made	Use sterile technique to prepare equipment and make cultures	
			Pathogens could form colonies on agar plates	Incubate plates at 25°C	
				Do not open plates once taped shut	
				Kill all microbes by autoclaving plates before disposing of them	

Method

- Prepare serial dilutions for each age of milk to give a dilutions of 10^{-2}, 10^{-4}, 10^{-6}, 10^{-8} and 10^{-10} by mixing 0.1 cm^3 fermented milk into 9.9 cm^3 medium at each transfer.

- For each dilution of each age of milk, spread 1 cm^3 of bacterial suspension on to the surface of solid nutrient agar in a Petri dish. Make three replicates for each dilution at each age.

- Tape the lid of each dish shut with two pieces of tape.

- Label the dishes on the base, with the dilution and age.

- Invert the Petri dishes and incubate them at 25°C for two days.

- For each age of milk, determine which dilution provides between 20 and 100 separate colonies in each plate.

- Count the colonies and calculate a mean.

- Autoclave the dishes prior to disposal.

Sample results

Time before use-by date / days	Milk age /AU	Number of colonies per plate				Suspension dilution	Number in undiluted sample (1 dp)	Log$_{10}$ number in undiluted sample (1 dp)
		1	2	3	Mean			
12	0	17	29	21	22.3	10^{-6}	2.2×10^{7}	7.3
9	3	11	8	13		10^{-8}		
6	6	98	84	96		10^{-8}		
3	9	81	69	65		10^{-10}		
0	12	96	95	74		10^{-10}		

- Complete the table of results.
- Plot a graph of the population growth over the previous 12 days. The milk age is the independent variable and is plotted on the horizontal axis. The vertical axis shows the number of bacteria in the undiluted sample, which can be plotted in various ways:
    - On linear graph paper with a log scale on the vertical axis, as below, plot the number of bacteria.
    - On linear graph paper, plot the log$_{10}$, as shown on p67.
    - On semi-log graph paper, plot the actual number on the vertical axis:

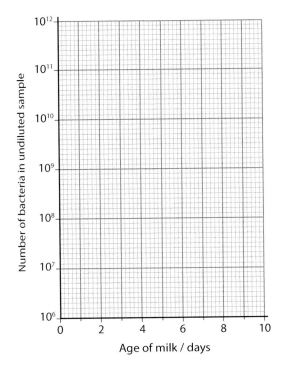

Bacterial growth curve

Further work

- Stain a sample of fermented milk with Gram stain and determine:
    - The shapes of bacteria
    - The proportions of the different shapes that are Gram-positive and Gram-negative.
- Plot population growth curves for different fermented milks, e.g. kefir, leben.

CH 5

Population size and ecosystems

Population growth is controlled by biological and environmental factors. Succession is the change of structure and species composition of communities over time. Energy passes along a food chain as biomass, but its loss limits their length. All organisms require carbon and nitrogen and micro-organisms play a key role in recycling them through the ecosystem. Agriculture, deforestation and pollution have drastically altered ecosystems and so agricultural practices must be carefully evaluated and altered to support the continuing increase in human population.

Topic contents

By the end of this topic you will be able to:

- Explain factors controlling the size of a population.
- Describe how to sample populations and assess their abundance and distribution.
- Understand the concepts of ecosystem, habitat, community, primary and secondary productivity.
- Describe the transfer of biomass through the ecosystem, shown in pyramids of biomass.
- Explain why herbivores have a lower secondary productivity than carnivores.
- Describe and explain the differences between primary and secondary succession.
- Understand that decomposers are key organisms involved in the cycling of nutrients.
- Describe the carbon cycle and the effects on it of deforestation and combustion.
- Understand the causes and effects of global warming.
- Explain the need for changes in agricultural practice.
- Describe the role of micro-organisms in the nitrogen cycle.
- Describe the importance of fertilisers and the ploughing and draining of agricultural land.
- Describe how the use of fertilisers can lead to water pollution.

Population growth

Factors controlling population size

There are factors in an ecosystem that affect all the organisms living in it and these factors constantly change. So ecosystems are dynamic:

- The intensity of energy flowing through the ecosystem varies.
- Biological cycles, such as the nitrogen cycle, vary the mineral availability.
- Habitats change over time as succession occurs.
- New species arrive and some species are no longer present.

As a result, the number of individuals in a **population** does not remain the same. The size of a population at a particular time is determined by:

- **Birth rate**, or natality, which also refers to hatching, reproduction by binary fission and all other ways that living organisms increase their numbers.
- Death rate, or mortality.
- **Immigration**.
- Emigration.

Birth and immigration increase population size, while death and emigration decrease it. When the combined effects of birth and immigration exceed those of death and emigration, the population size increases.

Population size

Different strategies for population growth are used by different species, depending on their characteristics:

- Fugitive species are species that are poor at competition: instead, they rely on a large capacity for reproduction and dispersal to increase their numbers. They invade a new environment rapidly, e.g. algae colonising bare rock; rose bay willow herb colonising soil cleared by fire.
- **Equilibrium species** control their population by competition within a stable habitat. Their usual pattern of growth is a sigmoid (S-shaped) curve called the one-step growth curve. It is seen, for example, when bacteria are put into fresh nutrient solution or when rabbits are newly introduced to an island.

The one-step growth curve

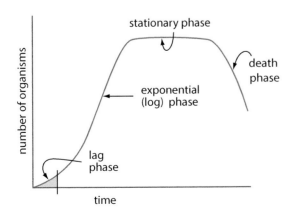

Graph showing changes in population growth

Exam tip

Choose your words carefully. Mammals are born but yeast and bacteria undergo binary fission; reptiles and birds hatch from eggs and flowering plants develop from seeds.

Key Terms

Population: An interbreeding group of organisms of the same species occupying a particular habitat.

Birth rate: The reproductive capacity of a population; the number of new individuals derived from reproduction per unit time.

Immigration: The movement of individuals into a population of the same species.

Equilibrium species: Species that control their population by competition rather than by reproduction and dispersal.

The one-step growth curve has four phases:

1. The lag phase. Initially, the population does not increase but then there is a period of slow growth. In bacteria, it may last from a few minutes to several days. It is a period of adaptation or preparation for growth, with intense metabolic activity, especially enzyme synthesis. In sexually reproducing organisms, such as rabbits, the lag phase represents the time for individuals to reach sexual maturity, to find a mate and gestate their young.

2. The exponential phase. As numbers increase, as long as there is no factor limiting growth, more individuals become available for reproduction. Bacterial cells divide at a constant rate and the population doubles per unit time. The cell number increases logarithmically and so the exponential phase is also called the log phase.

This rate of increase cannot be maintained indefinitely because **environmental resistance** sets in:

– There is less food available
– The concentration of waste products becomes increasingly toxic
– There is not enough space or nesting sites.

The population still increases, but more slowly, so the gradient of the graph decreases.

Environmental resistance includes all the factors that may limit the growth of a population. For bacteria in a flask these factors include:

– Available food
– Overcrowding
– Competition
– The accumulation of toxic waste.

However, in a less artificial situation other factors may play a part. For the rabbits on an island, these same factors apply but there are additional **biotic** factors including:

– Predation
– Parasitism and disease because increased population density allows infection to spread more rapidly
– Competition from other species for nesting sites and food.

Abiotic factors such as temperature and light intensity may also play a role in affecting population size.

3. The stationary phase occurs when the birth rate is equal to the death rate. The population has reached its maximum size, which is the **carrying capacity** for that particular environment, i.e. the maximum number of individuals that an area can support. The actual number depends on the resources available, e.g. more food increases the carrying capacity. The population is not absolutely constant, and it fluctuates around the carrying capacity in response to environmental changes, e.g. the number of predators.

4. The death phase. The factors that slow population growth at the end of the lag phase become more significant and population size decreases until the death rate is greater than birth rate and the graph has a negative gradient.

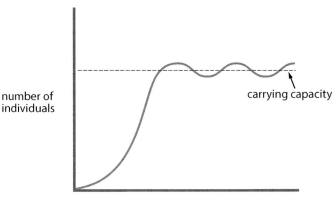

Graph showing carrying capacity

Key Terms

Environmental resistance refers to environmental factors that slow down population growth.

Biotic: A part of the environment of an organism that is living, e.g. pathogens, predators.

Abiotic: A part of the environment of an organism that is non-living, e.g. air temperature, oxygen availability.

Carrying capacity: The maximum number around which a population fluctuates in a given environment.

Environmental resistance and the carrying capacity

Predators are normally larger than their prey and tend to kill before they eat. The abundance of prey limits the numbers of predators and the number of predators controls the number of prey. A predator–prey relationship causes both populations to oscillate and these oscillations are regulated by negative feedback. In a classic experiment using data collected since 1920, the numbers of lynx and snow-shoe hares in parts of Canada have been estimated. Their numbers fluctuated around the carrying capacity for each because:

- A large number of lynx predate hares, so the hare population decreases.
- Then there is not enough food for the lynx, so lynx numbers go down.
- There is therefore less predation on the hares, so hare numbers increase again.
- There is more prey for lynx, so lynx numbers go up.

This cycle repeats at approximately 10-year intervals, and illustrates how population numbers at the carrying capacity can depend on the numbers in other species. An equivalent explanation would apply to other predator–prey relationships, e.g. owl and mouse numbers in a woodland, and herbivore–plant relationships, e.g. koalas and eucalyptus.

Calculating population increase from a graph

When a population increase is very large, for example in a population of bacteria in a test tube, the range of numbers is too great to plot a graph on a linear scale. A $\log_{10}$ scale is therefore used, in which each mark on the population scale is ten times the previous mark.

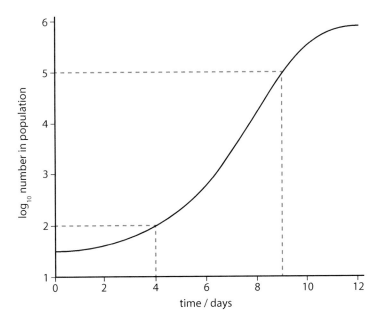

Bacterial growth curve

The rate of growth in the exponential phase =

$$\frac{\text{number of bacteria at day 9} - \text{number of bacteria at day 4}}{9 - 4} \text{ per day}$$

As a log scale is plotted, we cannot read the actual numbers at day 9 and at day 4. The actual number is the antilog of the number on the scale, i.e. the number for which the numbers on the axis are the logs.

∴ the rate of growth in the exponential phase = $\dfrac{\text{antilog 5} - \text{antilog 2}}{9 - 4}$ per day

Going further ▶

65 million years ago, non-avian dinosaurs became extinct. It is likely that they exceeded their carrying capacity because they ran out of food. Evidence shows an asteroid impact and excessive volcanic activity. They put so much dust into the atmosphere, that the consequent lack of sunlight killed the plants. Herbivorous dinosaurs, e.g. *Triceratops horridus*, therefore had no food and died. This meant that the carnivorous dinosaurs, e.g. *Tyrannosaurus rex*, had no food either, so they died.

Exam tip

Take care if you are plotting on a log scale. There is no zero, because each mark on the axis is one tenth of the mark above.

- In the example shown here, the log is a whole number:

  $\text{Antilog}_{10} 5 = 10^5 = 100\,000$ and $\text{antilog}_{10} 2 = 10^2 = 100$

  $\therefore$ the rate of growth in the exponential phase =

$$\frac{100\,000 - 100}{5} = \frac{99\,900}{5} = 19\,980 \text{ per day.}$$

- If the log is not a whole number, you can use your calculator to find the antilog. For example, to find the antilog of 6.7:
  - Press SHIFT log
  - Press 6.7
  - Press = to give the answer 5.011873×10^6 (6 dp)

Factors that regulate population increase

Some environmental factors have more effect if the population in a given area is larger. These factors affect a greater proportion of the population if the population is denser and so they are called **density-dependent factors**. They are biotic factors and include disease, parasitism and depletion of food supply, e.g. if a population is denser, parasites are transmitted more efficiently and a greater proportion of individuals are affected. Similarly, if the prey density is higher, the predator encounters prey more readily and eats a greater proportion of them.

The effect of abiotic factors in the environment does not depend on the population density. These are density-independent factors. The effect is the same regardless of the size of the population and is usually due to a sudden change in an abiotic factor, e.g. flood or fire. *Daphnia*, the water flea, undergoes population crashes when the temperature suddenly falls; honeybee populations are killed by neonicotinoid insecticides.

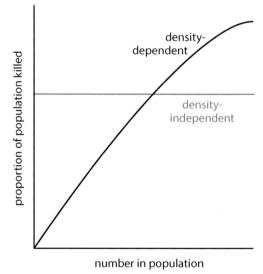

The effects of density-dependent and density-independent factors on populations

Populations fluctuate in numbers

In general, the balance between the birth rate and the death rate regulates the size of a population. However, populations fluctuate; they do not remain constant in size, although in equilibrium species these fluctuations are not usually large or erratic. The numbers in their populations fluctuate around a set point, which is the carrying capacity. The population size is regulated by negative feedback:

- If the population rises above the set point, a density-dependent factor increases mortality or reduces breeding to such an extent that the population declines.
- If the population falls below the set point, environmental resistance is temporarily relieved so that the population rises again.

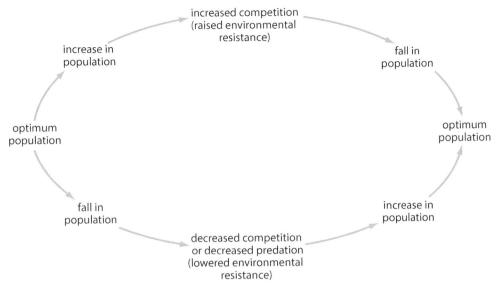

Negative feedback control of population

Knowledge check

Match the terms 1–4 with the statements A–D.

1. Stationary phase.
2. Lag phase.
3. Carrying capacity.
4. Density-dependent factor.

A. A period of preparation for growth.
B. The food supply is depleted as population density increases.
C. The maximum population size that can be supported by the available resources.
D. The birth rate of new individuals is equal to the death rate.

The abundance and distribution of organisms in a habitat

The study of species abundance and distribution is called biogeography. Alfred R Wallace was the first person to model biogeographic regions, and he defined six around the world. In his study of birds and vertebrates, he saw that rivers and mountain ranges marked the boundaries of many species' ranges. He saw different animals in similar habitats. This did not correspond with the prevailing explanation that organisms were all created to suit a particular environment, and it contributed to his understanding of natural selection.

A new habitat

When a new habitat is assessed, physical features, e.g. soil type and temperature, are described first because they determine the number and types of plants that live there:

- A soil derived from granite bedrock, e.g. Snowdonia, is more acid than one derived from chalk, e.g. the South Downs. Consequently, sundew grows on Snowdonia and half of all the native orchid species of the UK grow on the South Downs.

- If rainfall is very high throughout the year, e.g. parts of south-east Australia, tree ferns may be found, but in some very dry regions, e.g. south-western USA, cactuses are common.

- A very cold habitat, e.g. parts of Finland, may support only lichens but a warmer habitat, e.g. Guinea, may support a vast number of plant types.

The animals in a habitat depend on the plants, e.g. pandas in the wild are only found where bamboo grows. So in a new habitat, plants are described before animals.

Measuring abundance

The **abundance** of a species is a measure of how many individuals exist in a habitat. Animal abundance can be assessed by:

- Capture-mark-recapture experiments, using the Lincoln index calculation.

- Kick sampling in a stream and counting aquatic invertebrates.

WORKING SCIENTIFICALLY

The 'Wallace line' shows the boundary between animals in north-east Indonesia, which resemble Asian animals, and animals in New Guinea, which resemble those in Australia. This observation provided evidence for plate tectonics.

 Key Term

Abundance: The number of individuals in a species in a given area or volume.

Link In the first year of this course, you learned how to use the techniques quoted here to measure abundance of plants and animals to estimate the biodiversity of a habitat.

WORKING SCIENTIFICALLY

Salacia arenicola (the 'gobstopper plant': primates spend hours sucking the fruits for the sweet flesh inside) was discovered in 2013 in the Republic of Congo. It is vulnerable. Its locations are being investigated to define its distribution, to prevent its extinction.

Fruit of the gobstopper plant

Key Term

Distribution: The area or volume in which the organisms of a species are found.

The abundance of plant species can be measured by:

- Using a quadrat to calculate the mean the number of individuals in several quadrats of known area, to find the density, i.e. number / metre^2.

- Estimating percentage cover of a plant in which individuals are hard to recognise.

- Estimating percentage frequency.

Measuring distribution

The **distribution** of a species describes the area or volume in which it is found.

If a habitat is uniform, the positions of the outermost plants can be marked on a map and the area they surround can be measured. A small area may indicate that a species in under threat of extinction. Botanists use this technique to assess the distribution of threatened plant species. Mining companies and road building authorities can then be lobbied to protect specific sites and hence mitigate species loss.

If a habitat is not uniform, a transect is a useful technique for displaying the variation in organisms and its correlation with a changing abiotic factor. A transect is line along which abundance is assessed.

- A line transect shows the organisms that lie on a line, at measured intervals.

A footpath in Yorkshire. The red line shows where the tape measure was laid to make a transect.

The diagram shows a line transect across the footpath shown in the photograph:

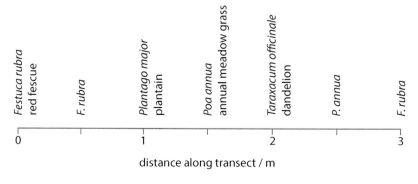

Line transect across the footpath

- A belt transect shows abundance data for a given area at measured distances along the transect. A quadrat is placed at each co-ordinate along the transect and readings may be taken for:
  - The density of chosen species
  - The percentage frequency of chosen species
  - The percentage area cover for all species.

The table shows the percentage cover across the footpath shown in the photograph:

Species	Common name	Percentage cover at distance along transect / m						
		0	0.5	1.0	1.5	2.0	2.5	3.0
Festuca rubra	red fescue	80	70					80
Holcus lanatus	London fog	20	30					20
Poa annua	annual meadow grass			70	60			
Plantago major	greater plantain			20	20			
Taraxacum officionale	dandelion				10	30	10	
Trifolium repens	white clover			10		20	40	

The kite diagram shows the percentage area cover of the species across the belt transect.

A footpath is most trampled in the middle and less so at the edges. A transect shows which species are more resistant to trampling, because they grow in the middle of the path. Both the line transect and the belt transect for the footpath shown in the photograph show that plantain and meadow grass are the least sensitive to trampling as they are found in the middle of the path. Red fescue and London fog are only found at the edges: they are more sensitive to trampling. The belt transect, however, being two-dimensional, provides more information about species distribution than the line transect, which is one-dimensional.

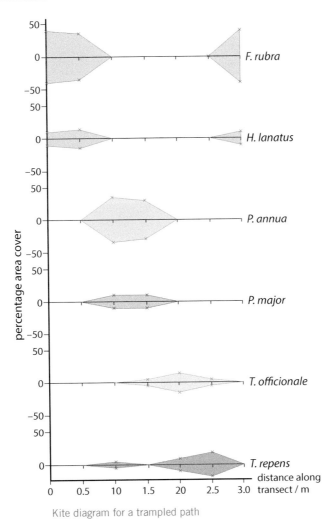

Kite diagram for a trampled path

⟨**Link**⟩ Transects across a rocky shore are described on p92.

▼ **Study point**

Kite diagrams are useful because they show clearly the abundance and distribution of species along a transect. But numbers are rounded when they are constructed and so accuracy is lost, in comparison with a table of data.

A transect is suitable for assessing plants or sessile animals but it is not suitable for motile animals, because they move. The distribution of animals is assessed by direct observation of individuals or their nests, faecal deposits or markings on vegetation.

Key Term

Ecosystem: A characteristic community of interdependent species interacting with the abiotic components of their habitat.

Ecosystems

A community comprises many species living and interacting together. The interaction of organisms with each other and with the non-living factors in their environment, such as the air, soil and water, comprises an **ecosystem**. The biotic and abiotic components of the ecosystem are linked by energy flow and by the cycling of nutrients.

- Ecosystems can be small, e.g. the human large intestine and its community of micro-organisms. Ecosystems can be very large. Seas, for example, cover about 70% of the Earth's surface and the Pacific Basin is the largest marine ecosystem. There are those who consider the whole planet to be one large ecosystem.

- Ecosystems can be temporary such as a puddle left after the rain or may last many millions of years, e.g. Lake Baikal in Siberia has existed for 25 million years.

Some examples are shown in the table:

Ecosystem	Example	Abiotic features	Characteristic organisms
marine	Pacific Ocean	aquatic; high mineral ion concentration	algae; Echinoderms, e.g. starfish
Arctic tundra	Canada, Siberia	temperature range between −50°C and +12°C; 15–25 cm rainfall per year; windy	low growing plants, e.g. moss, heather; reindeer, Arctic hare
temperate deciduous forest	Lady Park Wood, Monmouthshire	moist, warm	beech, oak, woodpecker, tawny owl
desert	Sahara, across 11 countries of north and central Africa	<25 cm rainfall per year; high temperature; high light intensity	cactus, camel

Energy and ecosystems

Energy is the ability to do work. No change happens unless energy changes occur. So the functioning of an ecosystem can be thought of as a sequence of energy changes, in which energy flows through the components of the ecosystem, subject to certain rules, described as the Laws of Thermodynamics.

There are many possible energy sources on Earth, e.g. geothermal, electrical and chemical.

- It is likely that the energy derived from an unequal distribution of protons allowed the non-living systems in the cavities of alkaline hydrothermal vents to make the transition into living systems.

- Early organisms used the energy released by chemical reactions to make carbohydrates by chemosynthesis. The electrons they need to reduce carbon dioxide or methane to sugar are derived from the oxidation of inorganic molecules such as hydrogen or hydrogen sulphide. Some Archaea and bacteria still do, but they tend to inhabit marginal ecosystems.

- The most significant energy source for ecosystems now is the light energy radiating from the Sun because light is the energy source for photosynthesis.

Biotic components of an ecosystem

Habitats

A habitat is an ecological or environmental area such as a fresh water stream or a temperate grassland, inhabited by a living organism. It provides the means of survival such as food, water, soil, appropriate temperature and pH. A habitat is not necessarily a geographical area, as part of one organism may be the habitat for another, e.g. the human duodenum can be the habitat of a tapeworm. A **microhabitat** is a very small area that differs from its surroundings and has the features that make it suitable for a particular species, e.g. the cabbage looper, which is a moth larva, feeds on the lower, but not the upper surface of leaves of cabbages and related crops.

Communities

The members of a species, living and interacting together in a habitat form a population. Populations interact to form a community. Community ecology studies the interactions of the species, related to their distribution and abundance, and their genotypic and phenotypic differences. It considers food web structure and predator–prey relationships.

Biomass transfer

The ultimate source of energy for most ecosystems is sunlight. Photosynthetic organisms convert sunlight energy into chemical energy which passes from organism to organism through a food chain. The study of the flow of energy through the ecosystem is ecological energetics. The energy available to a trophic level contributes to its biomass. Food chains can, therefore, be thought of as a means of transferring biomass.

Food chains

- Green plants, cyanobacteria and some Protoctista are called producers because they incorporate the Sun's energy into carbohydrates, which are the food, and therefore the energy source, for successive organisms in the food chain. They trap solar energy and synthesise sugars from inorganic compounds by photosynthesis.
- Only a small proportion of the total energy that reaches the plant as light is incorporated into the plant's tissues.
- Herbivores are primary consumers, i.e. they are animals that feed on plants. Carnivores are secondary, tertiary and higher consumers, i.e. animals that feed on other animals.
- Each of these groups operates at a feeding or trophic level with energy passing to a higher trophic level as material is eaten.
- Energy in the food consumed is incorporated into the molecules of the consumer.
- As energy is passed along the food chain there is a loss from the food chain at each level.
- As energy is lost at each trophic level, the energy flowing through the ecosystem reduces and ultimately the energy leaves the system as heat.

Decomposition

When producers and consumers die, energy remains in the organic compounds of which they are made. Detritivores and decomposers feed as saprobionts, i.e. they derive their energy from dead and decaying organisms, and they contribute to the recycling of nutrients.

- Detritivores are organisms, such as earthworms, woodlice and millipedes, which feed on small fragments of organic debris. This is detritus, the remains of dead organisms and fallen leaves.
- Decomposers are microbes such as bacteria and fungi that obtain nutrients from dead organisms and animal waste. They complete the process of decomposition started by detritivores.

Key Terms

Habitat: The place in which an organism lives.

Community: Interacting populations of two or more species in the same habitat at the time.

Trophic level: Feeding level; the number of times that energy has been transferred between the Sun and successive organisms along a food chain.

Biomass: The mass of biological material in living, or recently living, organisms.

Going further ▶

Plant parasites such as dodder and mistletoe are primary consumers even though they are plants, because they get their energy from their host, which is also a plant.

Exam tip

Take care with your use of words. Energy is not 'used', 'wasted' or 'destroyed'. It is 'lost from the ecosystem'.

Key Term

Saprobiont: A micro-organism that obtains its food from the dead or decaying remains of other organisms.

A detritus food chain can be summarised as: detritus ⟶ detritivore ⟶ decomposer.

Most of the energy entering a food chain is transferred through these detritus food chains, rather than being transferred between consumers.

Food chains and their length

A food web shows how the organisms in a community interact with each other through the food that they eat. A food chain is a linear sequence of organisms in a food web. A 'grazing food chain' with three consumers can be summarised as:

producer ⟶ **primary consumer (herbivore)** ⟶ **secondary consumer** ⟶ **tertiary consumer**

1st trophic level   2nd trophic level   3rd trophic level   4th trophic level

Energy is lost at each link along the food chain. After four or five trophic levels, there is not enough energy to support another one. The number of links in a chain is, therefore, normally limited to four or five, although their actual length may depend upon several interacting factors:

- The more energy that enters a food chain in the first trophic level, i.e. the more energy fixed in photosynthesis, the longer the food chain can be. So tropical food chains, which have high light all year, tend to be longer than Arctic food chains, which have much less light.
- If energy is transferred more efficiently between trophic levels, the food chain is longer.
- Predators and prey populations fluctuate and their relative abundance can affect the food chain length.
- Larger ecosystems can support longer food chains.
- Three-dimensional environments such as aquatic systems and forest canopies have longer food chains than two-dimensional habitats such as grasslands.

Photosynthetic efficiency

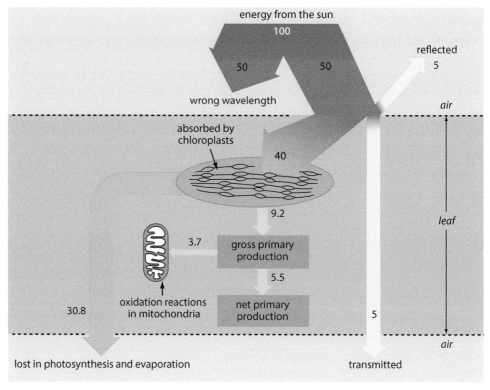

Fate of solar energy reaching the leaf of a crop plant (numbers are percentages of the total solar energy falling on the leaf)

▼ **Study point**

At a given light intensity, a leaf absorbs more energy per unit area if it is thicker, has a thinner cuticle, more chloroplasts, more chlorophyll or more grana.

- The energy flowing from one organism to another in the food chain originates as sunlight.

- About 60% of the light energy that falls on a plant may not be absorbed by photosynthetic pigments because it could be:

  - The wrong wavelength

  - Reflected

  - Transmitted straight through the leaf.

The diagram (opposite) shows the fate of 100 units of solar energy reaching the leaves of a crop plant per unit time, for a crop plant growing in in ideal conditions.

Photosynthetic efficiency (PE) is a measure of the ability of a plant to trap light energy:

$$PE = \frac{\text{Quantity of light energy incorporated into product}}{\text{Quantity of energy falling on the plant}} \times 100\%$$

In wild plants PE may be as low as 1%, but it is higher in crop plants, as they have been selectively bred for high productivity, e.g. for sugarcane, PE = 7–8%. The efficiency depends on a plant's genotype, and environmental factors such as light intensity and temperature.

Primary productivity

Gross primary productivity (GPP) is the rate of production of chemical energy in organic molecules by photosynthesis in a given area, in a given time, so its units are kJ m^{-2} y^{-1}. A substantial proportion of gross production is released by the respiration of the plant to fuel, for example, protein synthesis. The remains is **net (or nett) primary productivity** (NPP) and this represents the energy in the plant's biomass. This is the food available to primary consumers, or, in crop plants this represents the yield which may be harvested.

$$GPP - respiration = NPP$$

Both GPP and NPP are higher if plants have a high photosynthetic efficiency. Unlike crop plants, most plants are not highly selected for productivity and are not grown under ideal conditions, so, in contrast to the figures in the diagram above, a common figure for GPP is around 1% of incident radiation and for NPP, 0.5%.

Energy flow through food chains

Primary productivity is the rate at which producers convert energy into biomass. **Secondary productivity** is the rate at which consumers accumulate energy from assimilated food in biomass in their cells or tissues. Secondary production occurs in heterotrophs, including animals, fungi, some bacteria and some Protoctista.

Key Terms

Gross primary productivity: The rate of production of chemical energy in organic molecules by photosynthesis in a given area, in a given time, measured in kJ m^{-2} y^{-1}.

Net primary productivity: Energy in the plant's biomass which is available to primary consumers, measured in kJ m^{-2} y^{-1}.

Exam tip

If you are asked to calculate productivity, check that your answers are GPP ≈ 1% and NPP ≈ 0.5%.

Key Terms

Primary productivity: The rate at which energy is converted by producers into biomass.

Secondary productivity: The rate at which consumers convert the chemical energy of their food into biomass.

▼ Study point

A primary consumer converts about 10% of ingested energy into biomass. A secondary or tertiary consumer converts about 20%.

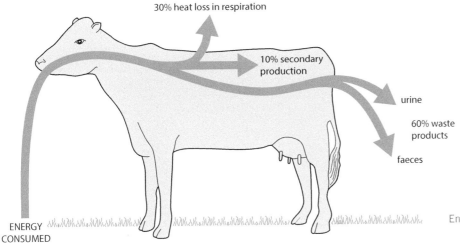

30% heat loss in respiration

10% secondary production

urine

60% waste products

faeces

ENERGY CONSUMED

Energy flow through bullock

Lion – usually a secondary consumer

As energy is passed along the food chain, there is a loss from the food chain at each level because:

- There is energy in molecules that are egested. Cows feed on plant material that contains cellulose, which is digested by their mutualistic micro-organisms. Its remains pass out of the body in faeces, which contain a high proportion of undigested material. This energy, however, is not lost from the ecosystem as it is available to decomposers. A carnivore's protein-rich diet is more readily and efficiently digested. Consequently, only about 20% of its energy intake is lost in waste products, compared with about 60% in herbivores.

- Energy lost as heat following processes fuelled by the energy generated in respiration, including muscle contraction.

- Energy remains in molecules in parts of an animal that may not be eaten, for example horns, fur and bones.

Herbivores, the primary consumers, have a conversion efficiency of about 10%, that is, for every 100 g of plant material ingested, only about 10 g is incorporated into their biomass. This means that only part of the NPP of the whole ecosystem is transferred to the primary consumers: herbivores do not eat all the vegetation available to them; for example, cattle grazing a field eat grasses and many other small herbaceous plants but not roots and the woody parts of plants. Carnivores are more efficient at energy conversion than herbivores, because their food is more quickly and easily digested, so they have a conversion efficiency of about 20%.

Fate of energy in consumers	% ingested energy in consumer	
	Primary	Secondary
Heat loss from reactions of respiration	30	60
In excreted and egested waste products	60	20
Secondary production	10	20

Calculating the efficiency of energy transfer

The efficiency of energy transfer between trophic levels is calculated using the following equation:

$$\text{efficiency of energy transfer} = \frac{\text{energy incorporated into biomass after transfer}}{\text{energy available in biomass before transfer}} \times 100\%$$

In a year-long study of a fresh water ecosystem, 1609 kJ m^{-2} in primary consumers (C1) was ingested by secondary consumers (C2). The detritus from the secondary consumers represented 193 kJ m^{-2} and 88 kJ m^{-2} was transferred to tertiary consumers (C3). The efficiency of transfer between primary and secondary consumers in this ecosystem can be calculated:

The data can be presented as a food chain:

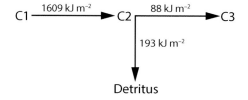

Using the equation: efficiency of energy transfer = $\frac{88 + 193}{1609} \times 100 = 17.5\%$ (1 dp)

This is close to the generally accepted figure of 20% for transfer between carnivores.

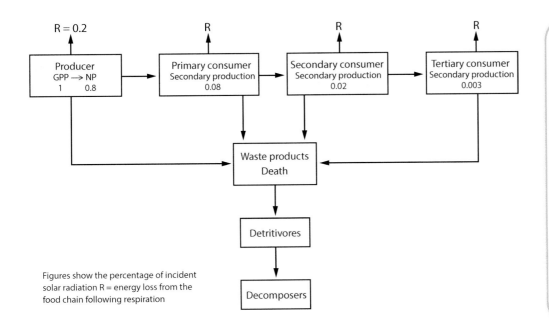

Figures show the percentage of incident solar radiation R = energy loss from the food chain following respiration

Energy flow through the grazing and detritus food chains of a grassland

Ecological pyramids

An ecological pyramid is a diagram that shows a particular feature of each trophic level in an ecosystem. Producers are drawn at the bottom and successive trophic levels are drawn above. The number of organisms, the energy or the biomass contained in each trophic level, is represented by a bar for each level. The area of each bar of the pyramid is proportional to the total number, energy or biomass at each trophic level. These are known as pyramids of numbers, energy and biomass respectively.

Although ecological pyramids are useful in describing ecosystems, they do not take account of the fact that some organisms operate at more than one trophic level at the same time. A human, for example, is an omnivore. When eating an egg sandwich, the human is:

- A primary consumer when eating the bread, which is made from wheat, a producer.
- A secondary consumer when eating the egg, which is produced by a chicken, a primary consumer.

The diagram below shows two different pyramids for the same ecosystem.

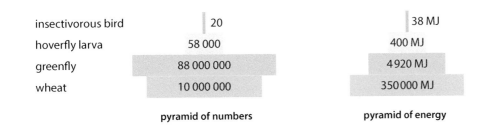

	pyramid of numbers	pyramid of energy
insectivorous bird	20	38 MJ
hoverfly larva	58 000	400 MJ
greenfly	88 000 000	4 920 MJ
wheat	10 000 000	350 000 MJ

Ecological pyramids

The pyramid of numbers

The pyramid of numbers is relatively easy to construct but:

- It does not take into account size of organisms.
- It does not recognise the difference between juvenile and adult forms.
- The range of numbers may be so large that it may be difficult to draw to scale.
- The pyramid, or part of it, may be inverted if one trophic level has more organisms than the previous trophic level, e.g. a single oak tree may have many thousands of caterpillars eating its leaves.

The pyramid of energy

The most accurate way of representing feeding relationships is to use a pyramid of energy, which shows the energy transferred from one trophic level to the next, per unit area or volume, per unit time. As material passes up through the food chain, energy is lost from the ecosystem so the area of the bars decreases accordingly. Since only some of the energy is passed on from one level to the next, energy pyramids are never inverted. Pyramids of energy make it easy to compare the efficiency of energy transfer between trophic levels in different communities.

The pyramid of biomass

Energy is incorporated into the macromolecules that make up the biomass of an organism. If the available energy is greater, more biomass can be supported, so a pyramid of biomass is closely related to the energy passing through the ecosystem. However:

- Pyramids of biomass are difficult to measure accurately as, for example, all the plants' roots must be included.
- They do not indicate productivity or the amount of energy flowing through the ecosystem.
- Pyramids of biomass may be inverted.
- In a pyramid of biomass, a trophic level may seem to contribute more to the next trophic level than it actually does. This is because many organisms contain structures with mass that will not transfer to the next trophic level, e.g. bones, beaks.
- Species with similar biomass may have different life spans. A direct comparison of their total biomasses is therefore misleading.

Inverted pyramid of biomass

In an aquatic ecosystem, phytoplankton are the major producers. A lot of energy flows through the first trophic level and the phytoplankton reproduce very quickly. Some are eaten immediately, leaving just enough to maintain the population. So their **standing crop**, i.e. the mass of individuals present at a given time is lower than the biomass of zooplankton, which eat them.

Data from Eniwetak coral reef in the Pacific Ocean		Data from the English Channel	
11	Secondary consumer		
132	Primary consumer	4	Zooplankton
703	Producer	21	Phytoplankton

Pyramids of biomass in g m^{-3}

Community and succession

Ecosystems are dynamic and subject to change. Organisms and their environment interact. A change in the environment affects the organisms, and a change in the organisms affects the environment. The change in community structure and species over time is **succession**. It may occur over decades, e.g. after a wildfire, or over millions of years, e.g. after a mass extinction. New species invade and replace existing ones, until eventually a stable community called the **climax community**, for example a mature woodland, establishes itself.

Primary succession

Primary succession is the sequence of changes following the introduction of species into an area that has not previously supported a community, e.g. bare rock or the site of a volcanic eruption. The sequence of communities, with the different species and structures, is called a **sere**, and a sere in a very dry environment is a **xerosere**. The stages of a sere are called **seral stages**. Each seral stage changes the environment and makes it more suitable for other species. When a new species immigrates, it may outcompete those that are there already and in this way, succession progresses.

In the UK, the most recent ice age lasted from about 110 000 to about 12 000 years ago. Bare rock eroded by receding ice has undergone a xerosere succession to mixed deciduous forest:

- The first organisms to colonise the bare rock are algae and lichens. These are **pioneer species** and form a pioneer community.

- The weathering of the rock, its erosion by lichens and the accumulation of dead and decomposing organic material leads to the formation of a primitive soil. Animals, e.g. mites, ants and spiders, can survive when there is enough food for them.

- Wind-blown spores allow mosses to grow and, as the soil develops, grasses and other small, herbaceous plants outcompete the mosses and become established. Well-dispersed species that make a lot of seed that can germinate in direct sunlight are favoured. Animals include nematodes, ants, spiders and mites.

- Tall grasses allow shade-tolerant species to become established and the community becomes more complex.

- As these plants and animals die and decay, the soil becomes thicker, with more minerals. The greater quantity of humus allows it to hold water more efficiently. As the soil builds up, deeper-rooted plants including shrubs, e.g. gorse and broom, and small trees, e.g. hawthorn, outcompete the herbaceous plants.

- The soil continues to deepen and increase in minerals and humus. Over a very long time, large trees such as oak and beech outcompete the shrubs and small trees to become established. These grow to form a stable, self-perpetuating climax community, which:

  - Has great species diversity

  - Has a complex food web

  - Is dominated by long-lived plants.

The species of the climax community depend largely on the climate, so the climax community is also called the **climatic climax** community. Animal diversity is at its highest and includes invertebrates, e.g. slugs, snails, worms, millipedes, centipedes, ants, and vertebrates, e.g. squirrels, foxes, mice, moles, snakes, birds, salamanders and frogs. But the tree canopy limits the intensity of light reaching the woodland floor and so plant diversity decreases slightly from its pre-climax state.

Key Terms

Succession: The change in structure and species composition of a community over time.

Climax community: A stable, self-perpetuating community that has reached equilibrium with its environment, and no further change occurs.

Primary succession: The change in structure and species composition of a community over time in an area that has not previously been colonised.

Pioneer species: The first species to colonise a new area in an ecological succession, e.g. algae, lichens and mosses in a xerosere.

▼ **Study point**

A xerosere sequence is pioneers → herbs and grasses → shrubs and small trees → large trees.

▼ **Study point**

Animals undergo succession, dictated by the plant types present at each stage.

The climax community is balanced, with equilibrium between:

- GPP and total respiration
- Energy used from sunlight and released by decomposition
- Uptake of nutrients from the soil and their return by decayed plant and animal remains
- New growth and decomposition, so the quantity of humus is constant.

bare rock colonised by algae, fungi and lichens

pioneer community of heather and mosses

herbs and low-growing shrubs

taller shrubs

oak/beech forest

birch and pine saplings

Succession

As a xerosere progresses, increases are seen in:

- Soil thickness and availability of water, humus and minerals
- Biomass
- Biodiversity
- Resistance to invasion by new species
- Stability to disruption by environmental challenges, e.g. abnormal weather.

Secondary succession

Secondary succession is the recolonisation of a habitat previously occupied by a community, but disturbed, for example, by fire or by tree felling. The area rapidly becomes re-colonised by a succession of organisms. The actual species depend on the conditions prior to the disturbance, e.g. soil thickness, mineral and humus content. Seeds, spores and organs of vegetative reproduction, such as bulbs, corms or rhizomes, may remain in the soil and dispersal of plants and migration of animals will assist in colonisation of the habitat. As the soil is fertile and there are organisms still present, a secondary succession has the same overall sequence as primary succession, but the succession is very rapid. Secondary succession is much more commonly observed and studied than primary succession. Common species that rapidly colonise are grasses and other herbaceous plants, heathers, brambles and birch trees.

Disclimax

Human interference can affect a succession and may prevent the development of the climatic climax community, e.g.:

- Grazing by sheep and cattle maintains grassland and prevent the shrubs and trees of a normal succession from growing.
- Farming of land removes all except deliberately introduced species and great effort is expended in excluding all others.
- Deforestation removes a community of large trees, and smaller trees may be replanted.

Key Term

Secondary succession: The changes in a community following the disturbance or damage to a colonised habit.

Going further ▶

A stable community maintained by human activity is called the anthropogenic climax or the disclimax, which is an abbreviation of 'disruption climax'.

Heather moors are subject to management to provide ideal conditions for game birds such as the red grouse. Much research has been done on the management of grouse moors and grouse breeding to support grouse shooting. The diagram shows a profile of the four growth phases in the life cycle of heather, *Calluna vulgaris*.

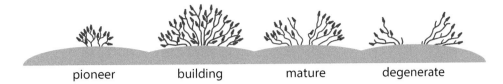

pioneer   building   mature   degenerate

Four stages of heather growth

Adult grouse feed mainly on young, succulent heather shoots. The main form of management is by burning, which is carried out about every twelve years, usually in the autumn or early winter. A secondary succession occurs:

- The pioneer phase supplies the best food for the adult grouse.
- The building phase provides the best shelter for nesting.

Without management the heather would pass through the mature to the degenerate phase and conditions would become unsuitable for the breeding of grouse.

Factors affecting succession

Migration

The arrival of spores, seeds and animals is vital for succession to progress. Immigrating non-native species may spread themselves widely, altering the community and the soil.

Competition

In all communities, organisms compete for survival. Plants compete for resources such as light, space, water and nutrients. Animals compete for food, shelter, space and reproductive partners.

- Intraspecific competition is competition between individuals of the same species. This type of competition is density-dependent since, as the population increases, there is more competition. When the population is denser, a greater proportion fails to survive. This has value to a population since organisms tend to produce more offspring than the habitat can support, so the numbers are regulated. Those organisms with alleles that make them best suited to the environment reproduce more successfully.

- Interspecific competition is competition between individuals of different species. Different species have some common needs, e.g. all fish species compete for dissolved oxygen in the water, but they also have different requirements. Each has its own **niche** in an ecosystem, i.e. it occupies a particular place and has a particular role within the community. The niche is a total description of an organism's way of life.

Competition operates at all the seral stages of a succession and illustrates how, for example, a herb may outcompete a moss at the end of the pioneer stage, but be outcompeted by a shrub as the soil thickens and more nutrients are available. In the long term, two species cannot occupy the same niche in a specific habitat. Whichever has the competitive advantage will survive.

The Russian scientist, Gause, cultured two species of *Paramecium,* with yeast as their food source. When grown separately under identical conditions, *P. aurelia* and *P. caudatum* showed typical one-step growth curves. When grown together, after a short time, the smaller, faster-growing *P. aurelia* outcompeted the larger, slower-growing *P. caudatum,* which eventually died. This counting of the population numbers of the *Paramecium* species cultured together and separately enabled Gause to formulate a general principle, known as the 'competitive exclusion principle', i.e. when two species occur in the same habitat, one will out-compete the other; two species cannot occupy the same niche.

Competition between two species of *Paramecium*

Facilitation

The association between individuals of two species is called **symbiosis**. There is a range of interdependence between the individuals. In some cases, the association is long term and the organisms are highly inter-dependent and in others, the association is loose.

To 'facilitate' means to enable something to happen. Ecology often looks at negative interactions such as predation and competition, but positive interactions, called facilitation, are important. They are increasingly significant as a succession progresses and as communities become more complex. Facilitation in an ecological community provides better resource availability and refuge from physical stress, predation and competition, e.g. in a xerosere:

Key Term

Mutualism: An interaction between organisms of two species from which both derive benefit.

- **Mutualism** is an interaction between species that is beneficial to both, e.g.:
  - The highly inter-dependent interaction between a fungus and an alga or cyanobacteria in lichens.
  - The relationship between flowering plants and their pollinators, in a climax community, e.g. bumblebees pollinating horse chestnut trees.
  - A bird eats insects off a deer so the bird feeds and the deer becomes insect-free, in a loose association.
  - Mycorrhizae, i.e. the close association between fungi and the roots of plants at any stage of succession.

Starlings perched on a red deer stag

- **Commensalism** is described as a loose interaction between organisms of two species in which one benefits and the other is unaffected, e.g.
  - A squirrel in a climax oak woodland is protected from predators and sheltered by an oak tree, and the tree is not affected.
  - Nurse plants are plants that make a canopy that protects individuals of other species. Seeds may germinate under the shelter of the leaves of a larger plant, which subsequently also provides soil moisture and nutrients, e.g. saguaro cactus seedlings, which may take up to 3 years to grow only 20 mm high, are shaded by shrubs. Initially the shrub is unaffected, but as the cactus grows, they may eventually compete with each other for light, minerals and water.

Over long periods of time, as the species in an interaction evolve, their relationship may change. An individual that is a parasite may evolve and provide its host with a useful molecule, so the relationship becomes commensal, e.g. bacteria in mammalian guts provide their host with vitamin K, which may also be obtained from the diet, so the mammal is not affected. But if the host does not acquire enough vitamin K from its diet, the relationship becomes mutualistic.

Recycling nutrients

Plants make their food in photosynthesis and all consumers acquire their food by eating producers or other consumers. Detritivores and decomposers break down the remains and the waste products of these organisms. Minerals return to the soil and plants take them up again. In contrast with energy, which is transferred in a linear fashion, minerals cycle between the biotic and abiotic components of the environment.

The carbon cycle

Carbon is a major component of all organic molecules, including carbohydrates, fats and protein. During the day plant photosynthesis converts carbon dioxide from the air into carbohydrate. All organisms return it to the air in respiration. The atmospheric concentration of carbon dioxide has fluctuated somewhat over millions of years but over the past few hundred years it has increased greatly. Two human activities have been the main cause of this increase:

- Burning fossil fuels releases carbon dioxide that was previously locked up in them, into the atmosphere.

- Deforestation has removed large quantities of photosynthesising biomass and so less carbon dioxide is being removed from the atmosphere.

The carbon cycle involves three major biological processes:

- Respiration – carbon dioxide is added to the air by the respiration of animals, plants and micro-organisms and by the combustion of fossil fuels.

- Photosynthesis – this takes place on so great a scale that it re-uses, on a daily basis, almost as much carbon dioxide as is released into the atmosphere by respiration.

- Decomposition – the production of carbohydrates, proteins and fats contributes to plant growth and subsequently to animal growth through complex food webs. The dead remains of plants and animals are then acted upon by detritivores and saprobionts in the soil, which ultimately release carbon dioxide gas back to the atmosphere.

Key Term

Commensalism: An interaction between organisms of two species from which one benefits but the other is not affected.

▼ Study point

Immigration, facilitation and competition all have significant roles in the relationships between species, as a community changes through a succession.

YOU SHOULD KNOW ›››

››› The role played by detritivores and saprobionts in nutrient cycles

››› How to draw a labelled diagram of the carbon cycle

››› The effects of deforestation and combustion in the carbon cycle

››› Causes and effects of global warming

▼ Study point

Energy moves along a chain but minerals cycle.

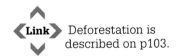 **Link** Deforestation is described on p103.

▼ Study point

If decay of biological remains is prevented by anaerobic or acidic conditions then organisms may become fossilised into coal, oil, natural gas or peat.

In aquatic ecosystems, carbon dioxide, as HCO_3^- ions, undergoes the same processes as described for terrestrial organisms. In addition, it is incorporated into magnesium and calcium carbonate in mollusc shells and arthropod exoskeletons, which sink after the animal's death. The carbonates become components of chalk, limestone and marble and are lost from the biosphere. But if geological processes expose them to the atmosphere they are eroded, releasing carbon dioxide back into the air.

The carbon cycle in air and in water

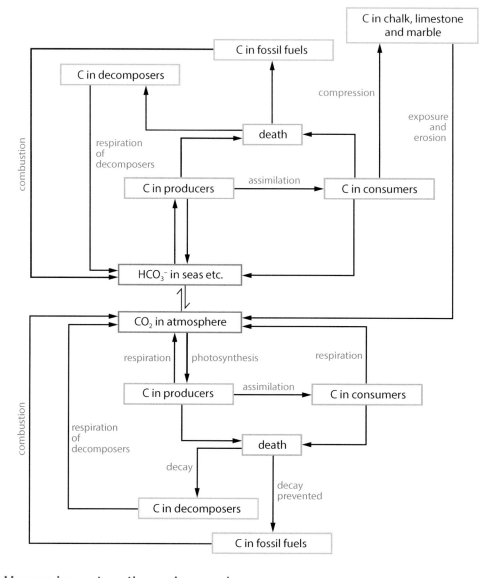

Human impact on the carbon cycle

Deforestation

Deforestation increases the carbon dioxide content of the atmosphere in two ways:

- The rate at which carbon dioxide is removed from the atmosphere by the process of photosynthesis is reduced by cutting down forests. On a global scale this is a massive reduction and contributes to global warming.

- When trees are cut down, they may be burned or left where they are cut, to decay. Both these processes release carbon dioxide into the atmosphere.

Deforestation in British Columbia

Climate change

Changes in global and regional climate patterns became noticeable in the second half of the 20th century, e.g. changes in average temperature, wind patterns and rainfall. The huge rise in atmospheric carbon dioxide shown in the table, and the rise in other greenhouse gases is thought to be the cause. The two major reasons for this increase in carbon dioxide are:

- Burning fossil fuels: this accounts for most of the increase, particularly from industrialised countries. The increase in industrialisation and in global transport have contributed to the steep, recent increase in greenhouse gas emissions.

- Deforestation: Nature Conservancy estimates that human activity has reduced global forest area by 40% of its pre-human cover. Forests affect the maintenance of the balance of carbon dioxide in the atmosphere.

This increase is important because carbon dioxide is a 'greenhouse gas'. It absorbs radiation from the Earth and if it accumulates in excess, it leads to global warming.

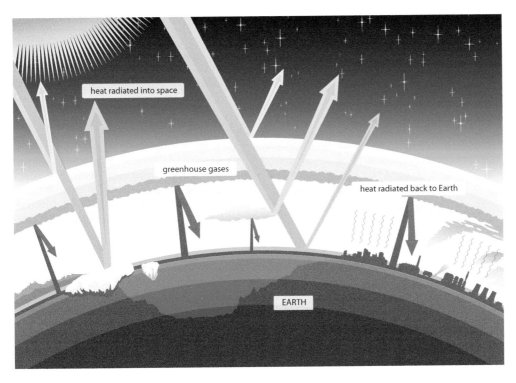

The greenhouse effect

The greenhouse effect

Greenhouse gases in the atmosphere are so-called because they are said to behave like the glass in a greenhouse. They include carbon dioxide, methane, nitrous oxide, CFCs, ozone and water vapour. These gases, like the windows of greenhouses, allow high-energy, short wavelength solar radiation to pass through to the Earth's surface. Much of this energy is absorbed by the Earth, which warms up and re-radiates lower energy, longer wavelength infra-red radiation. This is absorbed and trapped by the greenhouse gases in the atmosphere. These gases re-radiate the energy in all directions and the energy re-radiated back to the Earth's surface is absorbed. So the planetary surface and the atmosphere warm up. The greenhouse effect is a natural process, without which the average temperature on Earth would be about –20°C and too low to sustain life.

Going further ▶

Year	Atmospheric CO_2 concentration / ppm
1900	290
1960	315
1990	350
2015	398

Key Term

Global warming: The increase of average global temperature, in excess of the greenhouse effect caused by the atmosphere's historical concentration of carbon dioxide.

Going further ▶

International conferences have set targets for greenhouse gas reduction although these are not legally binding.

▼ Study point

A 'methane burp' is the release of methane into the atmosphere when methane clathrates (solid structures containing methane on the ocean floor) warm as sea temperatures rise. These and melting Arctic permafrost are likely to increase atmospheric methane.

Global warming

The enhanced greenhouse effect caused by high concentrations of greenhouse gases is called global warming. When scientists talk about the issue of climate change, their concern is about global warming caused by human activities. Even if greenhouse gas emissions stopped immediately, their influence would continue, because global processes take time to show an effect. Computer modelling indicates that the average global temperature would still increase overall by about 1.5°C above the average over the last 100 years. If the concentrations of greenhouse gases continue to rise at current rates, however, models predict that the average global temperature could increase up to 6.5°C within the next 50 years.

Some of the consequences of global warming may be:

- Melting of polar ice resulting in flooding in coastal areas.
- Increased frequency of extreme weather conditions such as droughts, hurricanes and cyclones.
- Increased frequency of forest fires.
- In tropical areas, decreased availability of water leads to the formation and expansion of deserts.
- Evolutionary adaptation is slow so as the climate changes, animals must move to find a more suitable environment. Plants can only move as fast as they are dispersed, and if this is not fast enough, they may be driven to extinction. Animals dependent on the plants will then become extinct and entire ecosystems may collapse.
- Fishing areas and crop-growing belts may move as the climate changes, e.g. peppers and pumpkins are now grown as crops in the UK whereas 50 years ago, it was not warm enough. Conversely, in 2012, the summer heat and drought reduced the soybean yield in the USA considerably below the average of the last 30 years.
- There may be increased crop yields as warmer temperatures allow more photosynthesis. But pest populations might also increase.
- World food production may decrease, with massive reductions in the grain crops of North America and Central Asia. There would be serious economic and political consequences.
- Increasing carbon dioxide in the oceans decreases the pH, which threatens many organisms, for example fish populations secrete mucus to protect their gills and then are unable to perform gas exchange; coral reefs have an external skeleton made largely of calcium carbonate, which is soluble in acid.

Global warming and farming

Agriculture is very vulnerable to the impacts of climate change and is likely to be affected through changes in temperature and the timing and quantity of rain. Extreme events, such as droughts and floods are likely to be more frequent. As global warming increases, fresh water will be critical to sustain food production and provide food for the world's growing population. Fluctuations in river flow are likely to increase and, based on observations of the River Nile, reduced water supply at critical times can halve crop yield.

Some problems for agriculture that are associated with climate change are described in the table:

Problem	Source	Altered farming practices to reduce gas emission
Carbon dioxide	Decomposition of soil organic matter	Improve soil quality by: • Conservation tillage: leave crop residues on the soil surface to reduce erosion, improve water use and add organic matter to the topsoil • Cover cropping: use, e.g. clover to cover soil to protect and improve it between crops. Enhances soil structure and adds organic matter to the topsoil • Crop rotation reduces pest numbers and mineral depletion
Methane	Digestive activities of farm animals used in the meat and dairy industries	• Reduce dietary intake of meat and dairy products • High-sugar grasses, oats, rapeseed and maize silage in cows' diets reduce the methane they release
	Decomposition in wet soils, e.g. in rice paddies	• Use rice varieties that grow in drier conditions • Select varieties that have a higher yield • Ammonium sulphate addition can favour non-methane producing micro-organisms in paddy fields in some conditions
Nitric oxide and nitrous oxide	Waterlogged and anaerobic soils	Improve drainage to remove water and aerate soil.
Low, fluctuating water supply	Low rainfall; high temperature	Use drought-tolerant crops, e.g. in Kenya, the use of drought-tolerant sorghum, millet and cowpea have increased yields significantly
Raised sea level	Cultivated land inundated by salt water	Salt-tolerant crops, e.g. a salt-tolerant potato; a salt-tolerant durum wheat yielding 25% more grain than the parent variety

Carbon footprint

Using an item generates greenhouse gases but producing the raw materials for its manufacture, its transport and its disposal may generate a great deal more. The **carbon footprint** is a way of measuring a contribution to the greenhouse gases in the atmosphere. It is defined as the equivalent amount of carbon dioxide generated by an individual, a product or a service in a year. It is defined in this way because carbon dioxide is not the only greenhouse gas. A molecule of nitrous oxide has 298 times the potential for global warming of one molecule of carbon dioxide and one molecule of methane has 25 times the potential. The effect of the emissions of these, and other greenhouse gases, are, therefore, often expressed as 'carbon dioxide equivalents'.

Crops absorb carbon dioxide from the atmosphere as they grow, but they also incur indirect sources of greenhouse gases, e.g.

- The production of farming tools
- The production of insecticides, herbicides, fungicides and fertilisers
- Farm machinery, powered by fossil fuels
- Transport of produce. Most crops are shipped hundreds of miles to processing plants before distribution.

To reduce the production of greenhouse gases, the '3Rs' have been publicised, i.e. reduce, reuse and recycle. They emerged from the first annual Earth Day celebration in 1970 and recommend that we:

- Recycle packing material
- Drive less
- Use less air conditioning and heating, e.g. by insulating, wearing suitable clothing
- Choose a diet to reduce animal protein, especially red meat, rice, because of methane-emitting paddies, foods transported long distance and heavily processed and packaged foods
- Avoid food waste; turn it into compost, if possible
- Plant trees in deforested regions.

Key Term

Carbon footprint: The equivalent amount of carbon dioxide generated by an individual, a product or a service in a year.

Going further ▶

The carbon footprint is one of several footprint indicators, modelled on the ecological footprint, conceptualised in the 1990s. Now there are also calculations for a water footprint and a land footprint.

Going further ▶

An atom of nitrogen takes about 2000 years to go around the cycle. So you probably have at least one nitrogen atom that belonged to the Buddha, Aristotle or Cleopatra.

Exam tip

Learn the generic names of some of the bacteria involved in the nitrogen cycle: *Nitrosomonas, Nitrobacter, Azotobacter* and *Rhizobium*.

Key Term

Nitrification: The addition of nitrogen to the soil, most commonly as nitrite (NO_2^-) and nitrate (NO_3^-) ions.

The nitrogen cycle

The nitrogen cycle is the flow of nitrogen atoms between organic and inorganic nitrogen compounds and atmospheric nitrogen gas in an ecosystem.

Living organisms need nitrogen to make amino acids, proteins and nucleic acids. Plants and animals are unable to use nitrogen gas. Instead plants absorb nitrates into their roots. The organic nitrogen compounds produced by plants are transferred through the food chain because primary consumers eat plants. The decomposition of plants and animals after death and of excreted and egested products of animals, releases the minerals back into the soil. Bacteria are the key organisms involved in the processes of the nitrogen cycle.

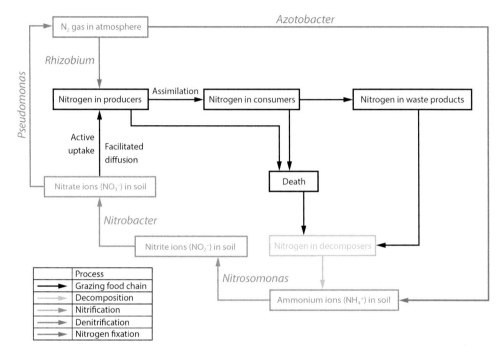

The biological processes in the nitrogen cycle

There are four main biological processes in the nitrogen cycle:

- **Ammonification**, also called putrefaction.

  Bacteria and fungi are decomposers. They secrete enzymes that decay dead organisms and animal products:

  – Proteases digest proteins into amino acids.

  – Deaminases remove ($-NH_2$) groups from amino acids and reduce them to ammonium ions (NH_4^+).

- **Nitrification**

  To nitrify means to add nitrites (NO_2^-) or nitrates (NO_3^-) to the soil. The ammonium ions formed in putrefaction are converted, in a process called **nitrification**, to nitrites, and then to nitrates. Various bacteria are involved, for example, *Nitrosomonas converts* ammonia to nitrite and *Nitrobacter converts* nitrite to nitrate:

$$NH_4^+ \xrightarrow{\quad Nitrosomonas \quad} NO_2^- \xrightarrow{\quad Nitrobacter \quad} NO_3^-$$

The nitrogen atom in the ammonium ion progressively loses hydrogen atoms and gains oxygen atoms and so the reactions are oxidations. *Nitrosomonas* and *Nitrobacter* therefore require aerobic conditions.

- **Denitrification**

  Denitrification is the loss of nitrate from the soil. Anaerobic bacteria, such as *Pseudomonas*, convert nitrate ions to nitrogen.

  $$NO_3^- \xrightarrow{\text{\textit{Pseudomonas}}} N_2$$

  This is a reduction reaction because oxygen is lost. It is, therefore, favoured by anaerobic soil conditions and so it is a particular problem in waterlogged soils.

- **Nitrogen fixation**

  Although 79% of the atmosphere is nitrogen, very few organisms can use it as they do not have the enzymes that break the triple bond between the atoms in nitrogen molecules. Geological processes releasing nitrates and ammonium ions are very slow. The biosphere therefore relies on several prokaryotic species possessing a group of enzymes that can reduce nitrogen molecules to ammonium ions, in the process called **nitrogen fixation**. The bacteria are the nitrogen-fixing organisms. ('Fixing' is an old-fashioned term meaning 'combining'.)

Azotobacter is a free-living nitrogen-fixing bacteria in the soil and accounts for most biological nitrogen fixation. Symbiotic nitrogen-fixing bacteria include *Rhizobium*, found in the root nodules of plants in the Fabaceae family, i.e. the legumes (peas, beans and clover).

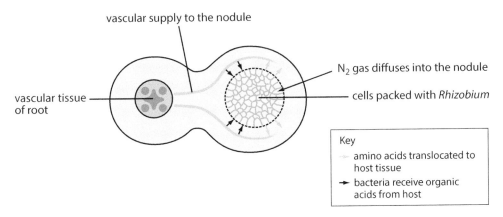

Key
- amino acids translocated to host tissue
- bacteria receive organic acids from host

Diagram of cross-section of root and root nodule of legume

Nitrogen gas diffuses into the legume root nodule and nitrogenase catalyses its reduction to ammonium ions, using energy from ATP. Nitrogen fixation requires a lot of energy so nitrogen-fixing organisms are often aerobic. But nitrogen fixation requires reduction reactions, and is poisoned by oxidising conditions. Root nodules contain a type of haemoglobin called leg-haemoglobin that binds molecular oxygen in the nodules and protects the reactions from oxidation. It makes the nodules pink.

Ammonium ions are converted to organic acids and then into amino acids for incorporation into bacterial proteins:

$$N_2 \xrightarrow{\text{Reduction by nitrogenase}} NH_4^+ \longrightarrow \text{organic acids} \longrightarrow \text{amino acids}$$

Some amino acids and ammonium ions are diverted into the vascular strand connecting the root nodule to the plant and in this way, the plant obtains nitrogen for its own metabolism.

Key Term

Nitrogen fixation: The reduction of nitrogen atoms in nitrogen molecules to ammonium ions, by prokaryotic organisms.

Going further ▶

The enzymes involved in nitrogen fixation include hydrogenase, ferredoxin and nitrogenase. Nitrogenase contains iron and molybdenum, although some bacteria have iron and vanadium instead and can fix nitrogen as low as 5°C.

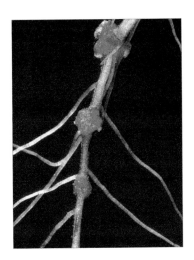

Frankia bacteria in nodules on *Alnus glutinosa*, black alder, contain leg-haemoglobin

‹Link› Plants without mutualistic nitrogen-fixing bacteria take up nitrates and ammonium ions from the soil by active transport and facilitated diffusion, as described in the first year of this course.

Knowledge check

Match the terms 1–4 with the statements A–D.

1. *Rhizobium.*
2. Ammonification.
3. Nitrogen fixation.
4. Denitrification.

A. The conversion of nitrates to nitrogen gas.
B. The decomposition of dead organisms.
C. The conversion of atmospheric nitrogen into nitrogen compounds.
D. A bacterium found in the root nodules of clover.

Going further ▶

The Haber process fixes vast amounts of nitrogen using 200 atmospheres pressure and 450ºC, so it is very energy intensive. Biological nitrogen fixation, by contrast, operates at ambient temperature and pressure.

The radicles of newly germinated legumes do not contain *Rhizobium*. In the soil, the newly emerged radicle and *Rhizobium* both secrete chemo-attractants. The radicle grows towards *Rhizobium* cells and the *Rhizobium* cells use their flagella to move towards the radicle. *Rhizobium* cells invade the cortex of the radicle and the large number produced when they replicate makes the swelling which is the root nodule. The nodules and the bacteria they contain enable leguminous plants to grow successfully even when soil nitrates are scarce. On the death of the plant, both the plant and the *Rhizobium* that the nodules contain are decomposed into ammonium compounds, which are released into the soil. The ammonium ions are nitrified to nitrites and nitrates, improving the soil quality further.

Non-biological processes have an impact in the nitrogen cycle:

- The application of agricultural fertilisers adds nitrogen to the soil
- Lightning adds a small amount of nitrogen to the soil
- Leaching of minerals removes nitrogen from the soil.

Human impact on the nitrogen cycle

As the human population increases, there is a need to produce increasing amounts of food. Plant breeding and genetic modification make a contribution to this, as does the use of pesticides. Treating soil and maintaining its structure are also vital for efficient food production.

Human activities can improve the circulation of nitrogen in agricultural soils:

- Ploughing fields improves soil aeration. This favours:
  - Aerobic organisms, such as free-living nitrogen fixers, enhancing the formation of ammonium ions in the soil.
  - Nitrifying bacteria and therefore enhances the conversion of ammonium into nitrites and nitrates.
  - Plant roots respire aerobically and generate ATP, which fuels their active uptake of minerals.
- Draining land allows air to enter the soil and so it reduces the anaerobic conditions, which favour denitrifying bacteria. In this way, the loss of nitrates is reduced.
- Artificial nitrogen fixation, e.g. the Haber process, converts nitrogen to fertilisers, essential to produce high volumes of good quality food. Fertilisers are largely compounds containing ammonium and/or nitrate ions, e.g. ammonium nitrate, and they augment the ammonium ions and nitrates produced by nitrogen-fixing and nitrifying bacteria, respectively.
- Large amounts of animal waste, e.g. chicken manure and cow dung, from stock rearing are used as 'brown' manure. The nitrogen and other nutrients it contains are essential for the growth of plants. It improves the soil structure so that the soil holds more nutrients and water, and is more fertile. It encourages microbial activity, which promotes the soil's mineral supply, improving plant nutrition. It is transported by road in large containers from areas where animals are farmed. Manure releases nitrogen compounds to the soil gradually, over a prolonged period.
- Slurry is a liquid made from manure and water. It is produced by more intensive livestock-rearing systems, where concrete or slats are used, instead of straw bedding. It is usually stored in a tank or lagoon before use. Slurries from intensive pig farming have an extremely unpleasant smell so they are usually injected into the soil, rather than being spread on the surface, as is brown manure. Herbivore manure, such as that from cows, has a milder smell than that from carnivores or omnivores because there is less protein in their diet and in their waste. Alternative feed for pigs is being developed to reduce this particular problem.

- Treated sewage sludge, known as 'biosolids', is a sustainable alternative to inorganic fertilisers.
- Planting fields of legumes, such as alfalfa or clover enhances nitrogen fixation. When the crop dies, it is ploughed back into the soil as 'green manure'. Its value is in its high nitrogen content.

Effects of fertilisers on habitats

Upland streams are oligotrophic, which means that they have very few minerals dissolved in them. As water flows over rocks, it dissolves minerals and their concentration increases. Water enriched with minerals is described as eutrophic. When the mineral concentration is so high that organisms die, the water is described as dystrophic. **Eutrophication** is the process of artificially increasing the mineral content of water as a result of nitrogen-containing fertilisers, especially nitrates, leaching from agricultural land.

In developed countries agriculture has become more intensive, providing ever-higher yields of crops from ever-smaller areas of land. However, this increase in food production has had a harmful effect on the environment. The increased use of nitrate-containing fertilisers has harmed both aquatic and terrestrial ecosystems.

- Problems caused by excess nitrate in soils: on agricultural land the increased use of fertiliser has reduced species diversity on grassland. Fertilisers increase the growth of grasses and plants such as nettles, which shade out smaller plants.
- Problems caused by nitrates leaching into rivers: the leaching of nitrates and phosphates from the surrounding land is a slow, natural process during which the concentration of salts builds up in bodies of water. In lakes and rivers, the mineral ions normally accumulate until equilibrium is reached, and their addition to the water is exactly counterbalanced by the rate at which they are removed. However, sewage and fertilisers are an additional source of these minerals and their leaching from the land into the water may result in eutrophication of lakes and rivers.

A eutrophic lake

▼ **Study point**

Draining land and ploughing fields ensures that anaerobic bacteria cannot compete with aerobic bacteria, and thus prevents denitrification.

YOU SHOULD KNOW

››› The effects of fertilisers on the environment

››› The process and effects of eutrophication

Key Term

Eutrophication: The artificial enrichment of aquatic habitats by excess nutrients, often caused by run-off of fertilisers.

Knowledge check

Identify the missing word or words.

Increased nitrate levels in lakes and rivers causes a massive increase in microscopic plants, referred to as an ....
This means .... is unable to penetrate to lower depths and plants are unable to carry out the process of .... and die.
The short-lived algae soon die and are decomposed by .... bacteria using up a lot of dissolved ....
Anaerobic bacteria then reduce nitrate to ....

Nitrate is highly soluble and is readily leached from soil and washed into rivers from surrounding land. It drains through the soil under gravity, as ground water but the rate increases in heavy rainfall and increases the concentration of nitrate in bodies of water.

- Nitrate is a fertiliser and algae respond so the first effect may be an algal bloom. The water becomes green and light is unable to penetrate to any depth.
- The plants in the deeper regions of the lake cannot photosynthesise and they die.
- There is a general decrease in animal species diversity, as they rely on the plants for food and shelter.
- The short-lived algae soon die and are decomposed by saprobiontic fungi. These are aerobic organisms and use a lot of oxygen creating a biochemical oxygen demand (BOD).
- The water in all but the very upper layers, which are exposed to the air, becomes deoxygenated, so that fish and other oxygen-requiring species die.
- Anaerobic bacteria in the water reduce nitrate to nitrite. They flourish and some species release gases with a characteristic smell, such as hydrogen sulphide.

To avoid the serious problem of high nitrate concentrations in waterways, farmers must comply with strict legislation to reduce the quantity of nitrate they release into the environment. They must:

- Restrict the amount of fertiliser applied to the soil.
- Only apply fertiliser at a time when the crops are actively growing, so that it is readily used and does not remain in the soil to be leached away.
- Leave a strip at least 10 metres wide next to watercourses. Nitrates can then not directly enter the water and when they do, it is over a longer period of time.
- Dig drainage ditches. The minerals concentrate in the drainage ditches, which undergo eutrophication, and this protects the natural watercourses. Water flows slowly in the ditches and they accumulate landfill. This has led to a local drop in invertebrate biodiversity and the water's high nitrate concentration reduces species diversity in the adjacent grassland. However, rare species have been reported growing in drainage ditches, in response to their unusual conditions.

Specified practical exercises

Assessing the distribution of organisms in a rocky shore transect

On a rocky shoreline, the most suitable sites for transects have exposed bedrock sloping towards the sea.

Line transect to identify species of algae on a rocky shore

Apparatus

- Ranging poles
- 1 m ruler
- Spirit level
- 10 × 10 gridded quadrat 0.5 × 0.5 m
- Alga identification guide

Method

1. Place a ranging pole at the top of the line chosen for the transect and another at the bottom. Run a tape measure horizontally between the two and, with the metre ruler, read the height of the tape measure above ground level every metre. Keep the tape measure taut and use the spirit level to ensure it is horizontal. If transect is very long or very steep, it should be worked on in sections.

2. At each site where the drop is measured, identify the alga growing there.

3. Avoid rock pools or fissures as these create microhabitats which introduce variability.

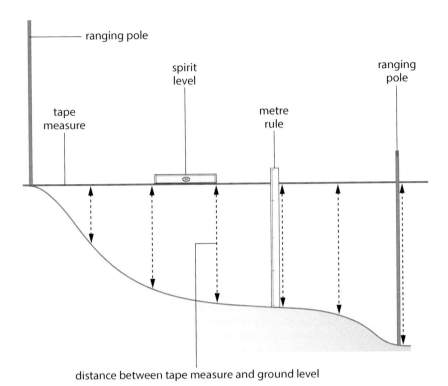

Constructing a shore profile and line transect

Sample results

The data can be displayed as a diagram showing vertical height and distance along the transect, with algae identified in the appropriate positions.

Distance along transect / m	Height above low tide mark / m	Height description	Common name	Biological name
0	5.4	very high tide	spiral wrack	*Fucus spiralis*
5	4.6	high tide mark	spiral wrack	*Fucus spiralis*
10	2.2	midshore (littoral)	rockweed	*Ascophyllum nodulosum*
15	2.1		bladderwrack	*Fucus vesiculosus*
20	1.7		bladderwrack	*Fucus vesiculosus*
25	0.9	low tide mark	serrated wrack	*Fucus serratus*
30	−0.2	very low tide (sublittoral)	kelp	*Laminaria digitata*

F. spiralis — *F. spiralis* — *A. nodulosum* — *F. vesiculosus* — *F. vesiculosus* — *F. serratus* — *L. digitata*

0   5   10   15   20   25   30

The height above low tide mark can be plotted to show the shore profile:

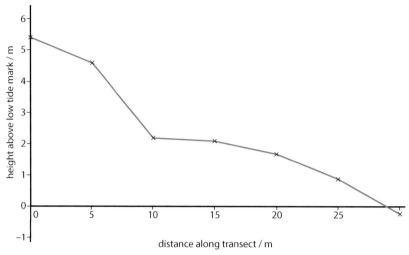

Shore profile

Belt transect to estimate distribution of algal species on a rocky shore

At chosen intervals along the transect, place the same corner of a quadrat at each co-ordinate and assess the abundance of algae in each:

(a) A rough estimate can be used, in which a numbered scale indicates the approximate area cover. This type of assessment, however, cannot be used quantitatively. There are several such scales in use, e.g. the ROFCASE scale:

% area cover	Score
1 or 2 individuals	Rare
A few scattered individuals	Occasional
<5	Frequent
5–30	Common
30–60	Abundant
60–90	Super-abundant
>90	Extremely abundant

(b) Make a numerical assessment of abundance. The data below are for percentage cover:

Sample results

Distance of quadrat along transect / m	Height above low tide mark / m	Biological name	Percentage area cover
0	5.4	*Fucus spiralis*	10
5	4.6	*Fucus spiralis*	80
		Ascophyllum nodulosum	15
10	2.2	*Fucus spiralis*	10
		Ascophyllum nodulosum	85
15	2.1	*Fucus vesiculosus*	68
		Ascophyllum nodulosum	20
20	1.7	*Fucus vesiculosus*	100
25	0.9	*Fucus vesiculosus*	36
		Fucus serratus	64
30	−0.2	*Laminaria digitata*	30

Kite diagrams

The data can be dislayed as a kite diagram: for each species, plot data points to the left and the right of the 0% line, with each data point representing half of the area cover observed. Join the data points to the left and those to the right.

The complete diagram is shown below, with kites for each species. This gives an easily visualised description of how the plants vary in distribution along the transect. The horizontal line at 30 m of the kite diagram for *Laminaria digitata* shows that beyond this distance, no more readings were taken.

The diagram shows a zonation of different species, from *Fucus spiralis*, the best adapted for exposure to air, at the very high water mark, to *L. digitata*, the least adapted for exposure to air below the low water mark.

By convention, kite diagrams illustrating aquatic habitats are drawn with vertical kites, whereas those for terrestrial habitats are generally drawn with horizontal kites.

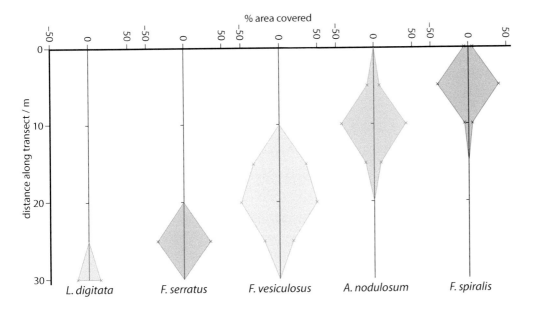

Kite diagram along a rocky shore transect

Risk assessment

Hazard	Risk	Control measure
Tripping on slippery rocks	Grazes, strains and sprains	Care where walking; use suitable footwear
Cut off by tide	Discomfort; drowning	Know times of high tide and leave site before tide comes in
Biting and stinging insects	Adverse skin response	Cover skin at all times; use insect repellent
Weather may be too hot, too cold, too bright	Hypothermia, overheating, sunburn	Wear appropriate clothing; use sunglasses; use sun cream

CH 6

Human impact on the environment

Human activity has caused massive habitat destruction. Each species has intrinsic value but more food is needed to support the increasing human population and agricultural production conflicts with conservation. Nine interacting global systems generate environmental stability. Their disruption is likely to bring sudden, catastrophic environmental change. The increased use of fossil fuels causing global warming is a grave concern. Humans recognise the obligation to take greater responsibility for the Earth. Informed, evidence-based political decisions must be made, based on knowledge gained using sound scientific principles.

Topic contents

By the end of this topic you will be able to:

- Explain how species become endangered and extinct.
- Describe how gene pools are conserved in the wild and in captivity.
- Understand the conflict between agricultural production and conservation, as illustrated by deforestation and overfishing.
- Describe how environmental monitoring may contribute to evidence-based political decision making.
- Describe the concept of planetary boundaries.
- Explain the status of each planetary boundary.
- Describe how technological innovations, such as the production of biofuels and seawater desalination may contribute to avoiding planetary boundaries.

Why species are at risk

The influence of humans has spread to every landmass on Earth, to all bodies of water on the planet's surface and to the atmosphere. Human artifacts occur throughout the solar system and, with the Voyager probes, beyond; however, only Earth's ecosystems will be discussed here. Human activities are altering ecosystems upon which they and other species depend. Massive destruction of habitats throughout the world has been brought about by agriculture, urban development, forestry, mining, and environmental pollution. A biodiversity crisis is occurring because species are increasingly deprived of the biotic and abiotic factors that they need. As a result, they reproduce less successfully and population numbers decrease. In some cases, this leads to extinction.

YOU SHOULD KNOW ›››

››› The reasons why species become endangered and extinct

››› Why conservation is important

››› Methods of conservation

Extinction

Extinction is a natural process that has been taking place since life originated. It is the current rate of extinction that underlies the biodiversity crisis. The fossil record shows that the normal 'background' rate of extinction is 10^{-6} y^{-1}. This means that each year, one species in a million becomes extinct. Evidence shows that humans have been responsible for death of megafauna, i.e. very large animals, such as the moa, a giant bird from New Zealand and *Megatherium*, the giant ground sloth, from South America, although climatic stresses may have contributed to their extinction. The extinction of the passenger pigeon in North America, in 1914, and the thylacine from Tasmania, in 1933, following their exposure to humans, is well documented. It is now estimated that human activity in tropical areas alone has increased extinction rates between 1000 and 10 000 times. Marine life has also been affected. About one third of the planet's marine fish species rely on coral reefs. Some data suggest that, at the current rate of destruction, all warm water coral reefs could have died by 2050.

WORKING SCIENTIFICALLY

It is suggested that the current wave of extinctions and the occurrence of novel chemicals mark a new geological era, the Anthropocene. This has yet to be accepted by the international bodies that determine such definitions.

Passenger pigeon

Thylacine

Geological evidence shows that five times in the history of life, the vast majority of species have been made extinct by a catastrophic change, such as low global temperature, shortage of dissolved oxygen in the oceans or ocean acidification. These are described as mass extinctions. The sixth appears to be underway now, caused by human activity.

Endangered species

The vast majority of Earth's earlier occupants, including the large and once dominant dinosaurs and tree ferns, have become extinct largely as a result of climatic, geological and biotic changes. At the present time, human activity has taken over as the main cause of species extinction. Many of the larger mammals such as mountain gorillas, giant pandas, tigers and polar bears are threatened.

The IUCN, the International Union for the Conservation of Nature, makes assessments of plants and animals and grades them according to their vulnerability to extinction. Species are rated depending on their numbers, rate of decline and distribution:

EX   Extinct (EX)

EW Extinct in the wild (EW)

Threatened:

    CR Critically endangered (CR)

    EN Endangered (EN)

    VU Vulnerable (VU)

NT Near threatened (NT)

LC Least concern (LC)

Many species are given the rating DD (Data deficient) or NE (not evaluated) and much research is urgently needed to identify and assess species at risk. Results are published in Red Data Lists. The African elephant, for example, is 'vulnerable' whereas the Sumatran elephant is 'critically endangered'.

The reasons species are threatened

Species become endangered or extinct for many reasons:

- Natural selection occurs when individuals less suited to prevailing conditions reproduce less successfully. Their numbers decrease, which may lead to their extinction. Human activities are causing habitats to change faster than new mutations allow species to adapt, and so they are driven to extinction at a faster rate than before humans had such influence over their environment.

- Non-contiguous populations. The total number of individuals in a species may suggest that numbers are sufficient to ensure the continuation of the species. But if groups are isolated from each other, they cannot interbreed and each group functions as a separate population. There may be too little genetic diversity in each to ensure a healthy population, leading to their extinction. Such is the case with black rhinos in Africa.

- Loss of habitat, e.g. by

  - Deforestation

  - Drainage of wetlands

  - Hedgerow loss. Hedgerows have separated fields for centuries. They provide a habitat for insects, nesting sites for birds and reptiles, food for many species and varying light intensity and water availability for diverse plants. Hedges act as wildlife corridors enabling reptiles, birds and mammals to move from one area to another, helping to maintain biodiversity. Their removal, often to accommodate the large agricultural machinery used in modern farming, has destroyed large areas of this specialised habitat. Herbivores and other consumers reduce in numbers with consequent reduction at higher trophic levels.

  - Farmers often sow crops in autumn rather than spring, which means that plants are an unsuitable height for the birds to build their nests. This has led to a decrease of many well-known birds, such as the skylark and lapwing.

  Deforestation is discussed on p103.

 Drainage of wetlands is discussed on p92.

Lapwing

- Overhunting by humans including:
  - Trophy hunting: countries that allow this, and charge for the privilege, claim that only old or sick animals are targeted.
  - Some traditional medical practices, such as the use of tiger body parts and of rhino horn.
  - The bush meat industry in which primates, among others, are killed for food.
  - Overfishing.
  - Agricultural exploitation.
- Competition from introduced species including domestic animals and the accidental introduction of organisms.
  - The dodo, in 17th-century Madagascar, was driven to extinction because rats, brought on European ships, ate the dodo eggs; similarly on the Galapagos Islands, rats have diminished native species, degrading this iconic site where Charles Darwin made significant observations.
  - The north American signal crayfish has invaded UK streams and rivers and the native crayfish, which is smaller, is being out-competed.
  - Red squirrels in the UK have declined due to habitat loss, and in many places, they are being out-competed by the north American grey squirrel.
- Pollution
  - Oil is shipped worldwide in supertankers, some of which are too big ever to enter a port. Accidental discharge of oil in to the sea occurs, such as when ships run aground, e.g. the Torrey Canyon disaster in the English Channel in 1967; the Exxon Valdez disaster off the coast of Alaska in 1989; the largest ever petroleum spill in fresh water was from a Royal Dutch Shell tank ship, Estrella Pampeana, at Magdalena, on the coast of Argentina, in 1999. It polluted the environment, contaminated drinkable water and killed plants and animals.

    Oil floats and prevents oxygenation of surface water. Animals that break through the surface are covered by a film of oil. Birds, for example, subsequently are chilled to death because their feathers clump together and cannot provide insulation. Oil washed up on beaches is ingested by shore-dwelling animals, which are poisoned by it.
  - PCBs are polychlorinated biphenyls. They are ingested with food. Because they are neurotoxins, carcinogens and hormone disruptors, their use was progressively banned in the UK between 1981 and 2000. They used to be manufactured in Newport, South Wales, until the 1970s and waste was dumped in a quarry west of Cardiff. They are still detected in wastewater from the site.

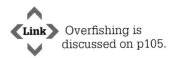

Link Overfishing is discussed on p105.

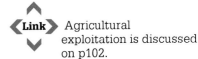

Link Agricultural exploitation is discussed on p102.

Going further ▶

The Mid-Wales Red Squirrel Project in the Tywi Forest in central Wales, has been supporting red squirrel populations since 2002.

▼ Study point

Species may become threatened or extinct because of habitat loss, overhunting, species introductions and pollution.

Key Term

Conservation: The protection, preservation, management and restoration of natural habitats and their ecological communities, to enhance biodiversity while allowing for suitable human activity.

Conservation

Conservation is the sensible management of the biosphere to maintain habitats and enhance biodiversity, while allowing human activity. It maintains genetic diversity, both in the wild and in captivity. Conservation is addressed in various ways, at local, national and international levels:

- Protecting habitats protects the species that live there and communities act as living gene banks. Official designation recognises local nature reserves, sometimes as small as a few hectares. There are larger, national nature reserves, such as the Gower coast, SACs (special areas of conservation), SSSI (sites of special scientific interest) such as the Brecon Bog in Carmarthenshire and other sites. They have varying levels of legal protection and may be managed and monitored by wardens.

- International co-operation restricting trade in e.g. ivory and whaling. International law allows some countries to practise 'scientific whaling', but many consider the term 'scientific' to be disingenuous and there are organised attempts to stop it from happening.

- Gene banks:
  - Endangered species are protected and entered into breeding programmes in specialised zoos and botanic gardens. Pandas are held in great affection and attempts by zoos to persuade them to breed, e.g. in Edinburgh Zoo, have much media publicity. Records of matings are kept so that genetic diversity can be increased by deliberate choice of parents.

Tian Tian asleep in her enclosure at Edinburgh Zoo

  - Sperm banks are used to store genes of economically important animals and of threatened species. Rather than moving animals, sperm samples can be sent around the world to use in breeding programmes in other zoos.

  - Seed banks maintain stocks of seeds of traditional varieties and of vulnerable species, in highly controlled conditions, often in liquid nitrogen. Although a plant from a 2000-year-old Judean date palm seed has been successfully grown, seeds degrade over time. So periodically, samples are thawed and germinated. Plants grown from them reproduce and another generation of seeds is collected. Seed banks in some countries have very high levels of protection as they are viewed as a potential source of food in case of catastrophic environmental degradation.

  - Rare breed societies maintain older, less commercial varieties for special characteristics, e.g. hardiness, wool production.

  - Species reintroduction: the red kite in mid-Wales, the chough (a member of the crow family) in Cornwall, the Arabian oryx in Israel and Jordan, the giant condor in California and the Przewalski horse in Mongolia have all been driven to the brink of extinction. Following successful breeding programmes, they have been reintroduced to their former habitats.

The Hebridean sheep is preserved and protected by the Hebridean Sheep Society

Going further ▶

Some reintroduced animals have no fear of predators or have no knowledge of how to forage or find shelter. Systems are being developed to train them before release, to improve their competitive abilities.

Przewalski horse and foal

- Education:
  - Global organisations, such as the World Wide Fund for Nature, mount public-awareness campaigns.
  - In the UK, the Countryside Commission, a government body promoting nature conservation, advises government and groups whose activities affect wildlife and their habitats. It produces publications, proposes ecosystem management schemes and establishes nature reserves.
- Legislation: the EU Habitats Directive has imposed a range of measures to protect habitats and enhance biodiversity throughout Europe, preventing overgrazing, overfishing, hunting of game, collection of birds' eggs, picking of wild flowers, and plant collecting.
- **Ecotourism** recognises that mass travel is harmful globally and to specific habitats. It aims to:
  - Contribute to conservation efforts.
  - Employ local people and give money back to local communities.
  - Educate visitors about local environment and culture.
  - Co-operate with local people to manage natural areas.

Why conserve?

Some ecosystems are particularly at risk and, as a result, the species they contain are vulnerable. The coral reefs and tropical rainforests are of particular concern as they have such large and complex communities. The conservation of species ensures the conservation of gene pools. They may contain genes that are potentially useful for the species itself and for future generations of humans. Reasons for species conservation include:

- Ethical reasons: each species represents a particular combination of genes and alleles adapted to a particular environment and it is considered that the uniqueness of each is intrinsically valuable.
- Agriculture and horticulture: plants and animals used in agriculture and horticulture have been developed from those in the wild. Selective breeding increases genetic uniformity, with the loss of rarer alleles. In the past, breeders may have neglected some important qualities, such as resistance to cold or disease. These need to be bred back into cultivated varieties, using the wild plants and animals as a gene bank. If habitats and the wildlife that live in them are threatened, it may no longer be possible.
- If the environment changes, some alleles will provide an advantage to the individuals that carry them, and those individuals will be selected for, preventing the extinction of the species.
- Potential medical uses: the antibiotics we use are derived largely from fungi but many other of our medicinal drugs are synthesised by plants. A historical example is quinine. Since the 16th century, it has been extracted from the bark of *Cinchona,* an Amazon rain forest tree, to treat malaria.

Logic dictates that there are potential drugs, not yet discovered, that could be of immense value. The extinction of any plant species before its chemical properties have been investigated could be an incalculable loss. We therefore have an obligation to protect habitats so that we can identify plants that make useful drugs, before they become extinct.

Cinchona

Key Term

Ecotourism: Responsible travel to natural areas that conserves the environment and improves the well-being of local people.

▼ **Study point**

Conservation methods include habitat protection, restricted trade, use of gene banks, species reintroduction, education, legislation and ecotourism.

▼ **Study point**

Species conservation is important for ethical reasons, for agriculture and horticulture, for potential medical products, for specific alleles useful in plant breeding and may save species from extinction in a changing environment.

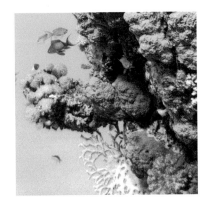

Coral reef

Tropical rain forest in New Zealand

Key Term

Monoculture: The growth of large numbers of genetically identical crop plants in a defined area.

Agricultural exploitation

Agriculture is essential to produce the quantity and quality of food required to feed the increasing human population. The way agricultural land is used, however, often conflicts with maintaining habitats and biodiversity. These opposing factors must be reconciled.

The efficiency and the intensity of food production are continually increased to meet the demand for food. Following World War II, more land was cultivated, fertiliser and pesticide use was increased and farms became more mechanised. These changes have increased in extent over the last 70 years and they have a number of environmental implications:

- Many hedgerows were removed to make larger fields to enable machinery to prepare the soil and harvest crops.
- The larger fields are used for **monoculture**, in which single crops, e.g. wheat or barley, are grown over a large area. With mixed crops there were many different microhabitats and so many different plants and, therefore, many different animals could live there. Monoculture provides only one habitat and so it reduces species diversity.

Combine harvesting wheat

Wheat growing in monoculture

- If the same crop is grown on the same plot year after year, yield progressively declines because:
  - The roots are always the same length so they extract the same minerals from the same depth of soil. Intensive cultivation has, therefore, hugely increased the use of inorganic fertilisers.
  - The same species is always susceptible to the same pests, which increase in number so more insecticides, herbicides and fungicides are used.
- Overgrazing by cattle causes grassland to become unsustainable. Their hooves compact the soil, driving out the air and preventing water draining through. Roots cannot penetrate the soil and so grass for grazing cannot grow.

In recent years, the views of government, farmers and consumers have changed. People value the countryside more, not only because it is a source of food, but also because it provides habitats for plants and animals, and places to visit for enjoyment.

EU and national schemes encourage farmers to manage their farms for biodiversity. Some land is given over to conservation and the farmers receive subsidies, i.e. money, to compensate them for reduced income due to lowered crop production.

▼ Study point

Farming has diminished biodiversity by hedgerow removal, monoculture, repeated growing of the same crop and by overgrazing of cattle.

Deforestation

Reasons for deforestation

Deforestation is widespread because timber is used extensively both as a building material and as fuel, as well as providing paper and packaging. Land is also cleared for farming, often to produce biofuels or grazing for the cattle destined for the meat industry. Specific high value trees, such as teak or mahogany, may be targeted, and felling and removing them damages many others in the process. New roads are built to provide a transport infrastructure for these activities, which contribute to the loss of forest cover.

Consequences of deforestation

- **Soil erosion**: tree roots bind soil together. Deforestation on the higher slopes of valleys allows heavy rain to sweep exposed topsoil down to the flood plains below. Topsoil is the fertile soil and what remains is not suitable for crop growth.

- Deforestation of uplands causes lowland flooding.

- Under normal conditions, on the lower slopes, plants, humus and leaf litter act as a sponge, soaking up heavy rainfall, and water is only gradually released into the soil. Trees transpire and return water to the atmosphere. After deforestation, there are no plants and water evaporates from the soil. This diminishes the quality of the soil:

  – Evaporation returns water vapour to the atmosphere more slowly than transpiration, so soil on deforested land becomes wetter. Water fills the soil's airspaces and so the oxygen available for roots decreases.

  – It takes longer for a wet soil to warm up than a dry soil, so these soils are also cold. Germination and root activity are reduced.

  – Cold, damp soil favours the growth of denitrifying bacteria, and so the soil loses its fertility.

YOU SHOULD KNOW ›››

››› Why deforestation occurs and its consequences

››› The importance of forestry management

Going further ▶

Paschalococos disperta, at 30 m high, was the tallest palm tree in the world before becoming extinct following deforestation on Easter Island.

Key Term

Soil erosion: The removal of topsoil, which contains valuable nutrients.

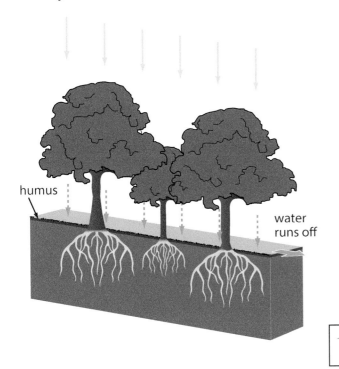

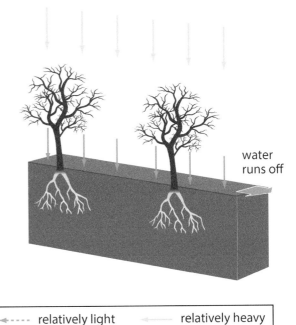

```
---- relatively light        ◄── relatively heavy
     flow of water                flow of water
```

Deforestation

- Less rainfall: water only returns to the atmosphere by evaporation from the soil, not by transpiration, and evaporation is slow. This accelerates desertification.

 Link The effects of deforestation on the carbon cycle are described on p84.

▼ **Study point**

The consequences of deforestation include soil erosion, lowland flooding, lower rainfall, habitat loss, biodiversity loss and an increase in atmospheric carbon dioxide.

Key Term

Coppicing: Cutting down of trees close to the ground and leaving them for several years to re-grow.

Going further ▶

The oldest coppiced tree in the UK is a lime tree in Gloucestershire, estimated to be 1000 years old.

Key Term

Selective cutting: Felling only some trees, leaving the others in place.

Trees are the right distance apart

- Habitat loss and reduction in biodiversity: it is estimated that at least 50% of the Earth's species live in the tropical rain forests, which cover only about 10% of the Earth's land area. Destruction of such natural habitats may lead to the loss of some tropical species. They may become extinct before their clinical properties have been investigated. It is estimated that every day, rainforest deforestation makes about 25 plant and animal species extinct.

- Effects on the atmosphere: as photosynthesising trees are cut down, the removal of carbon dioxide from the atmosphere by photosynthesis is decreased. Cut trees may be burned or left to decay, releasing carbon dioxide in the atmosphere.

Forest management

A traditional way of using a forest, since Neolithic times, is 'slash and burn'. A small forest area is cut and burned. People grow crops on soil fertilised with ash from the burned trees. When the soil is no longer fertile, people leave and the area regenerates. This is sustainable on a small scale but not on the large scale it occurs in rain forests now, e.g. in Brazil. As well as providing land to grow food, woodland and forests have been used as a source of timber for thousands of years. With careful management, it is possible to make use of this resource without destroying the ecosystem. Managed forestry involves sustainable replanting and regeneration.

Coppiced poplar trees

In Britain, **coppicing** has been used for thousands of years. A tree trunk is cut, leaving a 'stool' a few centimetres high. New shoots emerge from buds in the stool and grow into poles, which thicken over the years. The poles can be cut on rotation to produce timber of different widths. Coppiced plants can regenerate over long periods of time. A **long rotation time** increases sustainability as many years are left between harvesting adjacent areas of forest and a variety of habitats develop, favouring diverse wildlife.

Instead of removing all the trees in an area at one time, **selective cutting** can be used. This technique is valuable on steep slopes where the total removal of trees would leave the soil very vulnerable to erosion. Selective cutting also helps to maintain nutrients in the forest soil and minimises the amount of soil that is washed into nearby waterways.

With good forestry practice, land can be used efficiently:

- Planting trees the optimum distance apart. If they are too close, intra-specific competition occurs and the trees grow tall and thin, producing poor quality timber.

- Controlling pests and diseases so that trees grow well, producing high quality timber. Fewer trees need to be felled and best use is made of the land, reducing the total area required.

- Cutting a similar number of trees each year for long periods of time allows the forest ecosystem to be maintained. Habitats are left intact and species are able to live in the forest even though timber is being extracted.

Preservation of native woodlands

In prehistoric times the UK was largely covered with woodland. Domesday records show that in the 11th century, the UK was about 15% woodland. By the end of the 19th century, this had decreased to less than 5%. Since then, steps have been taken to increase forest cover, so that by 2015, the Forestry Commission reported that Wales was 15% forested, higher than the UK overall, at 13%. However, only 1% is natural or native woodland. It is essential that these native woodlands are preserved to maintain and enhance biodiversity. There is a need to plant more native species to provide a wide range of habitats for the great variety of species that live there.

An illustration of why it is so important to replant native trees is given by the experience in Sri Lanka. The government worried about forest depletion and so, in 1985, planted large areas of eucalyptus trees, which are mostly native to Australia. Their shallow roots absorb water very efficiently. Both the water content and the humus content of the soil fell and so biodiversity also fell. The eucalyptus trees are now being cut down.

Overfishing

The dramatic increase in the intensity and efficiency of commercial fishing has caused **overfishing** in many areas of the world. The Grand Banks, a sea area off the northeast coast of Canada, provides an example. The Grand Banks was among the most productive fishing areas in the world. In the 15th century, the huge numbers of cod are said to have prevented ships moving through the water. From the late 1950s, however, new equipment, such as trawlers, and new technology, such as radar, were used. So many fish were caught, that the cod population fell very steeply. To prevent them dying out entirely, in 1992, after 500 years, the Canadian government closed the area to fishing.

Nets with a very small mesh catch young fish before they have become sexually mature. This means that, as time goes by, there are fewer individuals left to reproduce and so the population size decreases. It may be harder for the remaining fish to find a mate and with a smaller number reproducing, the genetic diversity of the population decreases. Commercial fishing operates:

- Drift netting: pelagic fish live in surface waters. They swim into a net, suspended from floats, stretched between two boats. But with thousands of miles of nets, non-target species, e.g. dolphins and marine turtles, become trapped.

- Trawling: fish that live in deeper in water, the mid- and bottom-feeders, are caught by a large net which is dragged through the water, catching whatever swims into it. Equipment used in trawling has damaged the ocean bed, destroying the habitats of molluscs such as clams, and other organisms, putting their populations at risk.

Commercial fishing boat

Effects of overfishing on other wildlife

The overfishing of a particular species has implications for the rest of the food chain. For example:

- When trawlers spread their nets, they catch a fish called capelin. These are not eaten by humans but they are an important prey species for cod, so removing them from the sea has contributed to the decline of cod stocks.

- Since the 1980s, six countries, including Japan and Russia, have harvested Antarctic krill. These 50 mm long shrimp form swarms many miles across. They are primary consumers, eating phytoplankton. They are the main

Antarctic krill

food of whales and they supplement the diets of seals, penguins, squid and fish. The ecological balance in the Antarctic has already been upset by overexploitation of whales; heavy fishing of krill will badly affect the rest of their food web, including the remaining whales.

The reduction of fish populations by overfishing also damages the livelihoods of fishermen. A balance must be struck between catching enough fish to make a living, while ensuring that there are enough breeding fish remaining to replenish stocks.

Methods to regulate fishing and allow breeding stocks to recover

- The mesh size of the nets must be large enough that young fish can swim through and survive. This is supported by legislation that prevents selling fish below a certain size.

- Quotas can be set so that only a certain mass of fish may be brought to land. It appears, however, that more fish are taken than allowed by the quotas and they are thrown back, dead, into the sea. These 'discards' are recognised as a problem and EU legislation is progressively making them illegal.

- Exclusion zones prohibit fishing in defined areas at certain times of the year, allowing the fish to reproduce. The position of fishing trawlers can be monitored by satellite technology and fines imposed for non-compliance. In October 2015, for example, the EU banned cod fishing for 12 weeks in some North Sea fishing areas, during the crucial spawning period.

- Consumers may choose to eat only those fish certified by the Marine Stewardship Council, which ensures fish are taken from sustainable sources.

- Legislation controlling the size of fishing fleets.

- Legislation controlling the numbers of days spent at sea.

- Fish farming may reduce overfishing.

Fish farming

In the UK, trout and salmon are the most commonly farmed fish. They can be bred and grown to maturity in ponds, lakes and managed enclosures in estuaries, where predation is reduced and food supplies are maintained. For plankton-feeders, the growth of phytoplankton can be increased by the addition of artificial fertilisers to the water. Fish grow rapidly when they are reared in the warm waters discharged from factories. Fish are sometimes farmed in a pod, a large, steerable device, which can be moved, depending on prevailing ocean currents, on local water temperature and other abiotic factors.

Raising fish rather than pork, poultry and beef has advantages:

- Fish convert their food into protein more efficiently
- A greater proportion of fishes' bodies are edible
- Fish farming has a lower carbon footprint.

Fish farming in a lake

Fish farm in SE Asia

Fish farming, however, is the cause of many problems:

- Diseased fish: farmed salmon are often very densely stocked and so they can easily transmit disease. Huge doses of antibiotics are required to keep them moderately healthy. The pesticides used to control fish parasites, such as sea lice, are known to harm marine invertebrates, especially molluscs.
- Pollution: the ecological balance of the waterways may be upset. Eutrophication, for example, can result when fish excreta, waste food and fertiliser are carried into the water around the rearing pens.
- Escaped fish: farmed fish have been selected for very rapid growth. If they escape, they out-compete wild fish for food, habitat and mates. They also transmit parasites and other infections. Escaped farmed fish interbreed with wild fish and set up fast-growing colonies which can push wild fish to extinction.
- Resource use: farmed salmon, which are carnivorous, eat three times their bodyweight in fish feed, which is made from other fish. This is a poor use of resources from an environmental point of view.
- Environmental toxins, e.g. methyl mercury, PCBs, dioxins and pesticides, are more concentrated in farmed than wild salmon, although their concentrations are so small that their effects do not outweigh the health benefits of eating fish.
- Environmental degradation: the shrimp industry, in particular, has been blamed for the salinisation of soil and groundwater and the destruction of the mangroves that normally protect coastal communities from tropical storms.

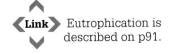

Link Eutrophication is described on p91.

▼ **Study point**

Fish farming results in several environmental problems: diseased fish, pollution, escaped fish, resource use and environmental toxins.

Sustainability and decision making

Environmental monitoring describes the quality of the environment. It establishes the current status and when done repeatedly, the pattern of the data can be used to identify trends and make predictions.

Going further ▶

The atmosphere normally gets colder with height, but in an 'inversion' it gets warmer a short distance above the Earth's surface instead. Air pollution gets trapped under this warm air.

- Air quality monitoring: the concentrations of air pollutants are measured because of the health effects associated with exposure to them. Air pollution is carried by the wind, so this must be taken into account when identifying sources of pollution.

- Soil monitoring looks at soil structure and density, its water-holding and drainage capacity, pH, organic particles, earthworms and other organisms, as implied by measuring enzyme activity and respiration rate.

- Water quality monitoring looks at several aspects of water including:

Going further ▶

Rivers and lakes are classified in terms of their quality in a system defined by the EU Water Framework Directive, the Environment Agency, the Countryside Council for Wales and the Scottish Environmental Protection Agency.

  - Chemical: in the last 30 years acid rain and greenhouse gases have made water monitoring essential because of their potential to do harm. Since the 1960s, the concentration of 'environmental oestrogens' in water supplies has increased, as a result of the use of oral contraceptives and of materials used in industry. They have a feminising effect on some aquatic organisms and it is suspected that they may influence humans as well, both in embryo development and in the timing of puberty.

  - Biological: many animals act as indicator species for water quality, e.g. brown trout in rivers indicate high quality water. The steep decline in salmon populations was an early indication of acid rain. Mosses can indicate heavy metal concentrations. Eels are used to study halogenated organic chemicals, which accumulate in their fat.

Exam tip

Be objective and scientific when you write about human impact on the environment, using clear and logical evidence-based arguments. Avoid sound-bites and clichés.

  - Microbiological: bacteria and viruses are monitored, especially in drinking water or water used for sports. Many sewage treatment plants do not sterilise the water they release. So water entering a river may look clean but could still have a very high bacterial count. Most will be harmless 'coliform' bacteria but if *E. coli* or coliform counts are high, the water will be tested for other, specific pathogens and treated.

Environmental impact assessments

Statistical analyses of environmental monitoring data, performed by dedicated software packages, mean that data collected can contribute to environmental impact assessments. These are documents that aim to predict environmental effects of a proposed project when activities risk harming the environment, e.g. road building. In the EU, environmental impact assessments were introduced in 1985. Among the topics they address are:

- A description of the project and site.

- Alternatives that have been considered, e.g. whether the fuel in a power station burning biomass, is local or not.

- A description of the environment, e.g. populations, fauna, flora, air, soil, water, human use, landscape and cultural heritage. Organisations with crucial local knowledge, such as the RSPB, have input into this.

- A description of significant effects on the environment, e.g. in a wind farm development, a significant impact may be collisions with birds.

- Mitigation, i.e. ways to avoid negative impacts are an essential aspect of planning. Evidence is required to show that reducing the impact of harmful activities has been considered. Countermeasures might be:

  - Wind farms to be built when birds are not nesting.

  - Proposed roads to avoid breeding ponds of endangered amphibians, e.g. the greater crested newt and the natterjack toad.

Natterjack toad

Different fields of study come together in planning new development. There is a need for engineers, mathematicians, economists, archaeologists, geologists, materials scientists and many others. The legal framework is such that planning must be done with a view to preserving habitats and enhancing biodiversity. Data collection is essential and decisions about how to proceed must be based on the most reliable evidence. Collaboration is required between practitioners in these different fields, who must extend their thinking skills into areas outside their own speciality.

Planetary boundaries

The concept of **planetary boundaries** was introduced in 2009 as a way of defining the 'safe operating space for humanity'. Nine global processes have been identified, that regulate the stability of planetary systems and the interactions of air, land and sea. Such systems are precarious and a tipping point may be reached, beyond which change to a factor and the response to it are not linear. At the tipping point, a small change has a very large and unpredictable effect on the environment. Numerical values have been estimated for upper and lower levels of these nine global systems, between which they could continue to operate without such extreme responses. Exceeding these limits is likely to result in sudden and catastrophic change to environmental conditions. The study of planetary boundaries is inter-disciplinary and its conclusions are based on vast amounts of research and data collection. It is hoped that political, economic and other decisions will take account of the knowledge these boundaries represent, to ensure sustainable development and support the health of the biosphere.

A circular graph is common way of displaying the status of planetary systems. Within the inner, green, circle, is the safe operating space. Between this and the outer, red, circle is an area of increasing risk, the 'zone of uncertainty'. Beyond the red circle, the values represent a high risk, the planetary boundary has been crossed and events are unpredictable.

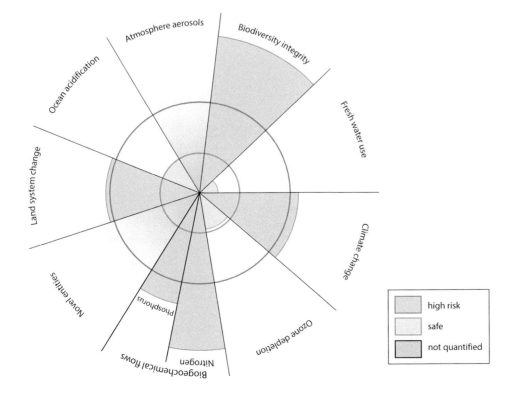

The status of planetary systems

YOU SHOULD KNOW ›››

››› The nine planetary systems

››› The status of each boundary

››› The Climate Change and Change in Biodiversity Integrity are the two core boundaries

››› Climate change may be partially offset using biofuels

››› That fresh water supply may be augmented by desalination

Key Term

Planetary boundary: Limits between which global systems must operate to prevent abrupt and irreversible environmental change.

▼ Study point

At the time of writing four planetary boundaries have been crossed: Climate Change, Change in Biosphere Integrity, Land-System Change and Biogeochemical Flows. This is likely to continue to change in response to human impact on global systems.

Going further ▶

In addition to the global planetary boundaries, regional boundaries have been developed for biosphere integrity, biogeochemical flows, land-system change and freshwater use.

1. The Climate Change Boundary

Link Some of the consequences of climate change for humans, plant and animals are described on p86.

Climate change is one of the two **core boundaries** and its planetary boundary has been crossed. Enough greenhouse gases have been added to the atmosphere that, even if emissions were to stop immediately, the average global temperature would still rise for decades or even centuries. The effects on wind, ocean currents, rainfall patterns and precipitation are being documented and a sea level rise of up to 7 m is predicted by 2100 unless drastic action is taken. Even if greenhouse gas emissions are so severely cut that the temperature rise is only 2°C, sea level is still predicted to rise by 2 m. The Kyoto Protocol, agreed in 1997, was the first major international agreement to address global warming, by setting targets for reducing the greenhouse gases in the atmosphere. Pressure groups stress that fossil fuel combustion must cease and that more political encouragement should be given to the development of alternatives. The **biofuel** industry has been developed to contribute to reduction in fossil fuel combustion.

Biofuels

The carbon dioxide that biofuels release on combustion has only recently been removed from the atmosphere. Growing more biofuel crops removes the carbon dioxide from the atmosphere again. This contrasts with fossil fuels, which release carbon laid down hundreds of millions of years ago.

Biofuels are made by biological processes such as anaerobic digestion of plant material or agricultural, domestic and industrial waste. They are useful in reducing the use of fossil fuels but growing plant material for them conflicts with the use of land for food production and it requires considerable irrigation. Because so much fossil fuel is used for transport, the International Energy Agency target is for at least 25% of the world's transport to use biofuels by 2050. First generation biofuels are made from the sugars and vegetable oils found in arable crops, which are easily extracted with conventional technology. Second generation biofuels are made from cellulose and lignin from woody crops, which are harder to extract.

There are social, economic and technical issues relating to biofuel production, in addition to the environmental issues:

- The 'food versus fuel' debate: land used to grow food has been turned over to crop production for biofuels, e.g. palm oil, so people have less food to eat or to export. An additional problem is that energy crops are often grown in monoculture.
- Carbon emissions: European bioethanol production and use reduce greenhouse gas production by 60–90%, compared with the production and use of fossil fuel.
- Sustainable biofuel production relies on sustainable planting and efficient technical systems.
- Deforestation to grow biofuel crops leads to soil erosion and biodiversity loss.
- Reduction in water availability, because biofuel crops require a large volume of irrigation water.
- Combustion of biodiesel produces more nitrous oxide (NO_2), a greenhouse gas, than fossil fuel.

Bioethanol

Bioethanol is the commonest biofuel and most car petrol engines are designed to use up to 15% bioethanol with petrol. Bioethanol is made by fermenting the carbohydrates in sugar or starch crops, such as maize, sugar beets, sugar cane and sweet sorghum. Brazil and the USA have developed bioethanol commercially, with 90% of the global production.

Bioethanol is made in several stages:

- Plant material is crushed and the stored starches are digested with carbohydrases to release sugars.
- Sucrose is crystallised out leaving molasses, which is rich in glucose and fructose.
- Glucose and fructose are fermented by yeast to produce a mixture containing ethanol.
- The mixture is heated by burning the fibrous waste, or bagasse, from the initial plant material and pure ethanol is distilled.

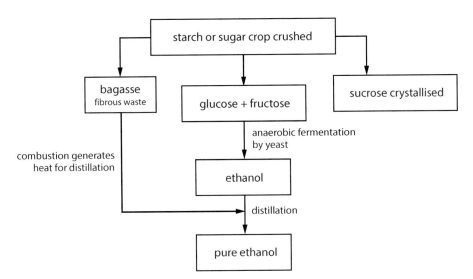

Flow diagram showing bioethanol production

Biodiesel

Biodiesel is the most widely used biofuel in Europe. It is made from vegetable oils including soya, rapeseed and palm oil and it can be made from algae. These crops are grown for their long chain fatty acid content. The fatty acids e.g. linoleic acid are reacted with an alcohol, commonly methanol, to produce methyl linoleate, also called 'methyl ester' or 'biodiesel':

methanol + linoleic acid $\longrightarrow$ methyl linoleate + water

Biodiesel contains less carbon and more hydrogen and oxygen than petrodiesel, so when it is pure, it produces about 60% of the carbon dioxide emissions. It also releases fewer carbon particles and less carbon monoxide, but produces more nitrous oxide. Pure biodiesel, B100, can be used in car engines. However, it is usually mixed with diesel to reduce pollution, e.g. B5 (5% biodiesel, 95% petrodiesel) is the most widely used in Europe.

Biogas

Biogas is a mixture of gases, comprising approximately 60% methane and 40% carbon dioxide. It is made by bacterial digestion of biodegradable waste materials, such as animal and human waste, or energy crops. It occurs in three stages, the last of which is an anaerobic process:

1. Macromolecules in the waste material are aerobically digested by amylases, proteases and lipases to sugars, amino acids, fatty acids and glycerol respectively.
2. Acetogenesis is an aerobic process that produces short chain fatty acids, especially ethanoic acid. Carbon monoxide and hydrogen gas are produced as the oxygen gets used up.
3. Methanogenesis is an anaerobic process: $C_6H_{12}O_6 \longrightarrow 3CH_4 + 3CO_2$.

   The solid material left over is dried and used as biofuel or fertiliser.

Going further ▶

Flueless fireplaces burn ethanol, but the heat output is less than electric or gas fires, and carbon monoxide poisoning is a danger.

▼ **Study point**

Diesel made from fossil fuel is called petrodiesel, to distinguish it from biodiesel.

Going further ▶

Biodiesel is a solvent and dissolves old deposits in the fuel tank and pipes. So if your car uses biodiesel, you have to change your engine filters more often.

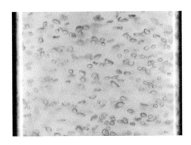

Photobioreactor containing algae for biofuel

Going further ▶

Ethanoic acid used to be called acetic acid, so its production is called acetogenesis.

Plastic tanks for small-scale biogas fermentation

Biogas installation processing cow manure on a farm

Biogas is made around the world in industrial installations and, notably in India and in China, in domestic digesters. 0.5 kg manure per day can provide enough fuel for the cooking needs of a family and the manure from one cow can produce the equivalent of 227 dm^3 of petrol each year.

Biogas is also produced through naturally occurring processes at landfill sites. It can be tapped off for use as a fuel. If it is left to escape into the atmosphere, it contributes to global warming, as methane and carbon dioxide are greenhouse gases.

2. The Biosphere Integrity Boundary

This used to be called the Biodiversity Loss and Species Extinction boundary but its new name was chosen to emphasise human impact on ecosystem functioning. It is the second core boundary and has also been crossed. While individual species may not be significant, together, their interaction produces 'ecosystem services' such as mineral recycling and absorbing carbon dioxide, on which the whole biosphere depends. Human activity has caused environmental changes to happen too fast for natural selection to produce organisms that are suitably adapted to their changing environment. In some cases, they fail to reproduce enough to maintain the species, which becomes extinct.

Habitat change can be observed worldwide but marine, tundra, coral reefs and coastal plains cause particular concern. Their biodiversity is demonstrably reduced.

Herd of wild reindeer in Arctic tundra

- Over 200 000 marine species have been described and this may only represent 10% of the total. By the year 2100, without significant changes, more than half of the world's marine species may be extinct. Oceans have been polluted with acid, oil, plastic and sewage effluence. The pesticide DDT has even been detected in Antarctic waters, illustrating that pollutants have global reach. Evidence correlates marine pollution with a decline in phytoplankton populations, for example in the Pacific Ocean. As these are the major ocean producers, profound changes at higher trophic levels occur. Much more data is needed to identify species at risk.

- Tundra ecosystems are altering as the temperature rises and species including herbaceous plants grow where, in the past, only lichens and mosses were seen.

- Coral reefs are biodiversity hot spots with up to 1 000 species per m². Coral reef bleaching continues to cause species loss, as warm temperatures cause the zooxanthellae that give the corals their colour, to leave. Their photosynthesis feeds the corals. In many cases they do not return, even if conditions improve, and so the corals die.

- Coastal communities are biodiverse. Mangroves, salt marshes and seagrass meadows absorb carbon dioxide up to 50 times more efficiently than the same area of tropical forest. But coastal plains become submerged as the sea level rises. Salt marsh and other coastal communities are flooded with salt water, and water loss by osmosis kills the plants, and, consequently, the animals of these communities.

Monitoring biodiversity in all these habitats is imperative. It will indicate which species should be prioritised in taking material for gene banks. Publicity will increase public awareness, which is seen as essential in reducing the human behaviour that has generated biodiversity loss.

Mangroves in the Florida everglades

▼ **Study point**

Climate Change and Change in Biodiversity Integrity are the two core boundaries.

3. The Land-system Change Boundary

Land-system change, for example deforestation, has occurred largely through the expansion of farming and raising livestock. Large forest areas have also been cleared for the growth of plants such as soya beans to make biofuels. Urban development is another significant factor. The use of land for biofuel crops is in direct conflict with the peoples' need to grow food. Similarly, land that could be used for food production for local communities is used to grow crops for export. The land-system boundary represents the misuse of land that results in too little food being produced. In 2010, the crossing of that boundary was predicted for 2050, but 2015 data suggests that it has already been crossed.

To reverse this boundary transgression and to provide enough food, farming should be concentrated into the most productive areas. A global reduction in meat consumption would reduce the land under cultivation, and there is a significant role for genetically modified plants. Using land for food, however, conflicts with using it for biofuel production.

Oil palm plantation

4. The Biogeochemical Flows Boundary

Biogeochemical flows refer to the cycling of minerals through the biotic and abiotic components of an ecosystem. Cycles have been described for many elements, such as carbon, sulphur, phosphorus and nitrogen. Mineral cycles are essential in maintaining the availability of elements in the ions that are absorbed and transmitted throughout a food web. Agricultural fertilisers have been used so intensively that the boundaries for both the nitrogen and phosphorus cycles have been crossed. This means that human activity has so disrupted these cycles that they are no longer self-sustaining. Polluting events that result from this, such as eutrophication, are further damaging ecosystems. To reduce fertiliser use, an aim for many years has been to transfer nitrogen fixing (nif) genes to crop plants but this has, so far, not been possible.

‹Link› The nitrogen cycle is described on p88. Eutrophication is described on p91.

5. The Stratospheric Ozone Boundary

90% of the world's ozone is in a layer in the stratosphere, 10–50 km up. Normally, ozone and oxygen are in equilibrium: $3O_2 \rightleftharpoons 2O_3$.

Halogenated hydrocarbons, such as the chloro-fluorocarbons (CFCs) alter the position of the equilibrium and favour the breakdown of ozone. Under the influence of ultra-violet light, CFCs release chlorine as free radicals and each one can break down 100 000 ozone molecules.

CFCs were widely used as propellants in spray cans, solvents, refrigerator coolants and in the manufacture of disposable food and drink containers. Their molecules are heavier than air but within 2–5 years climb to the stratosphere. Depletion of stratospheric ozone was observed first in the 1970s, when the concentration of atmospheric CFCs reached a tipping point. The ozone layer around the planet degraded but the thinning was so significant over Antarctica, that a 'hole' in the layer was detected each spring. Ozone absorbs uv-B, i.e. ultra-violet radiation with a wavelength 280–315 nm, and so the intensity of ultra-violet light at the planet surface increased. This range of wavelengths is strongly absorbed by DNA and so skin cancers and cataracts increased in number.

The use of CFCs in spray cans was banned in 1978. The Montreal protocols, agreed in 1987, banned the manufacture of CFCs and placed obligations on nation states to reduce their CFC use. The ozone layer continues to rebuild but it is feared that other atmospheric changes may make this a problem again in the future. However, for the present, this boundary has been avoided. It is the only one.

6. The Ocean Acidification Boundary

In the 17th century, the pH of the oceans was 8.16. Now it is 8.03. But as the pH scale is logarithmic, this represents a 30% increase in the concentration of hydrogen ions. Carbon dioxide from the air dissolves in bodies of water as hydrogen carbonate releasing a hydrogen ion. Hydrogen carbonate ions break down to carbonate ions and hydrogen ions. The increase in hydrogen ion concentration decreases the pH:

$$H_2O + CO_2 \longrightarrow H_2CO_3 \longrightarrow H^+ + HCO_3^-$$
$$HCO_3^- \longrightarrow H^+ + CO_3^{2-}$$

Low pH leaches calcium carbonate out of mollusc and coral shells and arthropod exoskeletons, softening them so organisms are vulnerable to physical and chemical attack:

$$CaCO_3 + H^+ \longrightarrow Ca^{2+} + HCO_3^-$$

Fish are particularly vulnerable to low pH, and their gill structure and functioning is damaged. Fish farms may have to consider changing species or re-locating to more suitable waters, if the problem is not addressed. The ocean acidification (OA) boundary has not yet been crossed, however, and reduction in fossil fuel combustion may prevent oceans undergoing catastrophic pH change.

7. The Freshwater Use Boundary

Liquid water is critical for the survival of living organisms. Many organisms, including the great majority of higher plants and most mammals, must have access to fresh water to live. The **fresh water** use boundary, below which organisms do not have enough regular fresh water to survive, is thought to be avoidable, but fresh water use must be reduced.

Fresh water occurs naturally on the Earth's surface in ice sheets, ice caps, glaciers, icebergs, bogs, ponds, lakes, rivers and streams and underground as groundwater in aquifers and streams. It comes largely from precipitation as mist, rain and snow.

Of all the water on Earth, 97% is saline. About 2.5% is fresh water, most of which is frozen in ice sheets. It is not uniformly distributed, e.g. the Antarctic ice sheet contains 61% of the fresh water on Earth.

Fresh water is not always potable (drinkable) as it carries materials from where it has been blown by the wind. In industrialised areas, it may be acidic; in coastal areas it may contain salts; in desert or dusty soils, it may contain sand and dust, e.g. rain falling in Brazil sometimes contains iron, blown from sandstorms in the Sahara.

Antarctic ice sheet

▼ Study point

There are three aspects to fresh water supply: its quality, volume, and timing. A change in one often leads to changes in the others.

Fresh water availability has decreased

Fresh water is not equally distributed, e.g. Canada has 20% of the world's fresh water supply; India has 10%, but has 30 times more people. About 12% of the global population does not have access to safe drinking water and the UN predicts that by 2025, 14% will have insufficient water. Water shortage damages sanitation, health, food production and global politics. The global supply has diminished to such an extent that one quarter of the world's rivers fail to reach the sea at some time of the year.

There are several reasons for the diminished supply of fresh water:

- Changing landscapes, e.g. draining wetlands, deforestation and soil erosion influence the flow of fresh water, which affects the water cycle.

- Agriculture consumes more fresh water than any other human activity. Strain on local fresh water resources damages ecosystems further. Agriculture may use water pumped from underground aquifers, but these will not be refilled in the foreseeable future.

- Increased demand as people move to warm climates, which have low levels of fresh water.

- Water pollution, e.g. eutrophication, makes water unsuitable for use. An example is provided by the cyanobacterium *Anabaena flos aquae*, which generates an algal bloom when the nitrate concentration is high. It secretes a neurotoxin called anatoxin-a, or VFDF (which stands for Very Fast Death Factor), lethal to birds and many mammals, including humans. Its presence makes water unavailable for use.

- Climate change:
  – As glaciers melt in increasing global temperature, fresh water availability initially increases, but there may be flooding. Then the availability decreases, causing drought.
  – The thermal expansion of ocean water raises sea levels, contaminating freshwater in coastal regions. Ground water becomes too salty for drinking and irrigation.
  – Semi-arid and arid areas are vulnerable because the rain, which is their water supply, occurs over a short space of time and evidence shows that these periods are already disrupted.

- Increase in population through increasing life expectancy.

- Increase in use, e.g. when a person in the USA has a shower, they use, on average, more water than a person in a developing country would use in a whole day.

Going further ▶

Fresh water extraction from rivers throughout central Asia has dried the Aral Sea, which used to be the world's the 4th largest lake. It now occupies only 10% of its former area.

A ship graveyard in what was the Aral Sea

Provision of fresh water

There are several methods for increasing the availability of fresh water, in addition to obtaining it from natural sources:

- Water conservation, e.g. it has been suggested that non-food crops, such as biofuels and cotton should not be irrigated.
- Water efficiency.
- Wastewater reclamation, for irrigation and industrial use.
- Urban runoff and storm water capture, recharging groundwater.
- Drip irrigation systems successfully reduce the water volume used for irrigation of food crops. This is because water containing fertiliser is dripped directly to the roots of individual plants, rather than being sprayed over a large area.
- **Desalination** is a technology that removes minerals from salt water producing fresh water and leaving salt and recycled wastewater. It is one of the few sources of water that does not depend on the rain. The process can have a high energy consumption, so the water it produces may be expensive. Desalinated water, however, may be a solution for some dry regions, though not if the places are poor, far from the sea or at high elevation. Yet these are the places with the biggest water shortages – 1% of the world's population is dependent on desalinated water.

Desalination plant in Lanzarote

Methods for desalination

- Solar stills distill seawater using heat from the Sun. They convert saltwater entirely into distilled water and do not produce air pollution or warm water discharges that endanger local lakes or river.
- Reverse osmosis: seawater is separated from fresh water by a fine, selectively permeable membrane. Water would be expected to move, by osmosis, from the fresh to the seawater, but pressure is applied from the seawater towards the fresh water and water is driven the other way. It moves from the seawater, across the membrane to the fresh water, i.e. against its water potential gradient. This is why the system requires energy. Most power stations use fossil fuels or nuclear power as their source of energy, and in some places, their waste heat is used to generate steam to drive desalination. The overall energy use is, therefore, comparatively efficient, making desalination a sensible method for making drinking water.

Key Term

Desalination: The removal of minerals from saline water.

Going further ▶

The membrane removes harmful contaminants but may remove minerals. Evidence from Israel, which has the highest use of desalination in the world, suggests a link between seawater desalination and iodine deficiency.

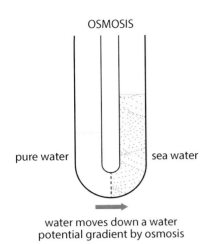

OSMOSIS

pure water   sea water

water moves down a water
potential gradient by osmosis

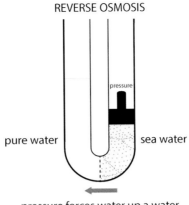

REVERSE OSMOSIS

pressure

pure water   sea water

pressure forces water up a water
potential gradient by reverse osmosis

Osmosis and reverse osmosis

Environmental problems from desalination:

- Brine (concentrated sodium chloride solution) is discharged: the potential environmental impact of such a high salt concentration is avoided by diluting the brine before returning it to the ocean.
- Brine is denser than seawater because of its high solute concentration. It sinks and damages the seabed ecosystem.
- Burning of fossil fuels to generate heat to run the system.

8. The Atmospheric Aerosol Loading Boundary

Atmospheric aerosols are the microscopic particles put into the atmosphere by combustion of fuels and by creating dust from digging and quarrying. Their concentration has doubled since before the Industrial Revolution.

Shanghai, China. Heavy air pollution is common in many Chinese cities

- They exacerbate respiratory problems, causing deaths from lung disease. An example is PM10 i.e. particulate matter of 10 μm diameter, which forms an aerosol and lodges in the lungs when inhaled. Its size makes it especially dangerous and it can lead to cancer. PM2.5, with a diameter of 2.5 μm, is small enough to transfer to the capillaries in the lungs and be carried around the body. It increases the likelihood of death from cardio-vascular disease. The main source of PM10 and PM2.5 is motor vehicles, especially those with diesel engines.
- Deposition on leaves reduces light absorption for photosynthesis and so crop yields are reduced.
- Sulphates in aerosols reflect sunlight, providing some cooling effect, but other particulates, e.g. soot, absorb sunlight and reradiate it, increasing warming. The balance of these two effects is not clear.

Aerosols are so variable that it has not yet been possible to suggest planetary boundaries for them.

9. The Introduction of Novel Entities Boundary

This was previously called the Chemical Pollution boundary. The name change reflects the increasing use of new technologies and materials. It encompasses organic pollutants, radioactive materials, nanomaterials and micro-plastics. Some chemicals are so toxic that their use has already been banned, for example, DDT and PCBs. It is estimated that there are 100 000 different manufactured chemicals in use in the world in millions of different products. They may interact with each other and produce additional harmful effects. As with aerosols, it is not possible to quantify the effects of these pollutants and their planetary boundaries have not been identified.

CH 7

Homeostasis and the kidney

Homeostasis is the term that describes the mechanisms by which the body maintains a constant internal environment. Mechanisms include thermoregulation, control of blood sugar level and osmoregulation. The chemical reactions that take place in living cells produce waste products that must be eliminated. The removal of waste made by the body's metabolism is excretion. In mammals the main organ of excretion is the kidney, which removes the waste product urea. The kidney also regulates the water content and the concentration of blood solutes. Kidney failure can be treated in various ways, including dialysis and transplantation. Animals show variations in their kidney structure that reflect the availability of water in their habitats.

Topic contents

By the end of this topic you will be able to:

- Explain the principles of homeostasis in terms of the stages involved in a feedback loop.
- Describe the structure of the kidney.
- Describe and explain the production of urine with reference to the processes of ultrafiltration and selective reabsorption.
- Explain the role of the loop of Henle and the counter-current multiplier mechanism in the reabsorption of water.
- Explain the role of the hormone ADH in osmoregulation.
- Describe some causes of kidney failure and methods of treating it.
- Explain why fish, birds, insects and mammals produce different excretory products.
- Explain how desert-living mammals have adapted to conserve water.

Homeostasis

The maintenance of a constant environment is homeostasis. The concept of the 'internal environment' dates from the 1850s, and it refers to the prevailing conditions within cells and within the body, in contrast to the external environment. The internal environment includes the tissue fluids that bathe cells, supplying nutrients and removing wastes and maintaining glucose concentration, pH, core temperature and solute potential.

Keeping the concentration of body fluids at a constant and optimum level protects cells from changes in the external environment. It ensures reactions continue at a constant and appropriate rate and allows cells to function normally, despite external changes. Body temperature, pH and water potential may alter but they fluctuate around a set point. The body is kept in a dynamic equilibrium – constant changes occur but a set point is resumed. Homeostasis is the ability to return to that set point. The endocrine system controls homeostatic responses, with hormones operating by negative feedback.

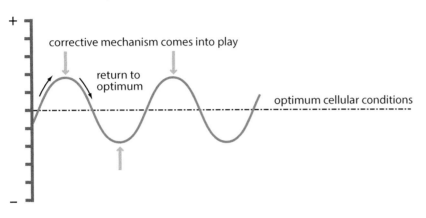

Set point

The control of a self-regulating system by **negative feedback** involves a series of stages, in which an output from an effector, i.e. a muscle or a gland, reduces the effect of a stimulus and restores the system to its original level:

- The set point for a factor is the norm at which the system operates.

- A receptor detects the level of the factor and its deviation from the set point.

- The receptor sends instructions to a co-ordinator or controller.

- The co-ordinator communicates with one or more effectors, i.e. muscles and glands, which make responses that are corrective procedures.

- The factor returns to normal, monitored by the receptor and information is fed back to the effectors, which stop making the correction.

Negative feedback

Going further ▶

Claude Bernard stated in 1856 *'La fixité du milieu intérieur est la condition de la vie libre et indépendante'*, which means 'The stability of the internal environment is a condition for being free-living and independent'.

Key Term

Negative feedback: A change in a system produces a second change, which reverses the first change.

▼ **Study point**

Insulin lowers blood glucose and glucagon raises blood glucose. (Remember: glucagon is produced when the glucose has gone.)

Examples of negative feedback systems include:

- Glucose concentration in the plasma. If glucose concentration increases above the set point, insulin is secreted, reducing the glucose concentration by converting it to glycogen and increasing the rate at which it is respired. If the level falls below the set point, glucagon is secreted, which results in glycogen being converted to glucose.

- If the body's core temperature falls below the set point, increased respiration generates heat, and constriction of superficial blood vessels allows the body to retain it. If the temperature rises above the set point, superficial blood vessels dilate, and heat radiates from the body, reducing its temperature.

Some systems in the body operate by **positive feedback**, in which an effector increases a change, i.e. movement away from the norm causes a further movement away from the norm:

- Oxytocin stimulates the contraction of the uterus at the end of a pregnancy. The contractions stimulate the production of more oxytocin, which increases the stimulus, i.e. the uterine contractions.

- When the skin is cut, the first stage of clot formation is that platelets adhere to the cut surface. They secrete signalling molecules, which attract more platelets to the site.

Excretion

Excretion is the removal of wastes made by the body. The mammalian body excretes several compounds, using four excretory organs:

Excretory compound	Metabolic process producing compound	Compound excreted in	Excretory organ
Carbon dioxide	Respiration	Expired air	Lungs
Water			
Urea	Amino acid breakdown	Urine	Kidneys
Creatinine	Muscle tissue breakdown		
Uric acid	Nucleic acid breakdown		
Urea	Sweating	Sweat	Skin
Bile pigments	Haemoglobin breakdown	Faeces	Liver

Water is a special case when considering how molecules are released from the body. It is needed by the body but it is:

- Excreted as a metabolic waste product of respiration

- Secreted, e.g. in tears and saliva

- Egested in faeces.

▼ **Study point**

Distinguish excretion from egestion (the removal of waste not made by the body) and secretion (the release of useful substances from cells).

◢ **Key Terms**

Excretion: The removal of metabolic waste made by the body.

Osmoregulation: The control of the water potential of the body's fluids by the regulation of the water content of the body.

The kidney

The kidney has two main functions:

- **Excretion** – the removal of nitrogenous metabolic waste from the body.

- **Osmoregulation** – the control of the water potential of the body's fluids (plasma, tissue fluid and lymph) by regulating the water content, and therefore the solute concentration.

Production of urea

Dietary protein is digested into amino acids, which are transported to the liver and then around the body, where they are assimilated into proteins. Any excess amino acids are deaminated in the liver and the amino group is converted to urea:

$$CHR.NH_2.COOH \longrightarrow O=CHR.COOH + NH_3 \longrightarrow O=C(NH_2)$$

amino acid $\qquad$ α-keto acid $\quad$ ammonia $\qquad$ urea

Other nitrogen-containing waste products can also be converted to urea, although a low concentration of creatinine is released in both sweat and urine. The urea is carried in the plasma to the kidneys and excreted in urine.

Structure of the kidney

Humans have two kidneys, one either side of the vertebral column. A tough renal capsule covers each kidney. Each receives blood from a renal artery and returns blood to the general circulation in a renal vein. The blood from the renal artery is filtered in the outer layer, the cortex, at the Bowman's or renal capsules. The medulla contains the loops of Henle and the collecting ducts that carry urine to the pelvis. The pelvis empties urine into the ureter and a ureter from each kidney carries urine to the bladder.

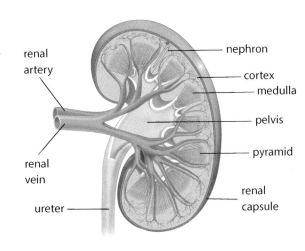

VS kidney

Fine structure

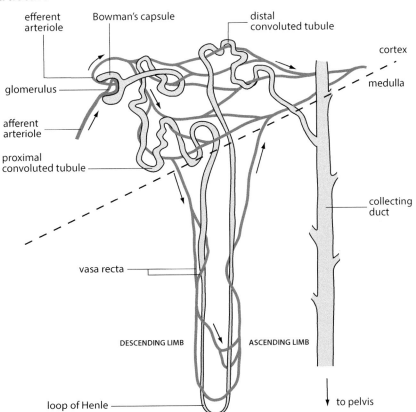

Structure of the nephron

YOU SHOULD KNOW ›››

››› The structure of the kidney

››› The location of the parts of the nephron

››› The names of the parts of the nephron

▼ **Study point**

The glomerulus and the proximal and distal convoluted tubules are in the cortex. The loop of Henle and the collecting duct are in the medulla.

Going further ▶

In an adult, a kidney is about 15 cm long, 6 cm wide and 4 cm thick. The ureter is about 30 cm long and 4 mm in diameter.

A thin section of a kidney observed in a microscope shows it is made of tubes. They are called nephrons or uriniferous tubules or kidney tubules. There are about a million, each about 30 mm long providing a large area for exchange.

A nephron is an individual blood-filtering unit. An afferent arteriole, which is a branch of the renal artery, brings blood to the nephron, and divides into about 50 parallel capillaries in the glomerulus, enclosed by the Bowman's or renal capsule. From there, the filtered blood is a carried by an efferent arteriole to:

- A capillary network surrounding the proximal and distal convoluted tubules.
- The vasa recta, a capillary network surrounding the loop of Henle.

The blood filtrate is diverted through the nephron and the collecting ducts of many nephrons join to carry urine to the pelvis and ureter.

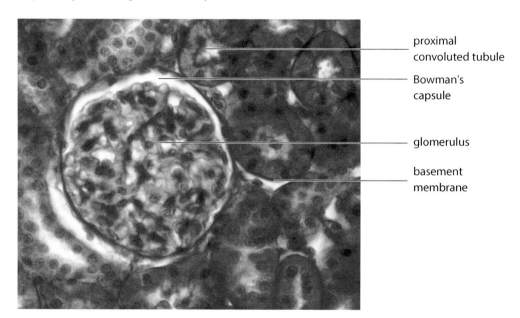

proximal convoluted tubule

Bowman's capsule

glomerulus

basement membrane

Thin section of kidney cortex seen in the light microscope

YOU SHOULD KNOW ›››

››› The ultrastructure of the Bowman's capsule

››› How ultrafiltration takes place

▼ Study point

High blood pressure is maintained in the glomerulus by the contraction of the heart and the afferent arteriole having a wider diameter than the efferent arteriole.

▼ Study point

The blood entering the glomerulus is separated from the space inside the Bowman's capsule by two cell layers and a basement membrane.

Ultrafiltration

The Bowman's capsule

Blood arrives in the capillaries of the glomerulus from the afferent arteriole. It has high pressure because:

- The heart's contraction increases the pressure of arterial blood.
- The afferent arteriole has a wider diameter than the efferent arteriole.

The blood entering the glomerulus is separated from the space inside the Bowman's capsule, called the Bowman's space, by three layers:

- The wall of the capillary is a single layer of endothelium cells with pores called fenestrae, about 80 nm in diameter.
- The basement membrane is an extra-cellular layer of proteins, mainly collagen and glycoproteins. It is a molecular filter and is the selective barrier, acting like a sieve, between the blood and the nephron.
- The wall of the Bowman's capsule is made of squamous epithelial cells called podocytes. Processes from each podocyte, called pedicels, wrap around a capillary, pulling it closer to the basement membrane. The gaps between the pedicels are called filtration slits. A way of visualising this is to think of a stool with short legs: the seat of the stool is the podocyte and the legs are the pedicels.

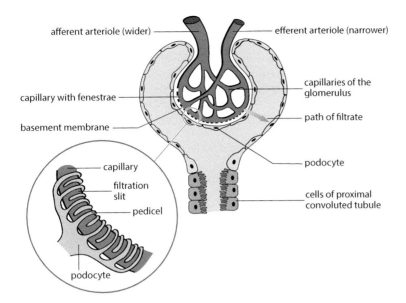

Structure of Bowman's capsule

The high blood pressure in the capillaries of the glomerulus forces solutes and water through the fenestrae of the capillaries, through the basement membrane and through the filtration slits between the pedicels into the cavity of the Bowman's capsule. Filtration under high pressure is **ultrafiltration**.

Key Term

Ultrafiltration: Filtration under high pressure.

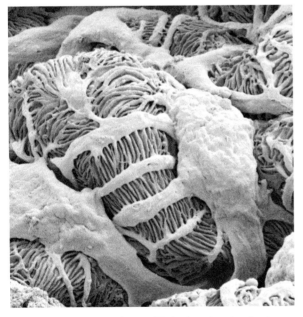

SEM of glomerulus showing capillaries in red and podocytes with their pedicels in beige

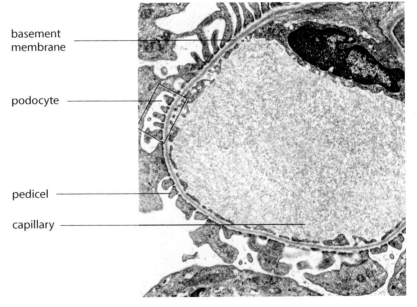

EM of section through a glomerular capillary

The solutes and water forced into the Bowman's capsule constitute the glomerular filtrate, which contains:

- Water
- Glucose
- Salts
- Urea
- Amino acids.

Molecules with RMM < 30,000 pass through the basement membrane easily. Molecules with RMM > 68,000 are too big to pass through. Blood cells, platelets and large proteins such as antibodies and albumin, therefore, remain in the blood. The glomerular filtrate, consequently, resembles plasma although it lacks large proteins. The blood that flows from the glomerulus into the efferent arteriole has a low water potential. This is because much water has been lost and there is a high protein concentration remaining.

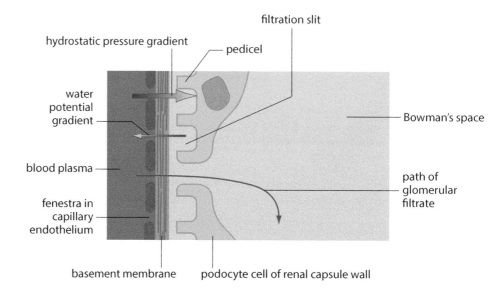

Ultrastructure of wall of glomerular capillary and renal capsule

Filtration rate

Of the blood that leaves the heart, about 20% goes straight to the kidneys. The rate at which fluid passes from the blood in the glomerular capillaries into the Bowman's capsule is called the glomerular filtration rate. It is determined by the difference in water potential between the two areas, i.e. the balance of their hydrostatic pressures and solute potentials. Together, the kidneys of an adult receive about 1.1 dm³ min⁻¹ and produce 125 cm³ min⁻¹ glomerular filtrate.

Blood volume entering kidneys per minute = 1.1 dm³

Volume of filtrate produced per minute = 125 cm³

∴ Blood volume leaving glomerulus per minute = 1100 – 125 = 975 cm³

$$\% \text{ filtered} = \frac{125}{1100} \times 100 = 11.4\% \text{ (1 dp)}$$

Selective reabsorption

The glomerular filtrate contains wastes that the body needs to eliminate, but also useful molecules and ions, including glucose, amino acids, sodium ions and chloride ions. **Selective reabsorption** is the process by which useful products are reabsorbed back into the blood, as the filtrate flows through the nephron.

Proximal convoluted tubule (PCT)

The proximal convoluted tubule is the longest and widest part of the nephron. It carries the filtrate away from the Bowman's capsule. The blood in the capillaries around the PCT reabsorbs all the glucose and amino acids, some of the urea and most of the water and sodium and chloride ions from the filtrate in the proximal convoluted tubule. The PCT has:

- A large surface area because it is long and there are a million nephrons in the kidney.

- Cuboidal epithelial cells in its walls. Their surface area is increased by microvilli, about 1 μm long, facing the lumen, and invaginations called basal channels in the surface facing the basement membrane and capillary.

- Many mitochondria, providing ATP for active transport.

- A close association with capillaries.

- Tight junctions between the cells of the proximal convoluted tubule epithelium. These are multi-protein complexes that encircle a cell, attaching it tightly to its neighbours. They prevent molecules from diffusing between adjacent cells or from the cell back into the glomerular filtrate.

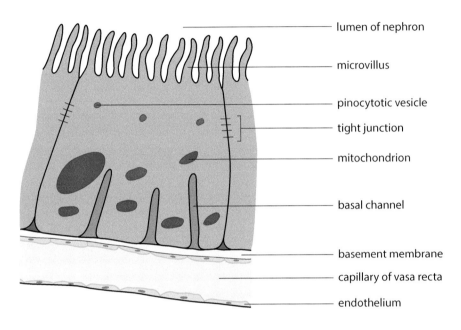

Cuboidal epithelium cell from of the proximal convoluted tubule wall

Selective reabsorption in the proximal convoluted tubule

- About 70% of the salts in the filtrate are reabsorbed to the blood. Some reabsorption is passive, although most uses active transport by membranes pumps.

- All the glucose and amino acids are reabsorbed to the blood by co-transport with sodium ions. A glucose molecule and two sodium ions bind to a transporter protein in the cuboidal epithelium cell membrane. They enter the cell by facilitated diffusion, dissociate from the transporter and diffuse across. Sodium ions are pumped into the capillary; glucose moves in by facilitated diffusion. Co-transport is called **secondary active transport** because active transport keeps the sodium ion concentration in the epithelial cell low, enhancing its diffusion into the cell, carrying in the glucose.

- About 90% of the water in the glomerular filtrate is reabsorbed to the blood passively, by osmosis, as reabsorbed ions lower the water potential of the blood.

- About 50% of the urea and small proteins in the glomerular filtrate is reabsorbed back to the blood by diffusion. So much water has been lost from the filtrate that their concentration there is high. As a result, the concentration gradient down which they diffuse is steep.

In summary, the filtrate has lost salts, water, urea, glucose and amino acids back to the blood. At the base of the proximal convoluted tubule, the filtrate is isotonic with the blood plasma.

Going further ▶

The PCT regulates the pH of the filtrate, by exchanging hydrogen carbonate ions, which increase the pH, with hydrogen ions, which decrease the pH.

 Link You learned about co-transport in connection with glucose absorption in the small intestine in the first year of this course.

Going further ▶

Not everything in urine comes from ultrafiltration. Tubular secretion is active transport into the filtrate from capillaries around the convoluted tubules and collecting duct of, e.g. creatinine, urea, some hormones and drugs, e.g. penicillin.

▼ **Study point**

All the glucose, some of the urea and most of the water and salts are reabsorbed from the glomerular filtrate back into the blood, leaving urine.

 Key Term

Secondary active transport: The coupling of diffusion, e.g. of sodium ions, down an electrochemical gradient providing energy for the transport, e.g. of glucose up its concentration gradient.

Going further ▶

Some water is reabsorbed at the distal convoluted tubule. Its other roles include regulation of filtrate pH; reabsorption of calcium, in response to parathyroid hormone, and of sodium in response to aldosterone.

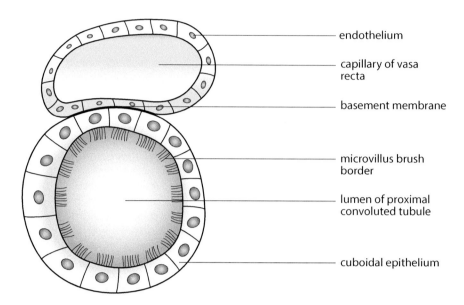

- endothelium
- capillary of vasa recta
- basement membrane
- microvillus brush border
- lumen of proximal convoluted tubule
- cuboidal epithelium

Proximal convoluted tubule and capillary

The glucose threshold

Glucose is an energy source and the body would be disadvantaged if it were lost. Under normal circumstances, the proximal convoluted tubule reabsorbs all the glucose that is present in the glomerular filtrate. If the concentration of glucose in the filtrate is too high, however, there may be too few transport molecules in the membranes of the proximal convoluted tubule cells to absorb it all. In that case, glucose will pass through into the loop of Henle and be lost in the urine. This may happen if:

- The pancreas secretes too little insulin (type I diabetes).
- The response of liver cells to insulin is reduced because insulin receptors in surface membranes are damaged (type II diabetes or gestational diabetes, which occurs in some women during pregnancy).

Reabsorption of water

A major challenge for terrestrial organisms is preventing dehydration. The body cannot afford to lose large volumes of water in urine; so much of it is reabsorbed back to the blood as the glomerular filtrate flows through the nephron. About 90% of the water filtered at the Bowman's capsule is reabsorbed into the blood from the proximal convoluted tubule. Some is reabsorbed back to the blood from both the distal convoluted tubule, in the cortex of the kidney, and from the loop of Henle in the medulla. About 5% is reabsorbed from the collecting duct.

The proximal convoluted tubule and the loop of Henle have 'stereotyped' functions, i.e. they always act in the same way, absorbing the same volume. But the distal convoluted tubule and collecting duct can reabsorb varying volumes of water in response to the body's needs, so they operate the fine control of the body's water content.

The mechanism of water reabsorption

The filtrate enters the descending limb of the loop of Henle and moves down into a hairpin bend and up into the ascending limb.

- The walls of the **ascending limb** are impermeable to water. They actively transport sodium and chloride ions out of the filtrate in the tubule into the tissue fluid in the medulla. A longer loop of Henle means that more ions can be exported into the medulla. The loops of Henle collectively concentrate salts in the tissue fluid, which therefore has a low water potential. As the filtrate climbs from the bottom of the hairpin, it contains progressively fewer ions. It becomes increasingly dilute and its water potential increases.

- The walls of the **descending limb** are permeable to water, and slightly permeable to sodium and chloride ions.
  - As filtrate flows down the descending limb, water diffuses out, by osmosis, into the tissue fluid of the medulla, which has low water potential. From there it moves into the vasa recta, i.e. the capillaries surrounding the loop of Henle.
  - At the same time, some sodium and chloride ions diffuse into the descending limb.

As the filtrate flows down the descending limb, it contains progressively less water and more ions and so, at the bottom of the hairpin, the filtrate is at its most concentrated, with the lowest water potential.

Having two limbs of the loop running side by side, with the fluid flowing down in one and up in the other, enables the maximum concentration to be built up at the apex of the loop. This mechanism is a **counter-current multiplier,** because flow in the two limbs is in opposite directions (counter-current) and the concentration of solutes is increased (multiplied). The solute concentration is even higher in the medulla.

- The collecting duct runs back down into the medulla, passing through the region of low water potential. Water therefore diffuses out of the collecting duct by osmosis, down a water potential gradient. The longer the loop of Henle, the lower the water potential in the medulla and the more water leaves the collecting duct by osmosis. The filtrate becomes more concentrated than the blood, i.e. it is hypertonic to the blood and by the time it reaches the base of the collecting duct, it is urine. The water is reabsorbed into the vasa recta, the blood capillaries surrounding the loop of Henle, and into the general circulation.

▼ **Study point**

The descending limb of the loop of Henle is permeable to water but much less so to ions; the ascending limb is permeable to ions but not to water.

Link The significance of the length of the loop of Henle and its relationship with the ion concentration in the medulla is discussed on p134.

Going further ▶

Removal of water makes urine hypertonic to blood. The energy to prevent water moving into it by osmosis is the energy expended in pumping ions out of the ascending limb.

▼ **Study point**

The vasa recta, the capillaries surrounding the loops of Henle in the medulla deliver nutrients to the medulla cells and carry water reabsorbed from the glomerular filtrate in the nephron.

Exam tip

The nephrons operate by ultrafiltration, selective reabsorption, secretion by active transport and osmosis. Make sure you know the locations of these processes in the nephron.

16

Knowledge check

Match the parts of the nephron 1–4 with the statements A–D.

1. Proximal convoluted tubule.
2. Glomerulus.
3. Ascending limb of loop of Henle.
4. Collecting duct

A. Secretion by active transport.
B. Selective reabsorption.
C. Ultrafiltration.
D. Osmosis

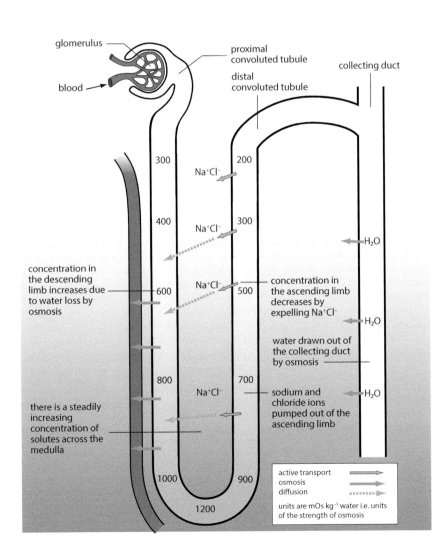

Counter-current multiplier

Going further ▶

Most of the body's water is lost as urine. Other losses are due to sweating, keeping exchange surfaces moist (e.g. alveoli), loss in faeces and loss in tears.

 Key Term

Antidiuretic hormone: Hormone produced in the hypothalamus and secreted by the posterior pituitary. It increases permeability of the cells of the distal convoluted tubule and collecting duct walls to water, increasing water reabsorption.

Osmoregulation

Osmoregulation is the homeostatic function that maintains concentrations of enzymes and metabolites, so that the reactions within cells occur at a constant and appropriate rate. To maintain the osmotic properties of their tissues and fluids, mammals must balance water gain with water loss. Humans gain most of their water from drinking and from food but about 10%, is 'metabolic water', i.e. water released from the body's reactions.

In the same way as other homeostatic mechanisms, osmoregulation operates by negative feedback: the hypothalamus, at the base of the brain, is the receptor, as its osmoreceptors monitor the solute potential of the blood. It is also the co-ordinator, as it signals the effector, the posterior lobe of the pituitary gland, to release stored **antidiuretic hormone**, ADH. This returns the system to normal if it deviates too far, by changing the behaviour of the walls of the distal convoluted tubule and the collecting duct.

Diuresis is the production of a large volume of dilute urine. A diuretic, such as alcohol, is a compound that causes the production of a large volume of urine. As its name suggests, antidiuretic hormone causes the production of a small volume of concentrated urine. It makes the walls of the collecting duct and distal convoluted tubule more permeable to water, so that more is reabsorbed from the filtrate back to the blood.

Negative feedback controls the volume of water reabsorbed. It restores the normal water potential if the blood is diluted or becomes more concentrated. A fall in water potential of the blood may be caused by:

- Reduced water intake
- Sweating
- Intake of large amounts of salt.

The reduced water potential is detected by osmoreceptors in the hypothalamus. Secretory granules carry ADH along axons from the hypothalamus to the posterior lobe of the pituitary gland, from where ADH is secreted into the blood stream. It is carried to the kidneys where:

- ADH increases the permeability of the walls of the distal convoluted tubule and the collecting duct to water.
- More water is reabsorbed from there into the region of high solute concentration, low water potential in the medulla.
- More water is reabsorbed from the medulla into the blood in the vasa recta.
- The water potential of the blood is restored to normal.
- The small volume of urine produced is relatively concentrated. Its concentration is close to the concentration of the tissues near the apex of the loop of Henle, and it is hypertonic to the body fluids.

The converse occurs if the water potential of the blood decreases, as a result of taking in a large volume of water. Less ADH is released by the posterior pituitary gland and so the permeability of the distal convoluted tubule and collecting duct walls decreases. Less water is reabsorbed to the blood, so its water potential is restored to normal. The body produces a larger volume of more dilute urine.

Going further ▶

Each minute, the kidneys produce about 125 cm³ glomerular filtrate and reabsorb 124 cm³, making about 1 cm³ urine or about 1.4 dm³ per day.

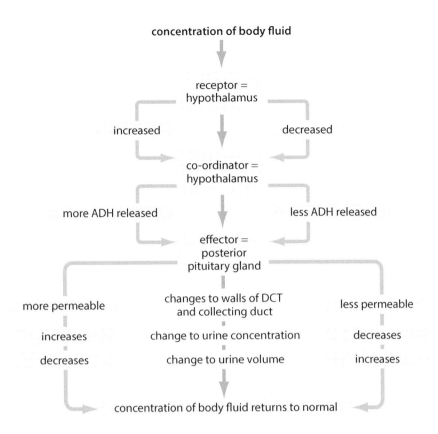

concentration of body fluid

receptor =
hypothalamus

increased   decreased

co-ordinator =
hypothalamus

more ADH released   less ADH released

effector =
posterior
pituitary gland

more permeable   changes to walls of DCT   less permeable
                        and collecting duct

increases   change to urine concentration   decreases

decreases   change to urine volume   increases

concentration of body fluid returns to normal

Negative feedback mechanism

Going further ▶

Reabsorption of water is crucial for maintaining the correct osmotic balance in the body. If no ADH were released, it is estimated that the body would dehydrate within 3 minutes.

ADH mechanism

Aquaporins are intrinsic membrane proteins with a pore through which water molecules move. There are thirteen types known and six of these operate in the kidney. The existence of many more types is suspected. In the walls of the distal convoluted tubule and collecting duct:

- ADH binds to membrane receptors.
- Adenyl cyclase catalyses the production of cyclic AMP, the second messenger.
- Vesicles containing aquaporins in the cytoplasm to move to and fuse with the cell membrane.
- Aquaporins are incorporated into the membrane.
- Water molecules move in single file through their pores into the cell, down a water potential gradient.

When intracellular cyclic AMP levels fall, the aquaporins are removed from the cell membrane and they accumulate again in vesicles.

▼ **Study point**

If the blood's water potential is too low, increased ADH increases water reabsorption. If it is too high, decreased ADH decreases water reabsorption. In both cases, water potential returns to normal.

17

Knowledge check

Identify the missing word or words.

To conserve water, osmoreceptors in the .... of the brain detect the solute potential of the blood, resulting in nerve impulses passing to the .... gland which releases .... into the blood stream. This hormone increases the permeability of the walls of the distal convoluted tubule and .... to water. The water is reabsorbed into the region of low water potential in the .... This results in the production of a volume of more urine.

▼ **Study point**

Ultrafiltration cannot occur if the blood pressure is too low. If the blood pressure is too high, damage to capillaries also prevents it.

▼ **Study point**

Kidney failure is treated with dietary restriction, drugs to reduce blood pressure and calcium and potassium ion concentration in the plasma, dialysis (haemodialysis and CAPD) and transplantation.

Kidney failure and its treatment

The major roles of the kidney are excretion and osmoregulation. If they fail, the body is unable to remove urea, so its concentration increases to toxic levels. The body is unable to remove excess water and so body fluids increase in volume and are diluted, compromising metabolic reactions.

The commonest causes of kidney failure are:

- Diabetes: high glucose concentration in the plasma results in the glomeruli losing protein, especially albumin, into the filtrate and causes some to proteins to link together, triggering scarring in a condition called glomerulosclerosis.
- High blood pressure: damage to the capillaries of the glomerulus prevents ultrafiltration.
- Auto-immune disease: the body makes antibodies against its own tissues.
- Infection.
- Crushing injuries, for example, in a road traffic accident.

The body can remain healthy with only one kidney. There may be slight loss of kidney function later in life but life span is normal. If both kidneys are compromised, however, treatments must reduce the concentration of waste products and control the volume of body fluids, to regulate solute concentration:

- Reducing intake of certain nutrients, in particular protein, to reduce urea formation and ions, e.g. calcium and potassium.
- Using drugs to reduce blood pressure:
  - Angiotensin-converting enzyme (ACE) inhibitors and angiotensin receptor blockers (ARBs) reduce the effect of angiotensin, a hormone that constricts blood vessels, increasing the pressure of the blood within.
  - Calcium channel blockers dilate blood vessels and reduce blood pressure.
  - Beta blockers reduce the effect of adrenalin, one effect of which is increased blood pressure as the heart rate increases.
- The concentrations of dissolved potassium and calcium ions are normally maintained by a balance of absorption in the small intestine and by selective reabsorption by the nephrons.
  - A high potassium concentration in the blood is treated with a combination of glucose and insulin. If untreated, it leads to heart arrhythmias so intravenous calcium is used in addition, to stabilise heart muscle membranes.
  - High calcium in the blood is correlated with increased risk of heart disease, kidney stones and osteoporosis. It is treated with bisphosphonates, which decrease the activity of osteoclasts, the cells that break down bone in its constant recycling. Calcium therefore accumulates in bone, and less circulates in the blood.
- Dialysis: the blood to be cleaned and a dialysis fluid are separated by a selectively permeable membrane. The dialysis fluid has the same water potential as the blood, but a low ion concentration and no urea. Inorganic ions, water and urea diffuse out of the blood across the membrane, down their concentration gradients. The dialysis fluid contains glucose at the normal concentration of the blood, so none diffuses out of the blood.
- Kidney transplant: a patient with end-stage renal disease may be offered a kidney transplant.

Haemodialysis

Haemodialysis uses a dialysis machine. Blood is taken from an artery, usually in the arm, and run through thousands of long, narrow fibres made of selectively permeable dialysis tubing. The fibres are surrounded by the dialysis fluid. The pores of the tubing let molecules in solution out into the dialysis fluid, but not the large proteins, blood cells or platelets. The blood and dialysis fluid run through the machine in opposite directions, enhancing diffusion out of the blood by a counter-current mechanism. The blood is returned to a vein. Heparin is added to thin the blood and prevent it clotting. A sensor in the dialysis fluid detects haemoglobin that would diffuse through if red blood cells were damaged. Patients routinely use a machine for several hours at a time, several days a week.

Going further ▶

Kidney functions not addressed by dialysis include erythropoietin production, so patients may become anaemic, and the conversion of dietary vitamin D into an active compound.

Thousands of fibres in a dialyser

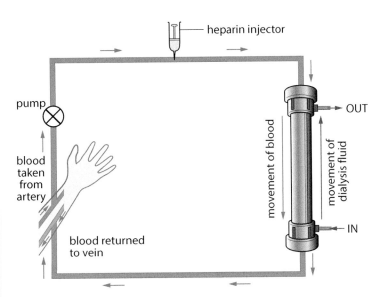

Haemodialysis

Going further ▶

Some people only have one kidney because they were born with only one, they had one removed, through illness or donation or both were damaged and they received a transplant.

Continuous ambulatory peritoneal dialysis (CAPD)

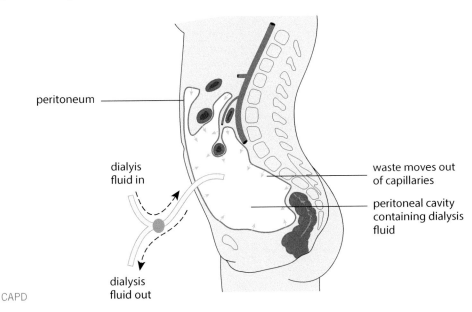

CAPD

This is 'ambulatory' because the patient can walk around, carrying on with normal activities while the dialysis operates. The patient drains a 1–3 dm^3 bag of dialysis fluid through a catheter in the abdomen, into the body cavity. The peritoneum is the membrane lining the body cavity and it has a rich supply of capillaries. It acts as the dialysis membrane and materials are removed from the blood in the capillaries into the dialysis fluid. After about 40 minutes, the fluid is drained from the abdomen, under gravity, into an empty bag. The process is repeated four times a day. Although dialysis removes wastes, retention of liquid is common and potassium ions accumulate in the blood. Patients therefore must drink very little and avoid foods such as bananas and tomatoes, which are rich in potassium.

Going further ▶

Rejection is recognised by an increase in serum creatinine and urea, a reduction in the volume of urine or the kidney may feel tender. It is confirmed with a biopsy.

18

Knowledge check

Complete the paragraph by filling in the spaces.

A patient with failing kidneys may have to restrict their .... and take drugs to reduce the potassium and calcium concentration in the blood and also their blood .... When kidneys fail, a patient's blood may be passed through a kidney machine, where it runs .... to a dialysis fluid on the other side of a selectively permeable membrane. If a patient is to receive a kidney transplant, the donor and recipient may have compatible .... groups.

Transplant

Donors and recipients

Donors may be living or have suffered brain stem or circulatory death. A kidney from a live donor generally works immediately and lasts longer. Deceased donor kidneys may take a few days or weeks to work and dialysis is used meanwhile. Some transplanted kidneys have survived over 30 years but most fail at some stage, and a patient returns to dialysis.

The donor and recipient must be compatible in their ABO blood group and in most of their HLA (human leucocyte antigens). Higher risk donors include those over 50 or those with high blood pressure or diabetes. Their donated kidneys have a higher failure rate than others, but a patient is still better off with a transplant than remaining on dialysis.

The transplant

The transplanted kidney is placed in the lower abdomen, in the groin and the renal artery and vein emerging from the transplanted kidney are attached to the iliac artery and vein, respectively. The circulation to the new kidney is then restored and when the kidney resumes a healthy pink colour and urine is seen emerging from the ureter, the ureter is joined to the bladder.

Immunosuppressants

A transplant recipient must take immunosuppressive drugs for the rest of their life, but even so, rejection may occur, and is commonest within the first six weeks. With a suppressed immune system, patients are more susceptible to infection, especially of the urinary tract. This can eventually damage the kidney and so long-term low-dose antibiotics may be used. Cytomegalovirus has infected 50% of the UK population. The donor kidney may infect an uninfected recipient and to avoid complications, anti-virals may be used. Immunosuppressants may also increase the risk of cancers, especially skin cancer and lymphoma.

Excretion and osmoregulation in different environments

Plants are producers and they make all the proteins that they need. Active transport and facilitated diffusion allow them to take up nitrate and ammonium ions from the soil, and plants in the Fabaceae family also obtain nitrogenous compounds from their mutualistic nitrogen-fixing bacteria. Plant cells combine ammonium ions with α-keto glutarate, making the amino acid glutamate; glutamate is converted into any other amino acid by **transamination** of other α-keto acids:

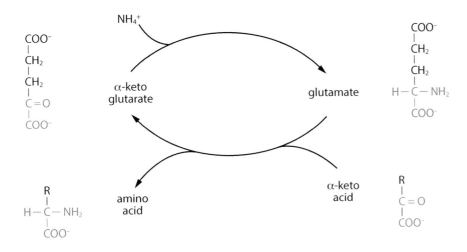

Making amino acids by transamination

Key Term

Transamination: An enzyme-catalysed reaction that transfers an amino group to an α-keto acid, making an amino acid.

Plants synthesise only the amino acids and proteins that they need, so they do not need to excrete nitrogen-containing molecules.

This is in contrast with animals. Animals are far less efficient than plants at transamination. They eat protein, and make the molecules they need from its constituent amino acids. They cannot store any they do not need, so these excess amino acids are deaminated and converted into another molecule, which is excreted. The environment in which an animal evolved determines the nature of the molecule it excretes:

- Most aquatic organisms, e.g. many freshwater fish and small organisms such as *Amoeba* excrete ammonia. It is highly toxic but extremely soluble in water. The large surface area of fish gills and of *Amoeba* allows ammonia to diffuse out rapidly and it is immediately diluted below toxic concentrations.

- Birds, reptiles and insects do not carry excess water and they convert amino acids into uric acid for excretion. Uric acid is almost insoluble in water and it is non-toxic. There is a large energy cost to its production but very little water is needed for its excretion. This is important in conserving water and allows these organisms to live in environments where there is a shortage of water or if their lifestyle requires them to be light enough for flight.

- Mammals excrete urea. Its production also requires energy but it is less toxic than ammonia and so tissues and body fluids dilute it below a toxic concentration and can tolerate it briefly. Desert-living mammals and those in aquatic habitats have adapted to their particular water availability and the concentration of their urine reflects this.

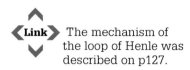 **Link** The mechanism of the loop of Henle was described on p127.

Going further ▶

The kangaroo rat makes very concentrated urine but has shorter loops of Henle than some animals making more dilute urine. Its high metabolic rate allows more active transport, creating a steep ion gradient in the medulla.

The loop of Henle

The longer the loop of Henle, the more the opportunity there is to pump ions into the medulla. The ion pumps in the ascending limb increase the concentration in the medulla above that in the loop of Henle. This low water potential in the medulla enhances water reabsorption from the descending limb and from the collecting duct, resulting in more concentrated urine.

There are two types of nephron, with their Bowman's capsule and loop of Henle in different positions:

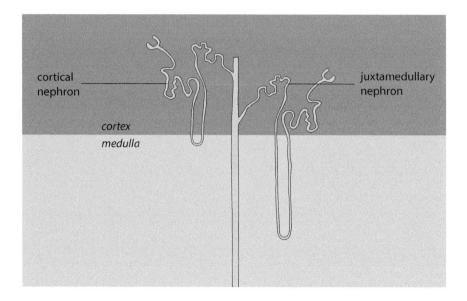

Kidney structure and water conservation

- Cortical nephrons have the glomerulus in the outer cortex and a short loop of Henle, which just penetrates the medulla, near its boundary with the cortex. In humans, most nephrons are cortical. Beavers and muskrats are animals that are not at risk of dehydration and their nephrons are mainly cortical, with very short loops of Henle. They make very dilute urine.

- Juxtamedullary nephrons have their Bowman's capsule closer to the cortex's boundary with the medulla. They have a long loop of Henle, which penetrates deep into the medulla. Mammals, such as Australian hopping mice, which live in very dry habitats, have a high proportion of juxtamedullary nephrons. They can therefore generate a very low water potential in the medulla and make very concentrated urine, conserving water very efficiently.

Mammal	Habitat	Relative length of loop of Henle	Urine concentration / mOsmol dm^{-3}
beaver	freshwater	very short	520
rabbit	mesic	short	3100
hopping mouse	desert	long	5500

The kidney tubule in mammals that evolved in habitats with different water availability

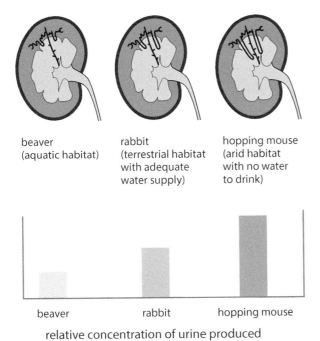

beaver
(aquatic habitat)

rabbit
(terrestrial habitat
with adequate
water supply)

hopping mouse
(arid habitat
with no water
to drink)

beaver   rabbit   hopping mouse

relative concentration of urine produced
(not to scale)

Key Term

Metabolic water: Water produced
from the oxidation of food reserves.

Exam tip

Adaptations to conserve water
may be anatomical, biochemical or
behavioural.

Metabolic water

Metabolic water is water produced from the breakdown of food and its respiration.
Xerocoles are desert animals, such as the kangaroo rat, that live in a very hot, dry habitat
and survive with little or no water. They do not drink and rely entirely on metabolic water.

Behaviour

Many desert animals remain underground
during the day, living in burrows, which are cool
and humid, reducing water loss by evaporation.
The hyrax is diurnal but remains in crevices
in rocks, in the shade. Scorpions, like many
desert animals, are nocturnal and are at less
risk of dehydration as less water evaporates
from their bodies at lower temperatures.

Rock hyrax

19

Knowledge check

Match the animals 1–3 with their
excretory product A–C.

1. Fish.
2. Insect.
3. Mammal.

A. Urea.
B. Ammonia.
C. Uric acid

Specified practical exercise

Kidney dissection

A butcher or supermarket will be able to provide suitable kidneys to dissect. These are likely to be lambs' kidneys, but any mammalian kidney will show the same structures as a human kidney.

Apparatus

- Kidney
- Chopping board
- Fine scalpel
- Fine scissors
- Fine forceps
- Seeker
- Glass rod
- Lens on a stand
- Microscope slide and coverslip
- Microscope

Method

1. Remove fat from the outside of the kidney. It can be pulled off by hand without damaging the kidney.

2. Place the kidney flat on the chopping board with the blood vessels and ureter emerging on the right.

3. Pierce the capsule of the kidney with the point of the scalpel and with the blade of the scalpel parallel with the bench surface, make small cuts, bringing the blade towards you each time. Rotate the kidney anti-clockwise after every few cuts, so that you cut right round the organ.

4. Extend the cuts through to the centre so that the kidney can be separated into two halves.

5. Identify the cortex at the outside of the kidney and the medulla further in. Note their colour difference.

6. Identify the pyramids and the pelvis. Use the seeker to follow the cavity of the pelvis into the ureter.

7. Place one half of the kidney on the chopping board with the cortex away from you. Lift the inner tissue of the medulla with forceps and cut through it with fine scissors, away from you, revealing white tubules, leading into the pelvis. Each of these is a major calyx.

8. Cut further towards the cortex to expose the finer branches. Each of these is a minor calyx. Use the lens to expose tubules as small as your eyes can see. The calices are continuous with the collecting ducts.

9. Place a small sample of cortical tissue on a microscope slide and macerate it in water with a scalpel and glass rod. Place a cover slip n the material and view under the microscope with a ×10 and a ×40 objective lens.

cortex

medulla

pelvis

inner tissue of medulla

renal capsule

Vertical section through a lamb's kidney

Risk assessment

Hazard	Risk	Control measure
Dissecting instruments are sharp	Can pierce or cut the skin	Care with use
Bacteria may be present on the surface of the kidney	Food poisoning	Wash hands immediately after dissection; care with disposal of dissected material.

CH 8

The nervous system

Detecting and responding to changes in the external environment enhances an organism's chance of survival. Structures that detect changes may be distant from those that respond, so organisms require a means of internal communication. Chemical communication involves hormones, secreted by endocrine glands into the blood. Physical communication is much faster and involves the passage of a nervous impulse from a receptor to an effector.

The mammalian nervous system is dual in nature. The central nervous system (CNS) co-ordinates and controls an animal's activities; the peripheral nervous system connects organs with the CNS. Reflex actions control many body functions and actions. Drugs can disrupt the normal communication between elements of the nervous system.

Topic contents

By the end of this topic you will be able to:

- Understand that the nervous system controls and co-ordinates actions by detecting stimuli, processing the information and initiating responses.
- Understand that stimuli are detected by receptors and responses are brought about by effectors.
- Describe the structure of the mammalian spinal cord.
- Contrast the simple nerve net system of the Cnidarian *Hydra* with the human nervous system.
- Describe the passage of nervous impulses in a reflex arc.
- Describe the structure of a motor neurone.
- Explain the transmission of a nervous impulse in terms of how an action potential is generated.
- Describe the factors affecting the speed of conduction of the nervous impulse.
- Describe the synapse as a junction between two conducting cells.
- Explain synaptic transmission by means of the neurotransmitter, acetylcholine.
- Explain how organophosphates and psychoactive drugs affect synaptic transmission.

The components of the nervous system

The nervous system:

- Detects changes, or stimuli, inside the body and in the environment.
- Processes and stores information.
- Initiates responses.

A **stimulus** is a detectable change in the internal or external environment of an organism that produces a **response** in that organism. Sensory receptors give an organism its senses. There are specialised sensory cells, such as pressure sensors in the skin, and in complex sense organs such as the ear and eye. Sensory receptors are transducers because they detect energy in one form and convert it into electrical energy. The electrical impulses travel along neurones and are called nervous impulses. They initiate a response in an effector, which may be a muscle or a gland.

Stimulus	Sensory receptor location	Sense
visible light	retina	sight
sound	inner ear	hearing
pressure	dermis of the skin	touch
heavy pressure	deeper in the dermis of the skin	pain
chemical	nose	smell
chemical	tongue	taste
temperature	dermis of the skin	temperature
gravity	middle ear	balance

The nervous system has two main parts:

- The **central nervous system** (CNS) comprises the brain and spinal cord. The CNS processes information provided by a stimulus. Both the brain and spinal cord are surrounded by tough, protective membranes called, collectively, the meninges. The structure of the spinal cord is shown in the diagram on p140. The white matter contains nerve fibres surrounded by myelin, which is fatty and so looks white. The grey matter has much less myelin and is largely the nerve fibres of relay neurones and the cell bodies of relay and motor neurones.

- The **peripheral nervous system** (PNS) comprises:

  - The somatic nervous system, i.e. pairs of nerves that originate in the brain or the spinal cord, and their branches. These nerves contain the fibres of sensory neurones, which carry impulses from receptors to the CNS, and motor neurones, which carry impulses away from the CNS to effectors.

  - The autonomic nervous system provides unconscious control of the functions of internal organs, e.g. heartbeat, digestion.

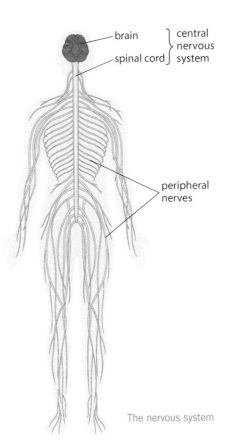

brain ⎫ central
spinal cord ⎬ nervous
⎭ system

peripheral nerves

The nervous system

Going further ▶

Mammals also sense pressure in blood vessels, ultra-violet light, electric fields, changes in humidity, magnetic fields, tissue damage and osmolarity of fluids.

Exam tip

A nerve is a bundle of neurones or nerve fibres.

▼ **Study point**

The ability to detect and respond to stimuli increases an organism's chance of survival.

20

Knowledge check

Identify the missing word or words.

The nervous system has two main divisions: the CNS, comprising the .... and .... and the peripheral nervous system. The latter is made up of the .... neurones that carry impulses to the CNS and the .... neurones that carry impulses away from the CNS.

The reflex arc

The simplest type of nervous response to a stimulus is a **reflex arc**. It is the neural pathway taken by the nervous impulses of a reflex action. An example of a reflex arc is the withdrawal reflex, when you immediately withdraw your hand if, for example, you place it on a hot object.

A **reflex action** is a rapid, automatic response resulting from nervous impulses initiated by a stimulus. The decision-making areas of the brain are not involved and the action is involuntary. Reflex actions are generally protective in function.

The elements of a reflex arc are stimulus $\rightarrow$ receptor $\rightarrow$ sensory neurones $\rightarrow$ relay neurone in CNS $\rightarrow$ motor neurone $\rightarrow$ effector $\rightarrow$ response. They can be identified in any reflex action. Two examples of reflex actions are given in the table below.

▼ **Study point**

The reflex arcs of some reflex actions do not have a relay neurone, e.g. the knee jerk reflex. The sensory neurone synapses directly with the motor neurone.

Going further ▶

If the sense receptor for a reflex arc is in the head, e.g. the retinal cells in the pupil reflex, the part of the CNS involved is the brain and not the spinal cord.

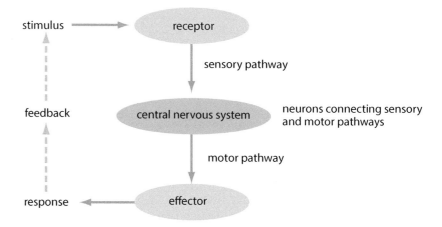

Main components of nervous control mechanisms

Stage of reflex arc	Withdrawal reflex	Pupil reflex
Stimulus	Heat	Light
Sensory receptor	Temperature and pain receptors in the skin	Photosensitive cells in retina
Sensory neurone	Sends impulse up the arm to the spinal cord	Optic nerve
CNS	Relay neurone in spinal cord transmits an impulse from a sensory neurone to a motor neurone	Brain
Motor neurone	Sends impulse to an effector, in this case a muscle	Carries impulses to muscles of iris
Response	Arm muscles contract and the hand is removed from heat source	Iris muscles relax or contract, altering the diameter of the pupil

Exam tip

Be prepared to label a diagram of a TS spinal cord and to show the direction of the nervous impulse in a reflex arc.

▼ **Study point**

The cell bodies of sensory neurones all lie together and appear as a swelling, called a ganglion.

The nervous pathways of the reflex arc

Nerve nets

Animals in the phyla that appear very early in the fossil record do not have nervous systems, e.g. the phylum Porifera, which includes sponges. Animals that appeared later have radial symmetry and their nervous system is a nerve net, e.g. the phylum Cnidaria, which includes jellyfish. Animals that appeared even later have bilateral symmetry and a central nervous system, e.g. Chordates, which includes humans.

A **nerve net** is the simplest type of nervous system. It is a diffuse network of cells that group into ganglia, but do not form a brain. There are two types of cell in a nerve net:

- Ganglion cells provide connections in several directions.
- Sensory cells detect stimuli, e.g. light, sound, touch or temperature.

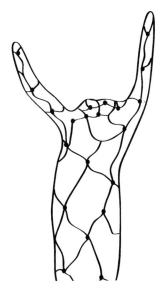

Hydra budding

The *Hydra* nerve net

Hydra, in the phylum Cnidaria, is the model organism for studying nerve nets because its nerve net:

- Has a simple pattern
- Is easy to manipulate in experiments
- Regenerates rapidly, e.g. when replacing a lost tentacle.

Hydra's nerve net is in its ectoderm, the outer of the two layers of cells of its body wall. The nerve net allows *Hydra* to sense light, physical contact and chemicals. In response, it can contract, perform locomotion, hunt and feed. So even without a brain, it shows complex movements and behaviour. It cannot, however, detect the direction of a stimulus, although a larger stimulus stimulates more cells and triggers a larger response.

The table shows contrasts between the *Hydra* nerve net and the human nervous system:

	Hydra	Human
Nervous system type	Nerve net	CNS
Number of cell types in nervous system	2	Many
Regeneration	Rapid	Very slow if at all
Myelin sheath	Absent	Present
Conduction speed	Slow – approx. 5 m s^{-1}	Fast – up to 120 m s^{-1}
Ability to regenerate neurones	Present	Absent

Going further ▶

Although sponges do not have nervous systems, they do have genes related to nerve formation. The formation of nervous systems is associated with the evolution of sodium channels in cell membranes.

WORKING SCIENTIFICALLY

Biologists use a model organism to investigate a particular phenomenon, e.g. *Drosophila* in genetics, tobacco in micropropagation. The insight the model organism provides is tested in other species.

Going further ▶

The human nervous system contains non-neuronal cells called glia. They support and protect neurones, including generating the myelin sheath.

Going further ▶

Vertebrate hormones, e.g. steroids occur in *Hydra*'s nerve net. The nervous systems of *Hydra* and vertebrates are so different that it is unlikely that the hormones have the same role.

YOU SHOULD KNOW ›››

››› The structure of a myelinated motor neurone

››› The transmission of the nerve impulse

››› What is meant by a resting potential

››› How a resting potential is established

››› What is meant by an action potential

››› The importance of the refractory period

Link Neurotransmitters have a role at synapses, which is described on p147.

Exam tip

The only neurone that you are required to know the structure of is the motor neurone.

21

Knowledge check

Identify the missing word or words.

Neurones which bring impulses from receptors into the CNS are called .... neurones, whereas those that carry impulses from the CNS to the effectors are called ....neurones. A neurone consists of a .... containing a nucleus and a long extension called the .... In vertebrates the extension has a fatty layer called a .... This acts as an .... and speeds up the transmission of impulses.

Neurones

Nerve cells or neurones are specialised cells adapted to rapidly carry nervous impulses from one part of the body to another.

There are three types of neurones, classified according to their function:

- Sensory – carry impulses from the sense receptors or organs into the CNS.
- Motor – carry impulses from the CNS to the effector organs, i.e. muscles or glands.
- Relay, connector or association – receive impulses from sensory neurones or other relay neurones and transmit them to motor neurones or other relay neurones.

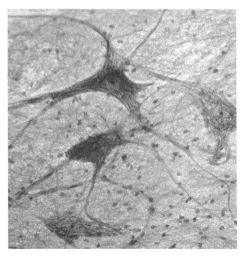

Photomicrograph of neurones

Structure	Function
Cell body / centron	Contains a nucleus and granular cytoplasm.
Cytoplasm	Granular; contains many ribosomes.
Nucleus	Holds DNA.
Nissl granules	Cytoplasmic granules comprising ribosomes grouped on RER.
Dendrite	Thin fibre carrying impulses towards cell body. A cell body may have several dendrites.
Axon	Thin fibre carrying impulses away from cell body. A cell body has only one axon.
Schwann cells	Surround and support nerve fibres. In vertebrate embryos, they wrap around the developing axons many times and withdraw their cytoplasm, leaving a multi-layered phospholipid myelin sheath.
Myelin sheath	Electrical insulator; speeds up the transmission of impulses.
Nodes of Ranvier	1 μm gaps in myelin sheath, where adjacent Schwann cells meet and where the axon membrane is exposed. They allow impulses to be transmitted rapidly.
Synaptic end bulb	Swelling at end of axon, in which neurotransmitter is synthesised.
Axon ending / terminal	Secretes neurotransmitter, which transmits impulse to adjacent neurone.

The structure of a myelinated motor neurone is shown below:

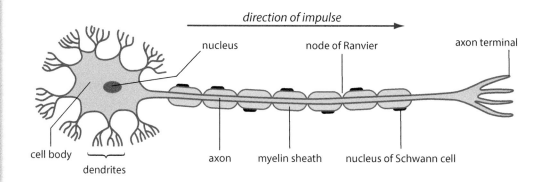

Mammalian motor neurone

The nervous impulse

Resting potential

A neurone is an excitable cell, which means it can change its resting potential, i.e. the potential difference across its cell membrane. Most cells are not excitable and so they cannot change their resting potential.

The potential difference across a cell membrane is about 70 mV. The membrane is more negative inside, so the resting potential is –70 mV. As there is a potential difference across the cell membrane, it is described as **polarised**. This resting potential results from the negative ions of large proteins, of organic acids such as pyruvate and of organic phosphates, e.g. ATP^{4-} in the cytoplasm, and from the uneven distribution of inorganic ions:

- The inside of a cell has both a higher concentration of potassium ions and a lower concentration of sodium ions than the outside. So the K^+ ions would tend to diffuse out and the Na^+ ions would tend to diffuse in. Some of the channels that allow the K^+ ions to diffuse out are open, whilst most of the channels that allow the Na^+ ions to move in are closed. This makes the axon membrane 100 times more permeable to K^+ ions, which therefore diffuse out faster than the Na^+ ions diffuse back in.

- Sodium–potassium exchange pumps pump K^+ ions back into the cell and Na^+ ions back out. They are trans-membrane proteins with ATPase activity that transport K^+ and Na^+ ions across the membrane against a concentration gradient, by active transport. They maintain the concentration and an uneven distribution of ions across the membrane.

The Na^+ ions are pumped out faster than the K^+ ions are brought in. The overall result is that the inside of the membrane is negative compared with the outside.

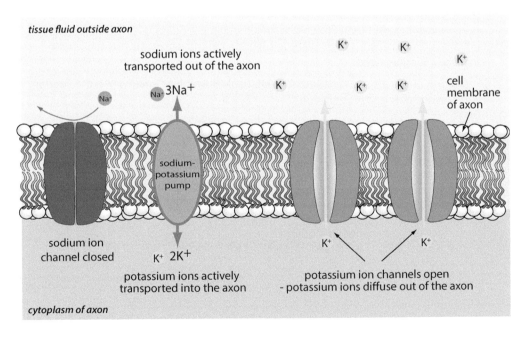

Distribution of ions at resting potential

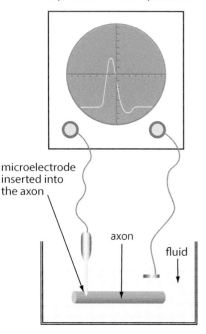

oscilloscope records passing impulse as an action potential

microelectrode inserted into the axon

axon

fluid

Detecting an action potential with an oscilloscope

 **Key Terms**

Action potential: The rapid rise and fall of the electrical potential across a nerve cell membrane as a nervous impulse passes.

Depolarisation: A temporary reversal of potential across the membrane of a neurone such that the inside becomes less negative than the outside as an action potential is transmitted.

Going further ▶

In a typical neurone, the action potential lasts about 1 ms. In skeletal muscle it lasts 2–5 ms and in cardiac muscle, it can last 200–400 ms.

▼ **Study point**

Nervous impulses are due to changes in the permeability of the axon membrane to Na^+ ions and then to K^+ ions, which change the potential difference across the membrane.

The action potential

An excitable cell is one in which the potential across the membrane can be altered. A nervous impulse is the transmission of a change in potential along a nerve fibre associated with the movement of sodium ions. The voltage change is very small but can be picked up through a pair of microelectrodes and fed into an oscilloscope. The oscilloscope trace is a graph showing how the voltage across the membrane changes with time. It is used to measure the magnitude and speed of transmission of the impulse and analyse the pattern of impulses generated in different parts of the nervous system and in different situations.

WORKING SCIENTIFICALLY

The model system for investigating the mechanism of transmission of nervous impulses was the squid giant axon. These axons are about 0.5 mm diameter, so an electrode could easily be inserted.

Neurones transmit electrical impulses along the cell surface membrane surrounding the axon. With a microelectrode inside an axon and another in the bathing solution, changes in potential read on an oscilloscope gave rise to the following Nobel Prize winning conclusions:

- The energy of the stimulus causes some of the **voltage-gated** sodium channels in the axon membrane to open. A voltage-gated channel is one which opens or closes in response to a particular voltage across the membrane. The sudden increase in the permeability of the membrane to Na^+ ions allows them to rapidly diffuse into the axon, down their concentration gradient. As a result, the negative charge of −70mV inside the axon rapidly becomes a positive charge of +40mV. This is the **action potential**. In this condition the cell membrane is **depolarised.** Once the potential inside the cell is +40mV, the sodium ion channels close, preventing further influx of sodium ions.

- The potassium channels open and K^+ ions diffuse out down their concentration gradient. The cell becomes less positive inside as more diffuse out and the membrane is **repolarised**.

- More K^+ ions diffuse out than Na^+ ions diffused in, so the potential difference across membrane becomes even more negative than the resting potential. This makes the membrane **hyperpolarised**.

- The sodium-potassium pump pumps K^+ ions back in and Na^+ ions back out, restoring the ion balance of the resting potential.

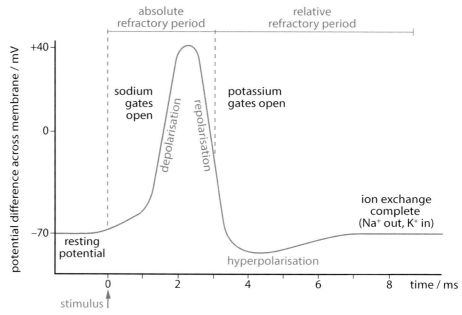

Action potential

How the action potential travels along an axon

At the site of an action potential, Na^+ ions diffuse into the axon and K^+ ions diffuse out. The resulting reversal of potential sets up local currents, as Na^+ ions move laterally through the axon. They depolarise the adjacent section of the membrane. This opens more voltage-gated sodium channels in those regions and more sodium ions flood in, depolarising the axon at this point. Sodium ions then diffuse further down the axon and in this way, a self-perpetuating wave of depolarisation spreads along the axon.

Meanwhile, at the site of the initial action potential, the sodium channels are inactivated and cannot open again until the resting potential has been re-established, so a new action potential cannot be initiated there. This is the **absolute refractory period**, and lasts about 1 ms. It ensures that the action potential is not propagated back in the direction from which it came, and the nervous impulse travels in one direction only.

For the next 5–10 ms, during the hyperpolarisation phase, if an impulse is strong enough, a new action potential may pass. This is the relative refractory period and occurs while the sodium–potassium pumps are restoring the resting potential.

Going further ▶

Nervous impulses travel at different speeds. Those controlling muscle position travel at about 120 m s^{-1}, pain impulses travel at 0.61 m s^{-1} and touch impulses at 76.2 m s^{-1}.

Key Term

Absolute refractory period: Period during which no new action potential may be initiated.

▼ **Study point**

In myelinated nerve fibres the speed of transmission is up to 120 m s^{-1}, whereas in non-myelinated nerve fibres it is about 0.5 m s^{-1}.

Passage of action potential along a non-myelinated nerve fibre

Properties of nerves and impulses

'All or nothing'

If the intensity of a stimulus is below a certain threshold value, no action potential is initiated. But, if the intensity of the stimulus exceeds the threshold value, an action potential is initiated. The action potential that is initiated is always the same size (+40 mV) and it remains that same size as it is propagated along the axon. No energy is lost in transmission.

An increase in the intensity of the stimulus does not give a greater action potential. Instead, the frequency of action potentials increases.

A nervous impulse is either initiated or not and it is always the same size. This is the '**all or nothing law**'. It allows the action potential to act as a filter, preventing minor stimuli from setting up nervous impulses, so the brain is not overloaded with information.

YOU SHOULD KNOW ›››

››› The 'all or nothing' law

››› The factors that affect the speed of conduction of a nervous impulse

▼ **Study point**

A stronger stimulus produces a greater frequency of action potentials as the intensity of stimulation increases but the size of the impulses is always the same.

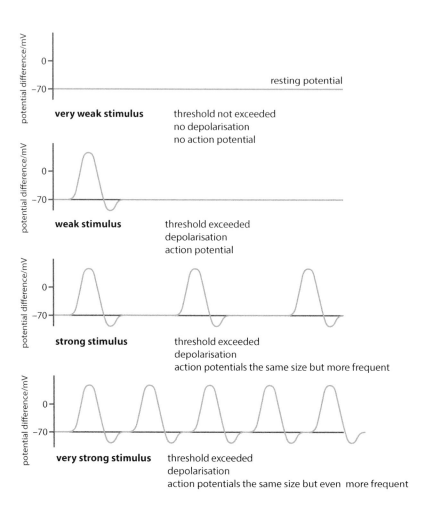

Effect of stimulus intensity on impulse frequency

Going further ▶

Dinosaurs were reptiles; there is evidence that some were warm-blooded, unlike modern reptiles, suggesting the dinosaurs could transmit nervous impulses rapidly.

▼ **Study point**

Myelinated axons use less ATP to transmit impulses than non-myelinated axons of equal diameter, because active transport of sodium ions occurs only at nodes, rather than along the whole length of the axon.

Key Term

Saltatory conduction:
Transmission of a nervous impulse along a myelinated axon, in which the action potential jumps from one node of Ranvier to the adjacent node.

Factors affecting the speed of conduction of the nerve impulse

The three major factors determining the speed of conduction of the nerve impulse are:

1. **Temperature** – ions move faster at higher temperatures than at lower temperatures, as they have more kinetic energy. So birds and mammals, the two warm-blooded taxa, transmit nervous impulses more quickly and have faster responses than all other groups of animals.

2. **The diameter of the axon** – the greater the diameter of the axon, the greater its volume in relation to the area of the membrane. More sodium ions can flow through the axon, so impulses travel faster. Human non-myelinated axons are 0.2–1.5 µm diameter. Such dimensions would allow only very slow transmission of impulses, especially at low temperatures. Some marine invertebrates evolved in cold habitats, with water temperature close to 0°C. To compensate, the squid, for example, has giant axons with a diameter of up to 1 mm; the earthworm, although living in warm habitats, has evolved giant axons up to 700 µm in diameter, which are associated with rapid escape responses.

3. **Myelination** – speeds up the rate of transmission by insulating the axon. Sodium ions flow through the axon, but a myelinated nerve fibre only depolarises where the resistance is low, i.e. at the nodes of Ranvier. The voltage-gated ion channels only occur at the nodes of Ranvier, so these are where sodium ions enter. The consequence is that the action potential appears to jump from node to node along the axon. This is **saltatory conduction**. The nodes of Ranvier are about 1 mm apart and so saltatory transmission is rapid, and much faster than conduction along a non-myelinated axon.

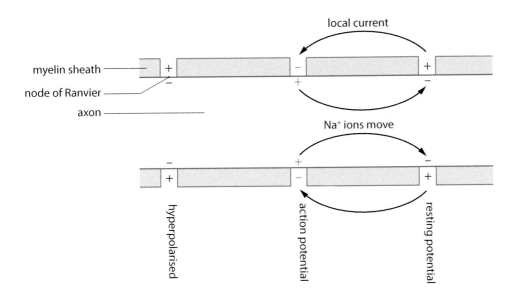

myelin sheath

node of Ranvier

axon

local current

Na⁺ ions move

hyperpolarised

action potential

resting potential

Saltatory conduction along a myelinated nerve fibre

Study point

Nervous impulses are faster if axons are wider, myelinated and at a higher temperature.

Going further ▶

The myelin sheath has evolved independently several times and occurs in Annelid worms, in some Crustaceans and in most vertebrates.

The synapse

Neurones are separated by synapses, which send the nervous impulse between neurones in one direction only.

There are two classes of synapse:

- The **electrical synapse** or gap junction is 3 nm across, small enough that an electrical impulse is transmitted directly from one neurone to the next.

- The **chemical synapse** is a 20 nm gap, too big for the nervous impulse to jump. Most junctions between neurones are chemical synapses. Branches of axons lie close to dendrites of other neurones but do not touch. The impulse is transmitted by a neurotransmitter, a chemical that diffuses across the synaptic cleft, from the pre-synaptic membrane of one neurone to the post-synaptic membrane of an adjacent neurone, where a new impulse is initiated.

YOU SHOULD KNOW ›››

››› The structure of a synapse

››› The role of the membrane and acetylcholine in synaptic transmission of nerve impulses

››› The function of synapses

››› The effect of organophosphates and psychoactive drugs on synapses

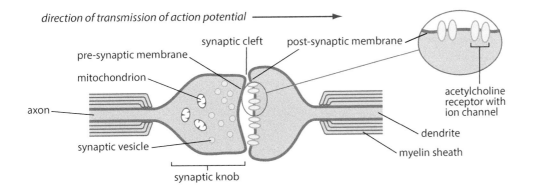

direction of transmission of action potential ⟶

synaptic cleft

pre-synaptic membrane

post-synaptic membrane

mitochondrion

axon

synaptic vesicle

synaptic knob

acetylcholine receptor with ion channel

dendrite

myelin sheath

A synapse

Key Term

Neurotransmitter: A chemical secreted in response to an action potential, which carries a chemical signal across a synapse, from one neurone to the next, where a new action potential is initiated.

Exam tip

This topic is synoptic in that it brings together several topics already encountered, including membranes, diffusion, enzyme action and mitochondria.

Link You learned about the co-operative binding of oxygen to haemoglobin in the first year of this course.

▼ **Study point**

Calcium ions enter through the pre-synaptic membrane to release acetylcholine. Sodium ions enter through the post-synaptic membrane to initiate a new action potential.

Going further ▶

Acetylcholinesterase has a high turnover number. Approximately 25 000 molecules of acetylcholine are hydrolysed each second, so acetylcholine is quickly removed.

22

Knowledge check

Place the statements 1–7 in the correct order:

1. Acetylcholine is released into the synaptic cleft.
2. Sodium ions diffuse into the post-synaptic neurone.
3. The synaptic vesicles fuse with the pre-synaptic membrane.
4. Calcium ions enter through the pre-synaptic membrane.
5. An action potential is initiated.
6. Acetylcholine attaches to a receptor site on the post-synaptic membrane.
7. Acetylcholine diffuses across the synaptic cleft.

Synaptic transmission

- The arrival of an impulse at the synaptic end bulb alters its membrane permeability, opening voltage-dependent calcium channels, so calcium ions diffuse into the end bulb, down their concentration gradient.
- The influx of calcium ions causes the synaptic vesicles to move towards and fuse with the pre-synaptic membrane. This releases the neurotransmitter acetylcholine, by exocytosis, into the synaptic cleft.
- The neurotransmitter diffuses across the synaptic cleft and binds to a receptor, an intrinsic protein spanning the post-synaptic membrane. The protein's sub-units have two receptor sites and the two acetylcholine molecules show co-operative binding when they attach.

When acetylcholine molecules bind with both of these sites, the receptor protein changes shape, opening a channel and sodium ions diffuse in, down their concentration gradient. The post-synaptic neurone is consequently depolarised and, if the membrane is depolarised enough, a threshold potential difference is reached and an action potential is initiated. But if insufficient acetylcholine is bound, the post-synaptic membrane will not be depolarised enough to exceed the threshold, and so an action potential will not be produced.

If acetylcholine were to remain in the synaptic cleft, it would constantly initiate new impulses in the post-synaptic membrane and impulses would not be distinct. This is prevented in three ways:

- Direct uptake of acetylcholine into the pre-synaptic neurone, so none remains in the synaptic cleft to bind to the post-synaptic receptor.
- Active transport of calcium ions out of the synaptic end bulb, so no more exocytosis of acetylcholine occurs.
- Hydrolysis of acetylcholine. After release, acetylcholine is quickly destroyed by the enzyme acetylcholinesterase, in the synaptic cleft. The products of this hydrolysis are choline and ethanoic acid. They diffuse back across the synaptic cleft into the pre-synaptic neurone, and re-form acetylcholine:

  – in the synaptic cleft: acetylcholine $\xrightarrow{\text{acetylcholinesterase}}$ ethanoic acid + choline

  – in the pre-synaptic neurone: acetyl CoA + choline $\longrightarrow$ CoA + acetylcholine

Energy is required to re-form these neurotransmitter molecules and for their exocytosis. Accordingly, there are many mitochondria in axon end-bulbs, in which ATP is generated.

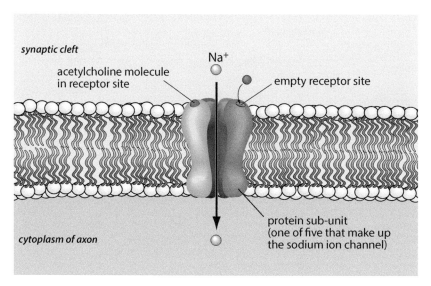

How the acetylcholine receptor works

Neurones transmit impulses in one direction only because:

- Repolaristion happens behind an action potential and so depolarisation could not happen at that point.
- Synaptic vesicles only occur at the end bulb of the pre-synaptic neurone.
- Neurotransmitter receptors only occur on the post-synaptic membrane.

Properties of synapses

- Transmit information between neurones.
- Pass impulses in one direction, generating precision in the nervous system.
- Act as junctions.
- Protect the response system from overstimulation, because the impulse is always the same size whatever the size of the stimulus.
- Filter out low-level stimuli: an action potential is only initiated when the depolarisation is large enough to reach a threshold value, about −55 mV. It can be built by:
  - Temporal summation, i.e. depolarisation builds up over time to reach the threshold at which an action potential is initiated.

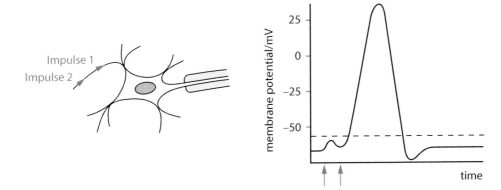

Temporal summation

  - Spatial summation, i.e. several pre-synaptic neurones synapse with the same post-synaptic neurone and all contribute to the growing depolarisation, which generates an action potential when it is large enough.

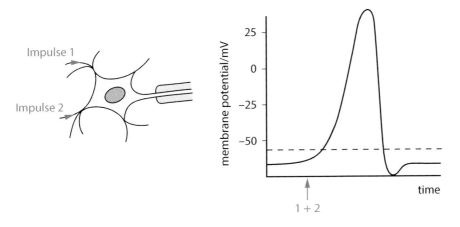

Spatial summation

Effects of chemicals on synapses

A drug is a molecule that has a physiological effect on the body when ingested, inhaled, absorbed or injected. Many drugs act at synapses and disrupt the normal functioning of neurotransmitters, producing abnormal patterns of nervous impulses. Many molecules have roles as neurotransmitters, including:

- Acetylcholine
- GABA
- Monoamines, e.g. dopamine, serotonin, noradrenalin
- Neuropeptides, e.g. endorphins

Different drugs affect the action of different neurotransmitters. They may be:

- **Sedatives**, e.g. alcohol inhibits the nervous system, creating fewer action potentials in post-synaptic neurones.
- **Stimulants**, also called agonists, e.g. amphetamines stimulate the nervous system by allowing more action potentials in post-synaptic neurones.

Mechanisms of drug action

There are various mechanisms of drug action, for example:

- Drugs may mimic the action of neurotransmitters. They may have the same shape and bind to the post-synaptic neurone in the same way, increasing the frequency of action potentials, e.g. the same group of atoms including a positively-charged nitrogen atom in nicotine and acetylcholine allow them both to bind to the same receptor. So nicotine increases the frequency of impulses. But, unlike acetylcholine, nicotine is not removed by hydrolysis so it continues to initiate impulses. The body may become habituated to nicotine, in which case, the nervous system only functions normally if nicotine is present. Then to achieve the desired effect more nicotine has to be taken. This is drug tolerance. If then no nicotine is taken in, impulses are not transmitted normally and the unpleasant symptoms associated with withdrawal are experienced. These are likely to be made worse because another effect of nicotine is to increase the release of the neurotransmitter dopamine in the brain, which generates pleasurable sensations.

- Drugs may prevent the breakdown of neurotransmitters, e.g. **organophosphates** (OPs) inhibit acetylcholinesterase. In the presence of OPs, acetylcholine is not hydrolysed and remains in the synaptic cleft, causing repeated firing of the post-synaptic neurone. Organophosphates are esters of phosphoric acid and are sometimes called phosphate esters. They can be inhaled, absorbed and ingested and have been implicated in many situations where long-term health has been damaged. They include:
  - Insecticides, e.g. malathion, dichlorvos
  - Herbicides, e.g. glyphosate
  - Nerve gases, e.g. sarin.

Nerve gases inhibit acetylcholinesterase at neuromuscular junctions, generating repeated, uncontrollable contractions of muscles. Where this occurs in antagonistic muscle pairs, it can break bones.

Psychoactive drugs

Psychoactive drugs act primarily on the central nervous system by affecting different neurotransmitters or their receptors, which affects the firing of neurones. This alters brain function and, consequently, perception, mood, consciousness and behaviour. They include therapeutic drugs such as Ritalin, Prozac and Paxil, and recreational drugs such as nicotine, alcohol, cannabis, cocaine, amphetamines, ecstasy, and heroin. The changes may be pleasant (e.g. euphoria) or advantageous (e.g. increased alertness or improved recall), and so many are abused, that is, used without medical supervision for reasons other than their original purpose. Dependence may develop, making the abuse difficult to interrupt.

Going further ▶

Alcohol appears to be a stimulant but is in fact a sedative because it suppresses activity in areas of the brain that contribute to self-control.

Going further ▶

It is said that nicotine is so powerful a drug that its addiction is harder to break than that produced by heroin.

▼ **Study point**

A neuromuscular synapse is between a nerve fibre and a muscle fibre. The post-synaptic membrane of the muscle fibre allows in influx of Ca^{2+} rather than Na^+, as in neurones.

Going further ▶

The use of nerve gases as biological weapons has been banned by international treaties, but evidence shows that they have, nevertheless, been used.

Exam tip

Caffeine is a stimulant that increases the metabolic rate in pre-synaptic cells. ATP production is increased, allowing more neurotransmitter synthesis.

Unit 3

1 Green vegetables, such as cabbages, are grown as food crops in many parts of the world. Where the conditions are suitable, such as in the UK, they are grown outdoors, but in places that have less suitable physical conditions, the crops are grown in greenhouses. A farmer planted 100 hectares with cabbages and noticed that in one of the fields at the edge of her land, the cabbage leaves were small and, in many cases, did not have the uniform green colour expected but were paler and, in places, yellow.

(a) (i) What term is used to describe plants with leaves that are yellow, and do not develop their normal green colour? (1)

(ii) The farmer knew that the plants in all the fields were genetically related and so the green colour is likely to have an environmental cause. What is a common cause of yellowed leaves in plants? (1)

(iii) How might the farmer have confirmed her suspicion? (1)

(b) The yield of a crop (Y) is the number of kg produced per hectare. The average actual yield (Ya) is the average yield over the previous 5 years. The table shows the data for the cabbage farm, measured in tonnes per hectare, where 1 tonne = 1000 kg.

Year	Yield (Y) / tonnes ha^{-1}	Average yield for the previous 5 years (Ya) / tonnes ha^{-1}
2005	71.9	73.1
2010	65.0	66.4
2015	62.1	63.9

Describe the trends of these findings and suggest an explanation for them. (3)

(c) The photosynthetic yield of plants determines whether or not they are useful as food crops. The molecules that plants use to grow are synthesised using light energy trapped in the light-dependent reactions and made from products of the light-independent reaction of photosynthesis.

(i) ATP is an important product of the light-dependent reaction. The energy to make it is made available by chemiosmosis, from a proton gradient between the thylakoid space and the stroma of the chloroplast. How is the proton gradient maintained? (3)

(ii) Identify the compounds X and Y in the diagram below, to show the names of intermediate compounds of the light-independent stage (1)

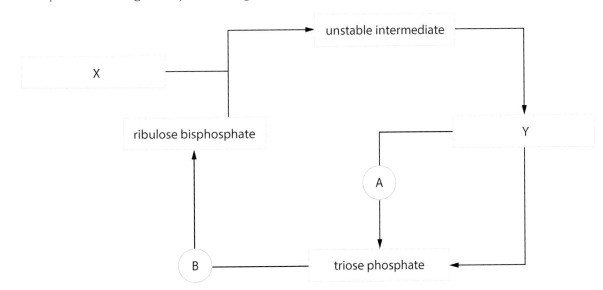

(iii) The reaction indicated as A uses about 17% of the ATP formed in the light-dependent reaction. How is the remaining 83% used? (1)

(iv) The arrow labelled B represents a sequence of reactions. Explain why this sequence of reactions is so important in the light-independent stage of photosynthesis. (2)

(Total 13 marks)

2 Respiration is a series of catabolic pathways comprising many different reactions in different locations within the cell. The pathways of aerobic respiration release chemical energy from respiratory substrates and make it available as ATP.

It is possible to disrupt these critical pathways with metabolic inhibitors, and deprive the cell of the essential energy it needs to function. 2,4-dinitrophenol (DNP) is one such respiratory inhibitor, although it has many commercial uses, including as a herbicide, a component of photographic developer, in wood preservative and in explosives. Inhibitors are useful in research because they can provide information about how cells function under normal conditions.

(a) (i) What is meant by the term 'catabolic pathway'? (2)

(ii) What is the advantage to the cell of having the electron transport chain membrane-bound, rather than being free in solution, as are glycolysis and the Krebs cycle? (1)

(b) The graph below shows the effect of the concentration of 2,4-dinitrophenol (DNP) on the production of ATP by an electron transport chain.

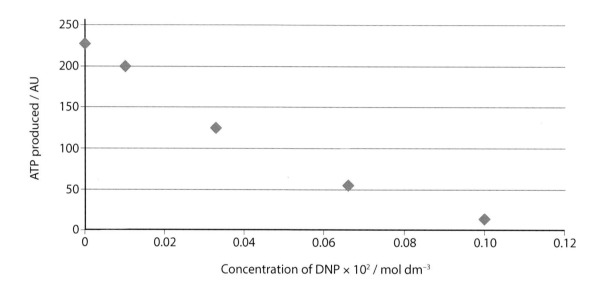

(i) Describe the effect of DNP on ATP production, as shown in this graph. (2)

(ii) DNP makes the inner membrane of the mitochondria permeable to protons. They leak across the entire membrane and the proton gradient is destroyed. How might this produce the results shown? (2)

(iii) For 60 years, some people have been using very small concentrations of DNP to lose weight. How might the action of DNP in mitochondria have this effect? (2)

(iv) Why has the use of DNP in diets caused great harm to many people, and, in a few cases, even deaths? (3)

(Total 12 marks)

3 Many species of bacteria can be cultured in the laboratory on growth media, either a liquid broth or solidified with agar. MRS agar is used to grow *Lactobacillus* bacteria. It contains:

- peptone
- egg extract
- yeast extract
- glucose
- sodium acetate
- polysorbate 80
- potassium hydrogen phosphate
- ammonium citrate
- magnesium sulphate
- agar

(a) (i) Why is MRS agar described as an 'undefined' medium? (2)

(ii) Explain the inclusion of glucose in MRS agar. (1)

(b) Gram-positive bacteria are susceptible to penicillin. They may acquire a plasmid carrying an allele conferring resistance to penicillin. Describe how a selective growth medium might allow these resistant bacteria to be cultured. (2)

(c) Bacteria were maintained in a liquid culture medium for 60 hours at optimum conditions of pH and temperature, with suitable concentrations of nutrients to permit unrestricted cell division.

(i) Describe how you would treat a sample taken from this culture at 40 hours and calculate the percentage of cells liable to be killed by treatment with penicillin. (8)

(ii) The table shows the number of bacteria / cm^{-3} at ten-hour intervals.

Penicillin was applied to the culture at 40 hours.

Time in culture / h	Number of bacteria / cm^{-3}
0	1.0×10^3
10	3.1×10^4
20	4.6×10^7
30	9.6×10^9
40	8.2×10^{10}
50	7.7×10^9
60	7.1×10^9

A microscopic analysis suggested that 87% of the bacteria would be killed by penicillin within 20 hours. Do the data support that suggestion? (4)

(Total 17 marks)

4 The sweet chestnut, *Castanea sativa*, is grown commercially using a traditional method of forest management called coppicing. At 15-year intervals, in late autumn when the trees are dormant, trunks are cut at the base to within a few centimetres of the ground, leaving a 'stool'. Buds in the stool produce new shoots, called poles, which grow a few metres in height, and their diameter increases over the following years. The poles can be cut at any time for use in, for example, fencing. If they are left for longer they grow thicker and can be used to make furniture. The timber is often used for outdoor construction because it is so durable and the chestnuts, which are the seeds, are used as food.

Sweet chestnut trees being felled
15 years after coppicing

An assessment, in August each year, of plants growing on the forest floor of coppiced sweet chestnut plantations gave the following data:

Year last coppiced	Mean % area cover on forest floor					
	Bramble	Willowherb	Soft rush	Grasses	Mosses	Leaf litter
2014	5.7	0	0	0	0.9	30.3
2012	6.5	0.9	3.0	0.8	1.0	15.7
2010	8.8	1.3	4.9	1.5	1.5	9.3
2008	8.1	1.8	7.2	1.2	1.4	8.8
2006	7.6	1.7	5.9	0.9	1.2	5.9
2004	6.9	1.4	3.4	0.8	0.9	0
2002	6.0	0.9	2.1	0.4	0.6	0
2000	5.1	0	0	0	0.2	26.6

(a) Why do the totals in each horizontal row not add up to 100% (2)

(b) Explain why the data show an abrupt change between 2000 and 2002. (1)

(c) (i) Describe the changes suggested by the data from 2002. (3)

   (ii) Suggest why the changes in plant community since 2002 may have happened. (4)

(d) Calculate the percentage change in soft rush between 2002 and 2008. (2)

(e) What is the name given to the changes that the data describe? (1)

(f) How might the data be different if this were a newly-seeded plantation where there had been no previous growth? (1)

(Total 16 marks)

5 The bromeliads, which include pineapples and some ornamental houseplants, are one of the more recent plant groups to have evolved. Most species are endemic to the Americas and *Pitcairnia feliciana* is the only one that is not. It is found on sandstone outcrops (inselbergs) of the Fouta Djallon highlands in Guinea, in tropical west Africa. Living on sandstone, it absorbs water through its leaves and nutrients from rainwater and even from its own dead tissues.

(a) The Guinea highlands are tropical. Why does this suggest that their biodiversity is likely to be high? (3)

(b) Although *P. feliciana* is only found in the uplands of central Guinea in small numbers, it does not have a threat rating, as research has not yet defined its abundance or precise distribution. What steps might be taken to ensure that this plant does not become extinct? (4)

(c) Guinea's central upland areas have been extensively deforested, threatening the habitats and, therefore, the survival of the plants that live there. Describe the likely effect of such deforestation on the quality of the remaining soil. (4)

(d) The data in the table shows causes of deforestation in Africa in the first decade of this century.

Land use	Percentage of deforested land	
	2000 – 2004	2005 – 2009
Subsistence farming	20	54
Intensive farming	20	34
Ranching / pasture	40	1
Logging	8	10

Data based on Mongabay.com

(i) How have patterns of farming changed in the period 2000–2009? (3)

(ii) The term 'intensive agriculture' often refers to monoculture of energy crops such as palm oil. Describe how palm oil is used as an energy crop. (3)

(e) Deforestation has changed the use of large areas of land and this change, worldwide, is thought to have caused the Earth to cross a planetary boundary. What is meant by the term 'planetary boundary'? (1)

(Total 18 marks)

6 The table gives data for the urine and plasma in five mammalian species that have evolved in different habitats. The Osmol is a non-SI unit that refers to the number of moles of a solute that contribute to the osmotic properties of a solution, and can be used as a proxy for concentration. A mesic habitat is moist and has an adequate supply of water.

Mammal	Habitat	Urine concentration (U) / mOsmol dm^{-1}	Plasma concentration (P) / mOsmol dm^{-1}	U/P
rat	mesic	2900		9.0
beaver	freshwater/land	520	306	1.7
human	mesic	1400		4.5
porpoise	marine	1800	360	5.0
camel	xeric	2800	350	8.0

Animals at the extreme: The desert environment. Based on Willmer et al (2000). *Environmental Physiology of Animals*. Oxford: Blackwell Science Ltd.

Measurements were made on the urine and plasma collected from well-fed, dehydrated animals, immediately before allowing them to rehydrate.

(a) Why were data collected from dehydrated animals? (1)

(b) How should data be collected such that it is reliable? (3)

(c) (i) Calculate the values for the plasma concentration in the rat and the human. (1)

(ii) Describe the values for plasma concentration, in comparison with those for urine concentration. (3)

(d) (i) The beaver, human, porpoise and camel are all large animals. For these large animals, describe the relationship between their urine concentration and the availability of water in the habitat. (1)

(ii) Describe the mechanism by which loop of Henle allows mammals make to urine that is more concentrated than their plasma. (6)

(iii) Rats and humans evolved in mesic environments but rat's urine is about twice the concentration of human urine. Suggest how the rat might achieve this? (1)

(Total 16 marks)

7

Myelinated nerves TEM

This false colour electron micrograph shows a transverse section through an axon.

(a) Identify the structures labelled A and C and describe a function of each. (2)

(b) (i) This electron micrograph is printed at a magnification of 50 000. Calculate the maximum thickness of B, using appropriate units and a suitable number of decimal places.

    (ii) How is structure A involved in the development of structure B? (2)

    (iii) Explain how you might use the image of structure B to estimate the length of a phospholipid molecule. (5)

(c) (i) The table below describes neurones of *Loligo*, the giant squid and of humans. Using the information given, describe factors that control the speed of a nervous impulse and suggest a possible mechanism for each. (6)

Neurone	Source	Approximate axon diameter / μm	Myelin sheath	Approximate speed of impulse / m s^{-1}
giant axon	squid	650	absent	20
motor neurone	squid	1	absent	4
sensory neurone ending in touch sensor	human	6	present	50
sensory neurone ending in temperature receptor	human	5	absent	20

    (ii) Using the data in the table below, suggest how temperature affects the speed of a nervous impulse. (3)

Neurone	Source	Mean environmental temperature / °C	Approximate speed of impulse / m s^{-1}
giant axon	squid	2	20
median giant fibre	earthworm	10	30

(Total 18 marks)

Sexual reproduction in humans

Reproduction makes new individuals that have broadly the same characteristics as their parents. Sex, in the biological sense, refers to the mixing of genes. Many organisms can perform reproduction without sex, e.g. bacterial binary fission, strawberries producing runners, sea anemones budding. In some cases, sex can happen without reproduction, such as when a bacterial cell takes up a plasmid with novel genes. In higher animals, such as humans, however, sex and reproduction must happen together. Two parents combine their alleles so the offspring have a new genetic combination. The placenta, a novel structure in mammals, has been of great evolutionary significance.

Topic contents

By the end of this topic you will be able to:

- Describe the structure and function of the male and female reproductive systems.
- Explain and the roles of mitosis and meiosis in spermatogenesis and oogenesis.
- Describe the preparation of the uterus for implantation.
- Describe fertilisation, zygote development and the processes leading to implantation.
- Describe the structure and functions of the placenta.
- Understand the roles of hormones in the menstrual cycle, pregnancy and birth.

The male reproductive system

The male reproductive system consists of:

- A pair of testes contained in an external sac, the scrotum
- The penis
- Ducts connecting the testes with the penis
- Accessory glands, i.e. a pair of seminal vesicles, a pair of Cowper's glands and the prostate gland. These glands secrete fluids that mix with the sperm to make semen.

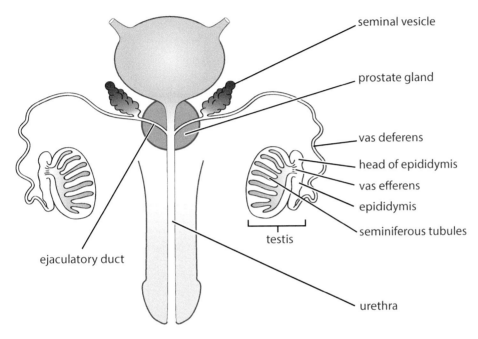

Male reproductive system

Each testis contains about a thousand coiled tubes called **seminiferous tubules**. The cells lining the seminiferous tubules undergo sperm formation and throughout the process, cells move towards the space, or lumen, running through the middle of the tubule. When spermatozoa reach the lumen, they move through the tubule and collect in the vasa efferentia. These are coiled tubes that carry sperm to the head of the epididymis. The sperm remain in the epididymis for a short time while they become motile, and then they pass into the vas deferens, during ejaculation.

The vas deferens carries sperm from the epididymis towards the penis. On the way, the seminal vesicles secrete mucus into the vas deferens. The mucus contains a mixture of chemicals including fructose, respired by the sperm for energy. The spermatozoa and seminal fluid move through the ejaculatory duct, which passes through the prostate gland, where zinc-containing prostate fluid is secreted. The secretions of these accessory glands are alkaline and they:

- Maintain sperm mobility
- Provide nutrients for the sperm, including fructose, their main energy source, amino acids and zinc ions
- Neutralise the acidity of any urine remaining in the urethra
- Neutralise the acidity of the vaginal tract.

The fluid emerging from the prostate gland is semen, a mixture of spermatozoa, seminal and prostate fluids. It is carried through the penis in the urethra.

Going further ▶

If you have studied Latin, you will know that 'vas' is a container. 'Efferens' means to carry away and 'deferens' means to carry down. The names of structures often describe their functions.

Histological examination of TS testis

Place a TS testis slide under the microscope. Using a ×4 objective, you will see seminiferous tubules, each surrounded by connective tissue and with a lumen in the centre. Move the slide so that a tubule with a clearly defined lumen is in the centre of the field of view. Using the ×10 and the ×40 objectives:

- Identify the germinal epithelial cells at the outer margin of the seminiferous tubule.
- Identify a sequence of cells along a radius of the tubule, if the plane of your section allows it. Moving inwards from the germinal epithelium identify primary spermatocytes.
- Secondary spermatocytes are rarely seen as they progress rapidly to spermatids, which can be seen further in towards the lumen.
- Adjacent to the lumen, spermatozoa with tails can be seen.
- Between the strands of developing spermatids are the Sertoli cells. They are columnar, very biochemically active and easily recognised by the large oval nucleus and dense nucleolus. They secrete a fluid, which nourishes the spermatids and protects them from the male's immune system.
- Groups of cells between the seminiferous tubules are called interstitial cells, or Leydig cells. They secrete testosterone, the male sex hormone, which has roles in sperm formation and maturation, as well as the development of male secondary sexual characteristics.

Use the eyepiece graticule and your calibration value to find the mean diameter of seminiferous tubules and of the lumen within.

23

Knowledge check

Link the appropriate terms 1–4 with the statements A–D.

1. Testis.
2. Prostate gland.
3. Seminiferous tubules.
4. Seminal vesicle.

A. Produces secretion containing fructose and supports sperm mobility.
B. Where spermatozoa are produced.
C. Contains seminiferous tubules separated by interstitial cells.
D. Produces a zinc-containing alkaline secretion.

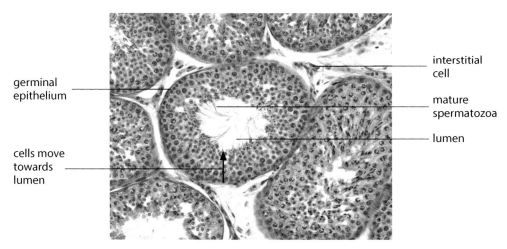

germinal epithelium — interstitial cell — mature spermatozoa — lumen — cells move towards lumen

Seminiferous tubules of human testis

The female reproductive system

There are two ovaries. Oocytes mature in follicles, which develop from cells in the germinal epithelium, around the periphery of the ovary. Mature follicles migrate to the surface of the ovary, from where a secondary oocyte is released at ovulation. The ovaries alternate each month in releasing the oocyte.

Cilia at the entrance of the oviducal funnel sweep the secondary oocyte into the oviduct, or Fallopian tube. The ciliated epithelial cells lining the oviduct convey the secondary oocyte to the uterus.

The uterus wall has three layers:

- The perimetrium is a thin layer around the outside.
- The myometrium is the muscle layer.
- The endometrium is the innermost layer. It is a mucous membrane, which is well supplied with blood. It is the layer which builds and is shed in a monthly cycle, unless an oocyte is fertilised, in which case, the embryo implants in the endometrium, establishing a pregnancy.

The uterus opens into the vagina through a narrow ring of connective tissue and muscle, the cervix. The walls of the vagina are muscular and open at the vulva.

Going further ▶

A blockage of an oviduct, e.g. as a result of infection, prevents the passage of the secondary oocyte to the site of fertilisation. Treatment usually involves microsurgery.

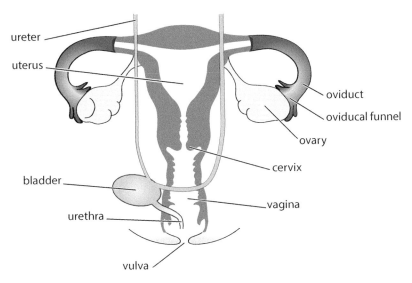

Female reproductive system

Histological examination of TS ovary

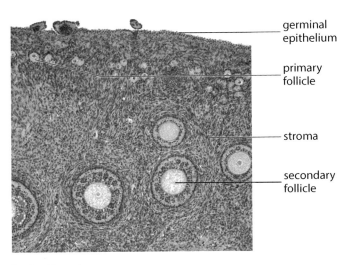

Section through ovary

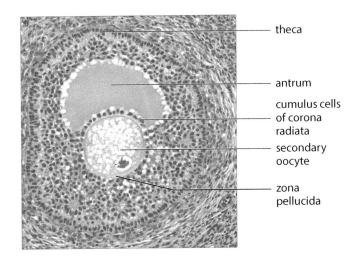

Developing secondary follicle

A section of ovary is oval or circular in outline, depending on the plane through which it was cut. Hold the slide up to the light and see if there is an apparently empty area, which is likely to be the antrum of a mature Graafian follicle. The Graafian follicle may be towards the centre of the ovary or, if the section was made soon after ovulation, at the edge of the ovary area, where the follicle has ruptured.

Going further ▶

The Dutch physician and anatomist Reinier de Graaf (1641–73) was the first person to describe the development of what we now call the Graafian follicle. He realised that it contained the oocyte.

- Place the slide under the microscope and, using a ×4 objective, identify any maturing Graafian follicles, by noting the fluid-filled antrum.

- The outer edge of the ovary comprises a germinal epithelium.

- Moving inwards, using a ×10 and a ×40 objective, identify primary follicles, which appear as clusters of cells, each with a larger primary oocyte in the middle.

- The oocyte in a Graafian follicle will only be seen if the plane of the section passes through it. If present, the dark-staining **zona pellucida** can be distinguished. If the first meiotic division has already taken place it is a secondary oocyte. If not, it is still primary oocyte.

- The secondary oocyte is surrounded by cumulus cells, which contribute to the **corona radiata**.

- The antrum may look opaque, despite being fluid-filled, as solutes tend to crystallise out during the preparation of the slide.

- The matrix of the ovary is called the stroma. It may be possible to see in it blood and lymphatic vessels.

Use your eyepiece graticule and calibration value to measure the diameter of a Graafian follicle.

Gametogenesis

The production of gametes in the sex organs is gametogenesis. Spermatogenesis is the formation of sperm in the testis. Oogenesis is the formation of secondary oocytes in the ovary.

The cells of the germinal epithelium of both the testis and the ovary undergo a sequence of mitotic and meiotic divisions to form haploid gametes. It is important that the gametes are haploid so that at fertilisation, the diploid number is restored and the chromosome number does not double in every generation.

Spermatogenesis

- Cells of the germinal epithelium are diploid. They divide by mitosis to make diploid spermatogonia and more germinal epithelium cells.

- The spermatogonia divide many times by mitosis and enlarge, making diploid primary spermatocytes and more spermatogonia.

- Primary spermatocytes undergo meiosis I, making secondary spermatocytes, which are haploid.

Going further ▶

Sperm production takes about 70 days. Sperm swim at 1–4 mm / minute. A man makes about 10^7 sperm / g testis / day. In the west, a sperm count of $2–4 \times 10^6$ /cm^3 ejaculate is considered normal.

- Secondary spermatocytes undergo meiosis II, making haploid spermatids.

- Spermatids mature into spermatozoa or sperm.

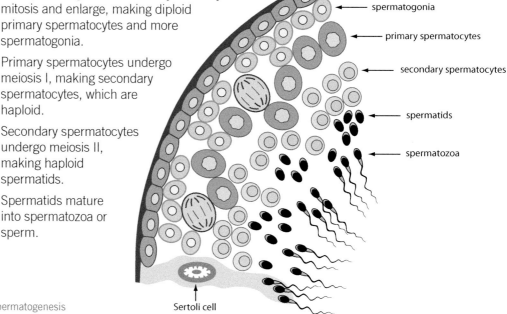

germinal epithelium cell
spermatogonia
primary spermatocytes
secondary spermatocytes
spermatids
spermatozoa

Spermatogenesis

Sertoli cell

Sperm structure

- The head contains a haploid nucleus, covered at the anterior end by a lysosome called the acrosome, which contains enzymes used at fertilisation.

- The middle piece is packed with mitochondria, which provide ATP for movement. They spiral around the microtubules, which extend from the centriole into the axial filament in the tail.

- The tail, or flagellum, makes lashing movements that move the sperm, although sperm are not motile until they have been modified in the epididymis.

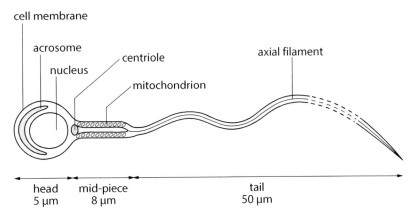

Diagram of a human sperm cell

Oogenesis

- Before birth, in the developing foetus, cells of the germinal epithelium of the ovary, which are diploid, divide by mitosis to make diploid oogonia, and more germinal epithelium cells.

- The oogonia divide many times by mitosis and enlarge, making diploid primary oocytes and more oogonia.

- The primary oocytes begin meiosis I but stop; a girl is born with millions of primary oocytes at prophase I in her ovaries.

- Germinal epithelium cells divide to form diploid follicle cells, which surround the primary oocytes, making primary follicles.

- From puberty onwards, hormones stimulate the primary follicles to develop further. Just before ovulation, a primary oocyte completes meiosis I, making a secondary oocyte, which contains most of the cytoplasm. The other product of meiosis I is a much smaller cell. It extrudes from the end of the secondary oocyte, and so it is called the first polar body. The secondary oocyte and polar body are both haploid.

- The primary follicle develops into a secondary follicle, which is called a Graafian follicle when it is mature. It migrates to the surface of the ovary where it bursts and releases the secondary oocyte, in a process called ovulation. Each month several primary follicles start to develop but normally, only one matures into a fully developed Graafian follicle.

- The secondary oocyte begins meiosis II but stops at metaphase II unless fertilisation takes place.

- After fertilisation, meiosis II is completed, making an ovum containing most of the cytoplasm. The other product of meiosis II is the second polar body.

- After ovulation the Graafian follicle becomes the corpus luteum (yellow body). If fertilisation occurs, it produces hormones, but otherwise, it regresses.

24

Knowledge check

Identify the missing word or words.

Spermatozoa are produced in the .... of the testes by a process called .... Spermatogonia divide many times to produce............................... These undergo meiosis and the products of the first meiotic division are haploid. These cells undergo the second division of meiosis, producing ...., which differentiate and mature into spermatozoa. The spermatozoa are protected and nourished by .... cells.

Exam tip

Make sure that you know the stages at which mitosis and meiosis take place in gametogenesis.

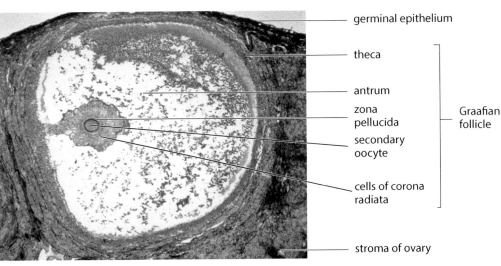

The Graafian follicle

25

Knowledge check

Identify the missing word or words.

Female gametes are produced in the ovary by a process called .... Two types of cell division are involved in their production. Oogonia divide by .... to form primary oocytes. These start to divide by ...., but only as far a prophase I until puberty, after which each month, one completes its first division to form a secondary oocyte. This is contained in a fully developed .... follicle, from which the secondary oocyte is released, in a process called ....

A clear glycoprotein layer called the **zona pellucida** surrounds the cell membrane of the secondary oocyte. The chromosomes of the secondary oocyte are at metaphase II. They are at the equator, attached to the microtubules that make the spindle apparatus. The periphery of the cytoplasm contains **cortical granules**, which are secretory organelles that prevent the entry of more than one sperm. Corona radiata cells surround the secondary oocyte and provide nutrients.

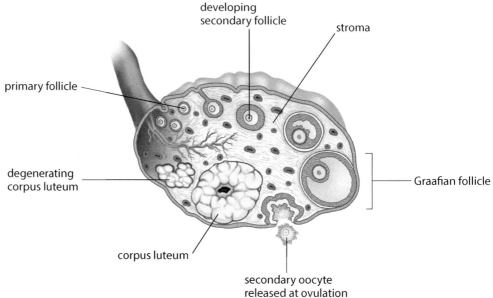

Illustration of an ovary

The flow diagrams below show that spermatogenesis and oogenesis follow the same principles, i.e. mitosis followed by meiosis. The equivalent names are given to equivalent stages. A major difference is that a primary spermatocyte produces four gametes but a primary oocyte produces one.

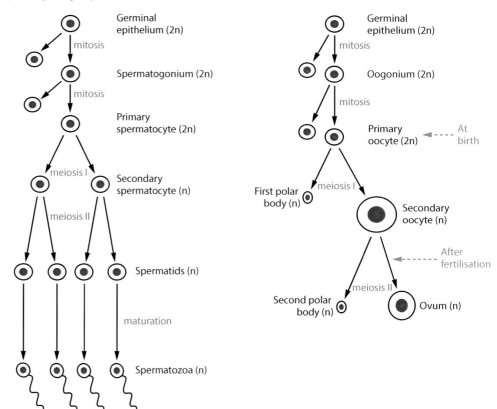

Flow diagrams of gametogenesis

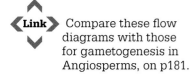 Compare these flow diagrams with those for gametogenesis in Angiosperms, on p181.

Sexual intercourse

Physical and psychological effects cause the arterioles entering the penis to dilate and the venules leaving to constrict. The build up of blood in spaces in the penis causes it to become erect. In sexual intercourse, the penis is inserted into the vagina, which is why it is described as an 'intromittent' organ. Movements of the penis result in contraction of smooth muscle in the walls of the epididymis, vas deferens and penis, which causes the ejaculation of semen into the vagina. The force of ejaculation is sufficient to propel some sperm through the cervix into the uterus, with the remainder being deposited at the top of the vagina. In men and women, the combination of physical and psychological events at their maximum intensity is an orgasm.

YOU SHOULD KNOW ›››

››› The process of fertilisation

››› The process of implantation

››› The structure and roles of the placenta

Fertilisation

Fertilisation is a process, not a single event. It may take many hours between sperm clustering around the oocyte and the completion of the fusion of the genetic material of the gametes.

- The sperm reach the secondary oocyte – it takes about 5 minutes from being deposited, for the sperm to respond to the oocyte's chemoattractants and swim through the cervix and through the uterus to the oviduct. The sperm remain viable for 2–5 days but are at their most fertile for 12–24 hours after ejaculation. If ovulation has recently taken place, there will be a secondary oocyte in the oviduct, although it only remains viable for about 24 hours unless fertilised. Despite millions of sperm being deposited, only about 200 reach the secondary oocyte in the oviduct.

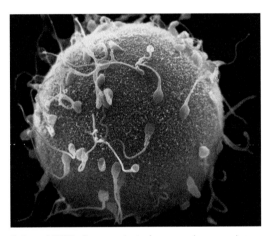

Coloured SEM of sperm clustering around a secondary oocyte

- **Capacitation** – sperm can only fertilise a secondary oocyte after a process called capacitation has taken place. This is the removal of cholesterol and glycoproteins from the cell membrane over the acrosome in the sperm head. Over several hours, the membrane becomes more fluid and more permeable to calcium ions.

- **Acrosome reaction** – the acrosome releases proteases, which digest the cells of the corona radiata.

  Then, on contact with the zona pellucida, the acrosome membrane ruptures and releases another protease, acrosin, which hydrolyses the zona pellucida around the secondary oocyte.

- Sperm head entry – the cell membranes of the secondary oocyte and sperm fuse and the head of the sperm sinks into the cytoplasm of the secondary oocyte. Following the entry of the sperm head, the secondary oocyte is called an ovum.

- **The cortical reaction** – this is the reaction of the oocyte that produces the fertilisation membrane, preventing **polyspermy**, i.e. the entry of additional sperm. When the sperm attaches to the secondary oocyte, the oocyte's smooth endoplasmic reticulum releases calcium ions into the cytoplasm. They make the cortical granules fuse with the cell membrane and release their contents of enzymes, by exocytosis. The zona pellucida is chemically modified and expands and hardens, making a fertilisation membrane, which is impossible for more sperm to penetrate.

Going further ▶

Internal fertilisation brings male and female gametes together without the risk of them dehydrating. It occurs in reptiles, birds and mammals. Its evolution is correlated with animals colonising land habitats.

Key Terms

Capacitation: Changes in the sperm cell membrane that increase its fluidity and allow the acrosome reaction to occur.

The acrosome reaction: Acrosome enzymes digest the corona radiata and the zona pellucida, allowing the sperm and oocyte cell membranes to fuse.

Cortical reaction: Fusion of cortical granule membranes with the oocyte cell membrane, releasing their contents, which convert the zona pellucida into a fertilisation membrane.

- Meiosis II – entry of the sperm also stimulates the completion of the second meiotic division of the ovum nucleus. It proceeds through anaphase II and telophase II, divides and expels the second polar body.

- Within about 24 hours, the first mitosis combines the genetic material of the parents to make the diploid cells of the embryo. The sperm chromosomes join the ovum chromosomes on the cell's equator. The cell is now a zygote, as the chromosomes have combined.

- This first mitotic division produces two cells, and until organs have developed at about week 10, this collection of cells is called an embryo.

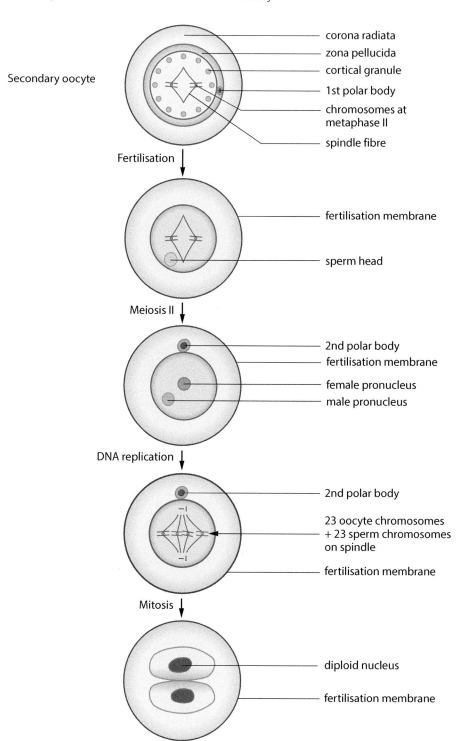

Sequence following sperm entry

Implantation

As the embryo moves down the oviduct, it divides many times by mitosis, in a sequence called **cleavage**. A solid ball of 16 cells, called a morula, forms within 3 days. Cleavage continues and the cells move in relation to each other. By 7 days, the ball of cells becomes hollow and is called a blastocyst. The cells around the outside of the blastocyst are **trophoblasts.** They divide to make an **inner cell mass** on one side. The blastocyst moves from the oviduct into the uterus.

After ovulation, the endometrium thickens and gets an increased blood supply to prepare it to receive an embryo. There is an 'implantation window' when the endometrium is receptive, between 6 and 10 days after ovulation. After about 9 days, protrusions from trophoblast cells of the blastocyst, called trophoblastic villi, penetrate the endometrium. The villi increase the surface area for the absorption of nutrients from the endometrium. The embedding of the blastocyst into the endometrium is **implantation** and 80% of blastocysts implant within 8–10 days.

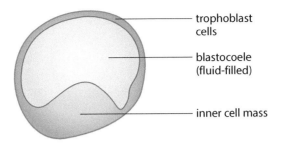

trophoblast cells

blastocoele (fluid-filled)

inner cell mass

Blastocyst at 7 days

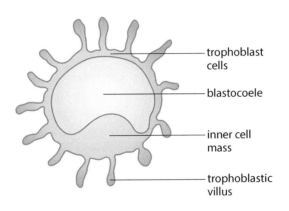

trophoblast cells

blastocoele

inner cell mass

trophoblastic villus

Trophoblast embedded in endometrium

Going further

Progesterone from the corpus luteum stimulates secretions in the uterus, which degrade the remains of the zona pellucida and provide nutrients, enhancing the blastocyst's growth and ability to implant.

Key Terms

Trophoblast: Cells forming the outer layer of the blastocyst.

Implantation: The sinking of the blastocyst into the endometrium.

Going further ▶

Of every 100 fertilised oocytes, fewer than 50 become a blastocyst, 25 implant and only 13 develop beyond 3 months.

Going further ▶

In an ectopic pregnancy, the blastocyst implants in the wall of the oviduct. The first a woman may know is an intense pain when the growing embryo ruptures the oviduct.

Going further ▶

About 70% of failures to implant are because the endometrium is not properly receptive and hormone treatment may help. In other cases, it is because of a problem with the embryo.

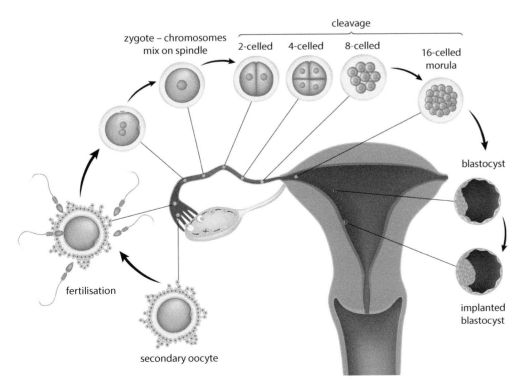

cleavage

zygote – chromosomes mix on spindle

2-celled   4-celled   8-celled

16-celled morula

blastocyst

implanted blastocyst

fertilisation

secondary oocyte

Summary of ovulation to implantation

The placenta

The **placenta** is about 22 cm long × 2 cm thick and weighs about 500 g. It connects the embryo and foetus to the uterus wall. It is made of tissues derived from the embryo and from the mother:

- The trophoblast develops into the chorion, an outer membrane surrounding the embryo. Cells of the chorion move into the trophoblastic villi and form the much larger **chorionic villi**. They acquire blood capillaries, which are connected to the umbilical artery and vein. These are the blood vessels that connect the embryo to the uterus wall through the umbilical cord.
- Projections from the endometrium between chorionic villi are the maternal tissues of the placenta.

The major roles of the placenta are:

- As an endocrine organ, producing hormones to support the pregnancy.
- Exchange between the mother's and foetus's blood, including nutrients, waste products, respiratory gases. Inter-villous spaces, called lacunae, containing the mother's blood, surround the chorionic villi. Chorionic villus cells have microvilli, giving a large area of contact with the mother's blood, for maximum exchange. The embryo's and the mother's blood do not make contact. The distance between them, through the walls of the chorionic villi, is only about 5 μm, a distance over which diffusion, facilitated diffusion, active transport, pinocytosis and osmosis operate effectively. The concentration gradient between the two circulations is maintained by a counter-current flow, enhancing exchange efficiency.
- A physical barrier between the foetal and maternal circulation:
  - Protects the fragile, foetal capillaries from damage by the higher blood pressure of the mother.
  - Protects the developing foetus from changes in maternal blood pressure.
- Providing passive immunity to the foetus: maternal antibodies cross the placenta and attack pathogens but do not attack the foetal cells, even though they carry the father's antigens, which are different from the mother's.
- Protection from the mother's immune system: the mother does not make an immune response against the foetus or placenta, even though they contain foreign genes, i.e. from the father. One reason is that cells of the wall of the chorionic villi fuse so there are no spaces between them, making a syncitium. Then, migratory immune cells, such as granulocytes, cannot get through to the foetal blood.

But the placenta does not always provide complete immunological protection:

- Some spontaneous abortions are equivalent to the rejection of a transplanted organ.
- Rhesus disease in a foetus is the destruction of its blood cells by antibodies made by a Rhesus negative mother against the blood cells of a Rhesus positive foetus. It is worse for each successive Rhesus positive foetus.
- In the 2nd trimester, some women develop pre-eclampsia, when they have very high blood pressure. One cause is an abnormal immune response towards the placenta.

▼ **Study point**

The trophoblast of the blastocyst, becomes the chorion and placenta; the inner cell mass forms the embryo and the amnion.

Link You learned how counter-current flow enhances exchange when you studied gas exchange in bony fish in the first year of this course.

Going further ▶

The fusion of cells in the chorionic villi is caused by the products of viral genes that have integrated into the mammalian genome in the evolutionary past.

Going further ▶

The mother's increased immune tolerance makes her more susceptible to infection when pregnant. Infections, e.g. influenza and malaria, are more serious then; however, vaccinations are still effective.

▼ **Study point**

The placenta's major roles are:
1. An endocrine organ
2. Exchange of materials
3. A barrier between the mother's and foetus's circulation
4. Passive immunity
5. Protection against the mother's immune system.

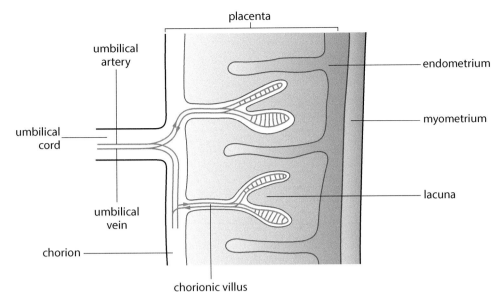

Diagram showing the structure of the placenta

Knowledge check

Match the terms 1–4 with their meanings A–D.

1. Capacitation.
2. Acrosome.
3. Cortical granule.
4. Chorionic villus.

A. Membrane-bound sac in sperm, containing enzymes facilitating fertilisation.
B. Projection from the chorion, containing capillaries where mother and foetal exchange occurs.
C. Vesicles in the secondary oocyte with contents that produce the fertilisation membrane.
D. Changes to sperm membrane allowing the acrosome reaction to take place.

Some micro-organisms, such as the *Rubella* virus, can cross the placenta from the mother's blood into the foetus's blood, as can some drugs, including nicotine and heroin.

The umbilical cord

The **umbilical cord** develops from the placenta and is about 60 cm long. It transfers blood between the foetus and the mother. The blood of the foetus comes to the placenta through the umbilical cord in two umbilical arteries. They are called arteries as the blood is coming away from the foetus's heart. This blood is low in nutrients and is deoxygenated. The blood exchanges materials with the mother's blood at the chorionic villi and returns to the foetus in a single umbilical vein. Returning blood is high in nutrients and is oxygenated.

▼ **Study point**

Arterial blood is usually oxygenated. The umbilical artery and the pulmonary artery are two arteries that carry deoxygenated blood. The converse is true for veins.

The menstrual cycle

Most mammals, such as pigs and dogs, have an oestrus cycle. This comprises a short period when they are fertile and sexually active or 'on heat'. In the absence of an implanted embryo, the endometrium is resorbed and an 'anoestrus' period follows.

In contrast, most primates (monkeys and apes, including humans) have a menstrual cycle, in which hormonal and physiological changes recur. In the absence of an implanted embryo, the endometrium is shed through menstruation. From the first period, which marks the beginning of puberty, until the menopause, about once a month, the endometrium detaches if a blastocyst has not implanted. The endometrium has a good blood supply, and it appears as bleeding, as it leaves the body through the vagina. The menstrual cycle is a system of positive and negative feedback operating between the events involving the brain, the ovaries and the uterus. In the graphs below, the cycle is shown to last 28 days, but it varies considerably between women and, for many, it varies from month to month.

The significant hormones of the menstrual cycle are:

- Follicle stimulating hormone (FSH)
- Luteinising hormone (LH)
- Oestrogen
- Progesterone.

YOU SHOULD KNOW ›››

››› The roles of FSH, LH, oestrogen and progesterone in the menstrual cycle

››› The sequences of events in the ovary and uterus

▼ **Study point**

Make sure you can spell 'luteinising'.

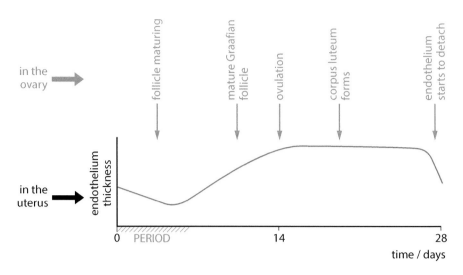

Events in the ovary and uterus in the menstrual cycle

Going further ▶

The menstrual cycle described in textbooks always lasts 28 days with ovulation on day 14. In fact, there is huge variation between people and for the same person at different times.

FSH and LH

The start of a cycle, day 0, is, by definition, the first day of a period. On day 0, the concentrations of all the relevant hormones in the plasma are low. Gonadotrophic releasing hormone (GnRH) is secreted by the hypothalamus and stimulates the anterior pituitary gland to secrete:

- FSH, which stimulates the development of primary follicles in the ovary. Only one matures. It forms a fibrous outer layer, the theca, and secretes fluid into a cavity, the antrum. A mature Graafian follicle is about 10 mm diameter. FSH stimulates the thecal cells to produce oestrogen.

- LH, which reaches its maximum concentration just before ovulation, on about day 12. The major role of LH is to induce ovulation: on day 14, its high concentration causes the Graafian follicle at the surface of the ovary to release the secondary oocyte. It has a positive feedback effect on FSH.

The remains of the Graafian follicle convert into the corpus luteum (yellow body), which secretes oestrogen and progesterone. These inhibit further secretion of FSH and LH and so their concentrations decrease.

▼ **Study point**

FSH stimulates Graafian follicle development. On day 14, the high LH concentration stimulates ovulation.

Going further ▶

Initially, LH stimulates thecal cells to secrete oestrogen. Oestrogen produces negative feedback on FSH production, which decreases. FSH concentration rises again because of positive feedback by LH.

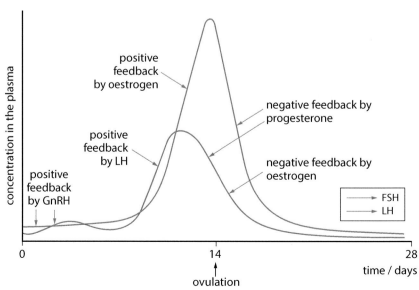

FSH and LH

Oestrogen and progesterone

As FSH concentration increases at the start of the cycle, it stimulates the production of oestrogen. As oestrogen becomes more concentrated in the plasma, it:

- Triggers the rebuilding of the endometrium, that was shed during menstruation
- Inhibits FSH secretion by negative feedback, which brings its own concentration down again
- Stimulates LH production by positive feedback.

The corpus luteum secretes oestrogen and progesterone. The progesterone maintains the newly rebuilt endometrium, so that if a secondary oocyte is fertilised, there will be suitable tissue in which the embryo can implant. But if there is no implantation, the falling concentrations of FSH and LH cause the corpus luteum to degenerate so progesterone production declines. The endometrium is no longer being rebuilt by oestrogen or maintained by progesterone, and so it is shed.

As oestrogen is low, it no longer inhibits FSH production, so the menstrual cycle restarts. This is the normal situation in modern humans. A secondary oocyte may be fertilised but fail to divide, implant in the wrong place or fail to implant. The data suggest that of those that are fertilised, 20–70% fail to establish a pregnancy. In most of these cases, it was because the blastocysts did not implant.

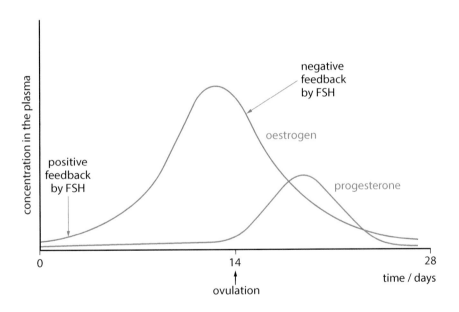

Oestrogen and progesterone

27

Knowledge check

Complete the paragraph by filling in the spaces.

At the start of the menstrual cycle, all the hormone concentrations are .... FSH increases and stimulates development of a .... LH concentration increases and stimulates .... on day 14. After ovulation, the Graafian follicle converts into the .... and secretes ....

WORKING SCIENTIFICALLY

Data provides evidence for mechanisms. The large range of figures given for failure to establish a pregnancy may be because data for this are hard to collect.

▼ Study point

A summary of hormone relationships:

- FSH up ∴ O up ∴ FSH down ∴ O down
- FSH up ∴ O up ∴ LH up ∴ ovulation ∴ progesterone up

▼ Study point

Main roles in the menstrual cycle:
FSH – follicle development
LH – ovulation
Oestrogen – rebuild endometrium
Progesterone – maintain endometrium

YOU SHOULD KNOW ›››

››› The roles of hCG, oestrogen and progesterone in pregnancy

››› The role of amniotic fluid

››› Hormonal changes associated with birth

Pregnancy

Pregnancy is defined as the time from the first day of the last period until birth, even though no embryonic development is possible until after ovulation. It lasts around 39 weeks.

The amnion

The embryo, later the foetus, develops and grows in the uterus, enclosed in the **amnion**, a membrane that is derived from the inner cell mass of the blastocyst. Initially the amnion is in contact with the embryo but in weeks 4–5, **amniotic fluid** accumulates and increases in volume for 6–7 months. The fluid is made by the mother initially, although from 4 months, the foetus contributes urine to it. The fluid pushes the amnion out, eventually as far as the chorion, the inner layer of the placenta.

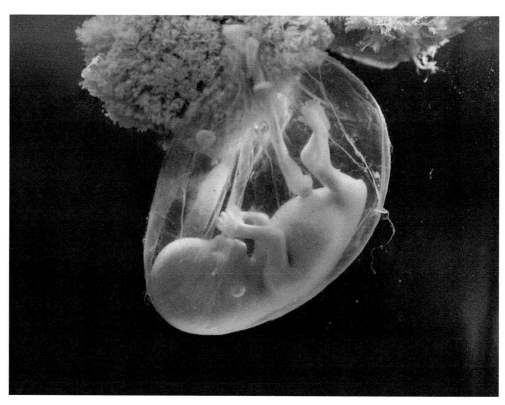

Model of human embryo surrounded by amnion

Going further ▶

There may be too little amniotic fluid if some has been lost through a ruptured amnion. The uterus is smaller than expected and the foetus may not move as much as normal. In cases of multiple pregnancies or gestational diabetes, there may be too much amniotic fluid. The foetus floats away as you apply pressure. The uterus is bigger than normal and the foetus's heart beat may sound muffled.

Amniotic fluid is 98% water, a solution of urea, salts, a little protein and a trace of sugar. It contains some foetal cells that have sloughed off the foetus. The foetus swallows about 500 cm³ of amniotic fluid each day and the volume of amniotic fluid present can be used to indicate that the foetus's swallowing reflex is normal. By the end of the pregnancy, about 1 dm³ remains. Amniotic fluid:

- Maintains the foetus's temperature
- Provides lubrication. In some cases, fingers and toes may become webbed if too little amniotic fluid circulates between them
- Contributes to lung development
- Allows movement so muscles and bones function before birth
- Acts as a shock absorber, protecting the foetus from injury from outside the uterus.

Stages of pregnancy

Pregnancy is classically divided into three trimesters. The 1st trimester includes conception, implantation and embryogenesis. All major organs are laid down. The 1st trimester has the highest risk of miscarriage and about 15% of women who know they are pregnant miscarry at this stage. But after week 10, the embryo is much less prone to miscarriage. It is about 30 mm long and is described as a foetus.

By the 3rd trimester, weeks 26–39, all major structures are complete; this is a period of growth. Fat is laid down, the foetus's mass increases three-fold and its length doubles. Development is such that at 28 weeks, more than 90% of premature babies survive. There is evidence that senses, such as hearing are developing. At 33 weeks, the foetus spends about 90% of its time asleep, while it is laying down millions of neurones each minute.

▼ **Study point**

The trimesters can be summarised as: 1st – all major organs laid down; 2nd – development; 3rd – growth.

Months	Trimester	Some early landmarks	
		Time	**Occurrence**
0–3	1st	22 days	2 brain hemispheres; heart beat.
		Weeks 5–6	Some electrical activity in the brain, but no evidence of conscious thought.
		Week 10	Eyes, mouth and ears visible; heart seen beating; Foetus makes involuntary movements.
4–6½	2nd	Week 18	Some foetuses open their eyes.
		Week 20	Movements of the foetus can be felt.

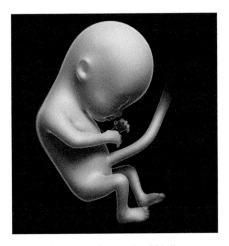

Model of 1-month embryo; it is about 5 mm long

Model of 4-month foetus; the CRL ('crown-rump length) is 150 mm

Going further ▶

At 28 weeks, a foetus may give a response to a repeated nursery rhyme. Psychologists interpret this as responding to a memory.

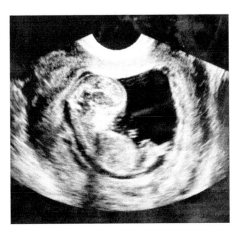

Ultrasound scan of a 3-month foetus

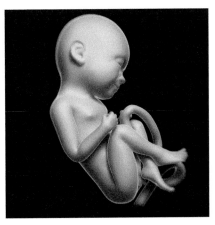

Model of 8-month foetus; the CRL is 300 mm

Going further ▶

A standard pregnancy test detects hCG in urine. It only works after 6 days, when the embryo is old enough to produce enough hCG, that having crossed the placenta, it is excreted in the mother's urine.

Placental hormones

- hCG – about 6 days after fertilisation, the embryo, at the blastocyst stage, begins to secrete human chorionic gonadotrophin (hCG). Following implantation, hCG is made by the chorion, the inner layers of the placenta. It is a glycoprotein and it maintains the corpus luteum in its secretion of progesterone, for the first 16 weeks of the pregnancy. Progesterone maintains the endometrium, which is essential for embryonic development, as it contributes to the structure of the placenta.

- Oestrogen and progesterone – secretion from the corpus luteum and then the placenta increases the concentration of oestrogen and progesterone in the plasma until the very end of the pregnancy. This inhibits the secretion of:

  - FSH, so no more follicles mature

  - LH, so ovulation is not possible

  - Prolactin, so no milk is made.

In addition:

- Progesterone inhibits oxytocin, so the myometrium and muscles in the milk ducts do not contract.

- Oestrogen stimulates the growth of the uterus to accommodate the foetus.

- Oestrogen stimulates the growth of the mammary glands, especially during the 3rd trimester, and increases their blood supply.

▼ **Study point**

The placenta secretes hormones, which influence the growth of the uterus and mammary glands and support the corpus luteum in progesterone secretion.

Hormones and birth

During the last three months of a pregnancy, the increase in oestrogen is greater than the increase of progesterone but just prior to birth, the progesterone concentration in the plasma declines. In biological terms, it is in the mother's interest to give birth as soon as the foetus can live independently. It is in the foetus's interest to delay birth as long as possible. They both release hormones supporting their conflicting needs but at 39 weeks, the mother's prevail.

At 39 weeks, foetal hormones that are transferred to the mother's bloodstream across the placenta decrease her plasma concentrations of progesterone. Throughout the pregnancy the mother's high concentrations of oestrogen and progesterone had inhibited the secretion of oxytocin and prolactin, but these are no longer inhibited and so:

- Oxytocin is secreted by the posterior pituitary gland. It causes contractions of the myometrium in the uterus wall. The contractions are mild initially but they stimulate the secretion of more oxytocin, by positive feedback and so the oxytocin concentration in the blood increases. Contractions, therefore, become stronger and more frequent. The myometrium contracts from the top down, so that the foetus can be pushed out through the cervix.

- Prolactin is secreted by the anterior pituitary gland and stimulates the glandular tissue in the mammary glands to synthesise milk. Milk is released when oxytocin causes the muscles around the milk ducts to contract. Prolactin secretion continues after the birth, as long as the milk is needed.

Going further ▶

The foetal adrenal glands secrete corticosteroids in response to immunological stress; the foetal lungs secrete prostaglandins when mature. These, and other, hormonal changes may be triggers for the onset of labour.

The table below summarises the main roles of hormones associated with sexual reproduction in humans:

	Hormone	Endocrine gland	Effects
FEMALES	FSH	Anterior pituitary	Stimulate development of the Graafian follicle
			Positive feedback on oestrogen early in cycle
			Negative feedback on oestrogen later in cycle
	LH	Anterior pituitary	Stimulate ovulation
			Positive feedback on FSH
			Stimulate conversion of Graafian follicle into corpus luteum
	Oestrogen	Theca; placenta	Rebuild endometrium
			Negative feedback on FSH
			Positive feedback on LH
			Secondary sexual characteristics
			Inhibit prolactin synthesis
			Inhibit oxytocin synthesis
	Progesterone	Corpus luteum; placenta	Maintain endometrium
			Negative feedback on FSH
			Negative feedback on LH
	Oxytocin	Hypothalamus, but stored in posterior pituitary	Contraction of smooth muscle in myometrium
			Contraction of smooth muscle in milk ducts
	Prolactin	Anterior pituitary	Milk synthesis
	hCG	Blastocyst; placenta	Maintain corpus luteum
MALES	FSH	Anterior pituitary	Sperm development
	LH (ICSH)	Anterior pituitary	Stimulate Leydig cells
	Testosterone	Leydig cells	Sperm development
			Secondary sexual characteristics

Sexual reproduction in plants

The flowering plants or Angiosperms are the most successful of all terrestrial plants. The flower is the organ of reproduction and usually contains both male and female parts. In Angiosperms the female part, the ovule, is never exposed but is enclosed within a modified leaf, the carpel. A key feature of the success of flowering plants is their relationship with animals. Pollen grains have no power of independent movement and have to be transferred to the female part of the flower to ensure fertilisation. Flowering plants have evolved the strategy of attracting animals, particularly insects, to their flowers, feeding them and exploiting their mobility to transfer pollen from flower to flower. Some plants are pollinated by the action of wind.

Topic contents

By the end of this topic you will be able to:

- Describe the basic structure and functions of the parts of the flower.
- Describe how the male and female gametes develop.
- Compare an insect- and a wind-pollinated flower.
- Explain what is meant by pollination and describe how cross-pollination results in far greater genetic variation than self-pollination.
- Describe the process of double fertilisation in flowering plants.
- Describe the development of the fruit and seed.
- Explain the differences between a fruit and a seed.
- Describe the structure of a maize fruit and a seed of the broad bean.
- Describe the requirements for germination and how food reserves are mobilised from the food store to the embryo plant.
- Describe the role of gibberellin in seed germination.

Flower structure

Not all plants have flowers. Those that do are called Angiosperms, and the flower is their reproductive structure. Most flowering plants are diploid. Meiosis takes place within the reproductive tissues and produces haploid spores which contain the gametes. Male spores are pollen grains, produced in the anther. The female spore is the embryo sac, produced in the ovule, in the ovary.

Many species have flowers that are hermaphrodite, i.e. the flowers contain both male and female parts. Flowers of different species have a great variety in their appearance, but similar patterns can be seen. A flower is four sets of modified leaves arising from the receptacle at the base of the flower:

- The outermost ring of structures is the calyx, which comprises the sepals. They are usually green and protect the flower in the bud, although in some flowers, e.g. lilies, the sepals are coloured.

- Inside the sepals is the corolla, a ring of petals. These range from absent to small and pale green to large and brightly coloured. There may be a nectary at the base, releasing nectar which is scented and attracts pollinators, such as insects.

- Inside the petals are the male parts of the flower, the stamens. Each stamen consists of a filament supporting an anther which produces pollen grains. The filament contains vascular tissue, which transports sucrose, mineral ions and water to the developing pollen grains. The anther usually contains four pollen sacs arranged in two pairs, side by side. When mature, the pollen sacs dehisce, which means they open and release the pollen.

- In the centre of the flower are one or more carpels, which are the female parts of the flower. Each carpel is a closed structure in which one or more ovules develop. The lower part of the carpel, surrounding the ovules, is the ovary and at its tip bears the style, which ends in a receptive surface, the stigma.

YOU SHOULD KNOW ›››

››› The names of the parts of the flower and their functions

››› The meaning of the term pollination

››› The differences between insect- and wind-pollinated flowers

▼ Study point

The stamen has two parts: the filament and the anther; the carpel has three parts: the stigma, style and ovary.

Going further ▶

The features of flowers of different species allow them to be classified into families. A tulip has flower parts in multiples of 3 so it is in a different family from a wild rose, with 5 petals, or a sweet pea, which has also has 5 petals, but of different sizes, with some fused together.

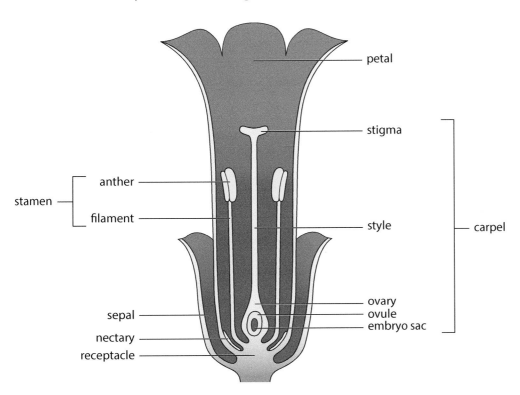

Diagram of generalised insect-pollinated flower

Cherry flower

Cross-pollination by insects or wind

Pollinators such as bees are attracted by large, coloured petals, by scent and nectar. They use their long tongues to reach the sugary nectar at the base of the petals. As the bee enters the flower, the anthers brush against its thorax and legs leaving the sticky pollen behind. When the bee enters another flower, it brushes some of the pollen against the ripe stigma, and cross-pollination has taken place.

Wind-pollinated flowers do not need the bright, scented petals that attract insects. Their anthers hang outside the flower so that the wind can blow away the small, smooth and light pollen. The feathery stigmas hang outside the flowers and provide a large surface area for catching pollen grains that are blown into their path.

Insect-pollinated flowers	Wind-pollinated flowers
Colourful petals, sometimes with nectar guides	Petals usually absent or small, green and inconspicuous
Scent and nectar (mainly sucrose)	No scent or nectar
Anthers within the flower	Anthers hanging outside the flower
Stigma within the flower	Large, feathery stigmas hang outside flower
Small quantities of sticky, sculptured pollen	Large quantities of smooth pollen
Produces larger pollen grains	Produces smaller pollen grains

Wind-pollinated flowers may be very small and grouped together in an inflorescence, e.g. *Plantago*, the plantain.

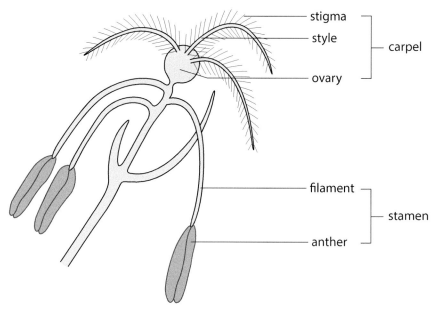

Diagram of generalised wind-pollinated flower

Inflorescence of *Plantago media*, the hoary plantain. Older flowers are at the top so mature carpels are above mature stamens. Pollen cannot fall on to the carpels and so they are wind-pollinated.

Exam tip

Be prepared to label a diagram of a flower and explain the functions of its parts.

Gamete development

Development of the male gamete

In the pollen sacs of the anther, diploid pollen mother cells undergo meiosis. Each forms a tetrad, containing four haploid cells, which become four pollen grains. The tapetum, a layer of cells around the pollen sac, provides nutrients and regulatory molecules to the developing pollen grains. It has a significant role in the formation of the pollen cell wall, which is tough and resistant to chemicals. It resists desiccation, so the pollen grains can be transferred from one flower to another without drying out. Ultra-violet light cannot penetrate the pollen cell wall, so the DNA in pollen that is carried at high altitude is protected from mutation.

Inside the pollen grain the haploid nucleus undergoes mitosis to produce two nuclei, a generative nucleus and a tube nucleus. The generative nucleus produces two male nuclei by mitosis.

When the pollen is mature, the outer layers of the anthers dry out, causing tension in lateral grooves. Eventually **dehiscence** occurs in which the tension pulls the walls of the anther apart and the edges of the pollen sacs curl away. An opening called the stomium exposes the pollen grains and they are carried away by insects or the wind.

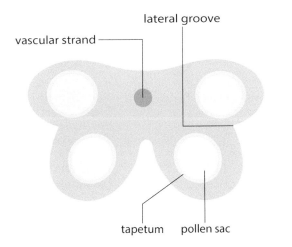

TS anther

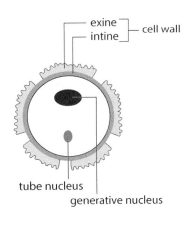

Mature pollen grain

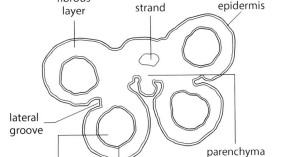

TS lily anther

Diagram of TS lily anther

YOU SHOULD KNOW ›››

››› The structure of the pollen grain and ovule

››› The names of the nuclei in the pollen grain and pollen tube

››› The names of the nuclei in the embryo sac

Going further

The high sucrose content of pollen grains contributes to their resistance to desiccation. The cell walls contain sporopollenin, one of the most chemically resistant compounds known.

Key Term

Dehiscence: The opening of the anther, releasing pollen grains.

Exam tip

The male nuclei inside the pollen grains, not the pollen grains themselves, are the male gametes.

Link See p193 for a description of how to draw an anther and calculate the anther's size and the drawing's magnification.

Development of the female gamete

- The ovary contains one or more ovules.
- In each ovule, a megaspore mother cell, surrounded by cells of the nucellus, undergoes meiosis making four haploid cells.
- Three disintegrate.
- The remaining cell undergoes three rounds of mitosis, producing eight haploid nuclei, one of which is the female gamete.
- Two of the haploid nuclei fuse to make a diploid nucleus called the polar nucleus.

These nuclei are in the embryo sac, surrounded by the nucellus, a layer of cells which provide nutrients. Around the nucellus, are two layers of cells called the integuments. A gap in the integuments is called the micropyle. As with the formation of the male gamete, the type of cell division that directly produces the female gamete is mitosis, not meiosis.

Pea pod with immature ovules

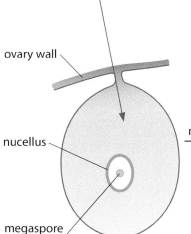

Embryo sac development

Immature ovule

Embryo sac in mature ovule contains:
3 antipodals (haploid)
2 synergids (haploid)
1 oosphere (haploid)
1 polar nucleus (diploid)

▼ **Study point**

The female gamete is also called the oosphere. Words that begin with 'o' or 'oo', including ovary and oocyte, often refer to females.

Link In the formation of the female gamete in mammals, three of the four products of meiosis disintegrate. See p163.

▼ **Study point**

The ovule consists of the outer integuments surrounding the nucellus and an embryo sac containing seven nuclei.

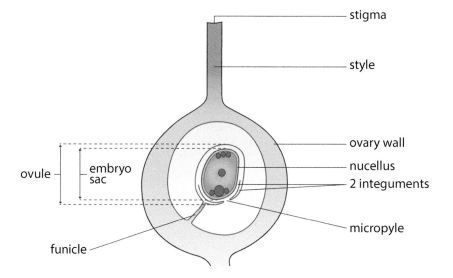

Diagram of mature ovule within the carpel

Ovary structure

The photograph shows a section through the ovary of *Narcissus tazetta* and the diagram is a low-power plan of the section.

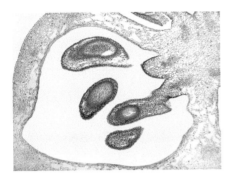

TS ovary of *Narcissus tazetta*

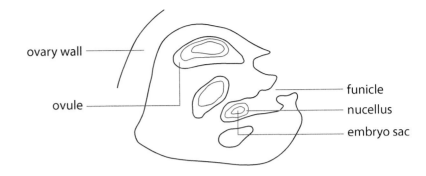

Diagram of *Narcissus tazetta* ovary

ovary wall

ovule

funicle

nucellus

embryo sac

The flow diagram summarises the formation of the nuclei in the mature pollen grain and in the embryo sac:

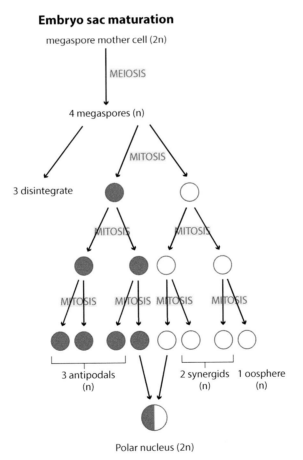

Pollen grain maturation

pollen cell mother (2n)

MEIOSIS

4 immature pollen grains in tetrad (n)
Each nucleus

MITOSIS

Generative nucleus (n)   Tube nucleus (n)

MITOSIS

2 male nuclei (n)

Embryo sac maturation

megaspore mother cell (2n)

MEIOSIS

4 megaspores (n)

MITOSIS

3 disintegrate

MITOSIS   MITOSIS

MITOSIS   MITOSIS   MITOSIS   MITOSIS

3 antipodals (n)   2 synergids (n)   1 oosphere (n)

Polar nucleus (2n)

Cell division and gamete development

181

Key Terms

Pollination: The transfer of pollen grains from the anther to the mature stigma of a plant of the same species.

Protandry: The stamens of a flower ripen before the stigmas.

‹Link› The importance of new genetic combinations in evolution is discussed on p224.

28

Knowledge check

Link the appropriate terms 1–4 with the statements A–D:

1. Anther.
2. Petal.
3. Sepal.
4. Stigma.

A. Attracts insects.
B. Where pollen grains are produced.
C. The receptive surface for pollen.
D. Encloses and protects the flower in the bud.

Self- and cross-pollination

Pollination is the transfer of pollen grains from the anther to the mature stigma of a plant of the same species. Pollination brings the pollen grains, containing the male gametes, into contact with the female part of the flower, which can result in fertilisation.

- Self-pollination: the pollen from the anthers of a flower is transferred to the mature stigma of the same flower or another flower on the same plant.
- Cross-pollination: most Angiosperms use cross-pollination, in which pollen is transferred from the anthers of one flower to the mature stigma of another flower on another plant of the same species.

The genetic implications of self- and cross-pollination

The two forms of pollination have very different genetic consequences.

Self-pollination leads to self-fertilisation, which results in inbreeding:

- Self-fertilised species depend only on independent assortment and crossing over during meiosis and on mutation to bring about genetic variation in the genomes of their gametes. Consequently, they display less genetic variation than cross-fertilised species, which also combine gametes from two different individuals.
- There is a greater chance of two potentially harmful recessive alleles being brought together at fertilisation.
- An advantage of inbreeding is that it can preserve those successful genomes that are suited to a relatively stable environment.

Cross-pollination leads to cross-fertilisation, which results in outbreeding:

- Outbreeding combines gametes from two individuals, in addition to events in meiosis and mutation, and so it generates more genetic variation than inbreeding.
- Outbreeding reduces the chance of producing harmful allele combinations.
- Outbreeding is of great evolutionary significance because, in the struggle for survival, some genomes are more successful than others. It may allow a species to survive in a changing environment, as there are always likely to be some members of a population with a suitable combination of alleles.

Ensuring cross-pollination

Different flowering plant species employ a variety of methods to ensure that cross-pollination takes place. These include:

- Dichogamy, i.e. the stamen and stigma ripening at different times. In **protandry**, e.g. daisy, the stamens ripen first. In protogyny, e.g. bluebell, which is rarer, the stigma ripens first.
- The anther is below the stigma so pollen cannot fall on to it, e.g. the pin-eyed primrose.
- Genetic incompatibility, e.g. in red clover. Pollen cannot germinate on the stigma of the flower which produced it.
- Separate male and female flowers on the same plant, e.g. maize.
- Separate male and female plants, e.g. holly.

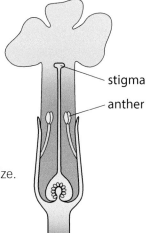

stigma

anther

Pin-eyed primrose

Double fertilisation

Fertilisation is the process in which a male gamete fuses with a female gamete to produce a zygote. In flowering plants, the female gamete is in the ovule, protected within the ovary. The male gamete is the nucleus contained in the pollen grain. The male gamete is delivered to the female gamete by a pollen tube.

- When a compatible pollen grain lands on the stigma, it germinates in the sucrose solution secreted by the stigma and produces a pollen tube.

- The pollen tube nucleus is at the tip of the tube, with the two male nuclei behind.

- The pollen tube grows out of the pollen grain through a gap in the cell wall, called a pit, and down the style, up a gradient of chemoattractants, e.g. GABA from the ovule. The pollen tube nucleus codes for the production of hydrolases, including cellulases and proteases, and it digests its way through the tissues of the style. The products of digestion are used by the growing pollen tube.

- The pollen tube grows through the gap in the integuments, the micropyle, and passes into the embryo sac.

- The pollen tube nucleus disintegrates, presumably having completed its function of controlling the growth of the pollen tube.

- The tip of the pollen tube opens, releasing the two male gametes into the embryo sac.

- The male and female gametes are haploid. One of the male gametes fuses with the female gamete, the oosphere, to form a zygote, which is diploid.

- The other male gamete fuses with the diploid polar nucleus to form a triploid nucleus. This triploid nucleus is the endosperm nucleus. When it subsequently divides repeatedly by mitosis, it generates the endosperm tissue, which takes over from the nucellus in providing nutrition for the developing embryo.

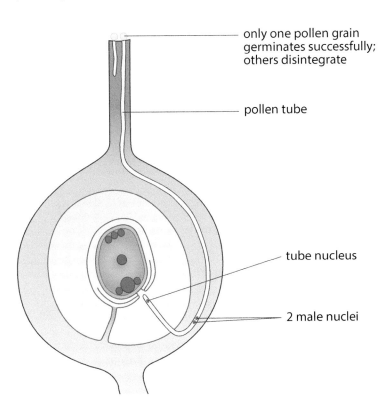

only one pollen grain germinates successfully; others disintegrate

pollen tube

tube nucleus

2 male nuclei

Growth of the pollen tube

Two fusions have occurred, one to form the zygote and one to form the endosperm. This is **double fertilisation**, a process unique to flowering plants.

 Key Term

Fertilisation: The fusion of a female and male gamete, producing a zygote.

‹Link› GABA also acts as a neurotransmitter in the mammalian brain. Neurotransmitters are described on p150.

29

Knowledge check

Link the appropriate terms 1–4 with the statements A–D:

1. Micropyle.
2. Zygote.
3. Triploid endosperm nucleus.
4. Embryo sac.

A. Structure which contains the oosphere and polar nuclei.

B. The gap in the integuments through which the pollen tube enters.

C. The result of the fusion of one of the male gametes and the female oosphere nucleus.

D. The result of the fusion between a male gamete and the polar nucleus.

Key Terms

Fruit: A structure developing from the ovary wall, containing one or more seeds.

Seed: Structure developed from a fertilised ovule, containing an embryo and food store enclosed within a testa.

▼ Study point

Note the correct spelling of radicle.

Going further ▶

A true fruit develops from the ovary wall, with seeds inside. A strawberry is a 'false fruit' because the part we eat develops from the flower receptacle. Its true fruits each contain one seed and are outside.

Development of the fruit and seed

After fertilisation, the **fruit** and **seed** develop. The seed develops from the fertilised ovule and contains an embryonic plant and a food store.

- The diploid zygote divides by mitosis, becoming an embryo which consists of a plumule (the developing shoot), a radicle (the developing root) and one or two cotyledons (seed leaves).
- The triploid endosperm nucleus develops into a food store, providing food for the developing embryo.
- The outer integument dries out, hardens and becomes waterproof, with deposits of lignin. It becomes the seed coat or testa. The micropyle remains as a pore in the seed.
- The ovule, comprising the embryo, endosperm and testa, becomes the seed.
- The funicle, or stalk, of the ovule becomes the funicle of the seed. It attaches to the seed at the hilum.
- The ovary becomes the fruit. In some species, such as a cherry, the ovary wall becomes sweet, juicy and pigmented. In others, such as the almond, the ovary wall becomes dry and hard.

Structure of the fruit and seed

The ovary of a broad bean flower contains several ovules. After fertilisation, the ovary elongates into a pod, which is the fruit. The ovules mature into seeds, which are the broad beans.

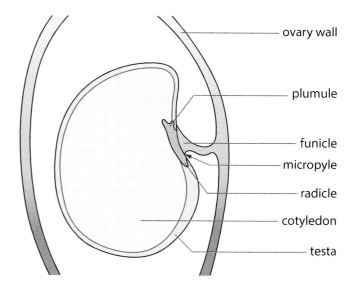

ovary wall
plumule
funicle
micropyle
radicle
cotyledon
testa

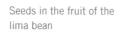
Seeds in the fruit of the lima bean

Diagram of a bean seed

The broad bean is classed as a dicotyledon, which means its seeds have two seed leaves or cotyledons. The embryo lies between them. The plumule is the part of the embryo that will become the shoot, and the radicle is the part that will become the root. The endosperm, which was the food store for the early embryo, is absorbed into the cotyledons, so the broad bean has a 'non-endospermic' seed.

Maize is classed as a monocotyledon as it has only one cotyledon. Typically of cereal grains, the endosperm remains as the food store, so maize seeds are 'endospermic'. The cotyledon remains small and does not develop further. Like other grasses and cereals, the testa of a maize seed fuses with the ovary wall so maize has a one-seeded fruit.

Seeds become **dormant**. Their water content falls below 10%, which reduces their metabolic rate. They can survive long periods in this state and will not germinate until conditions are suitable.

Going further

The oldest seed that has grown into a viable plant was a narrow-leafed campion, from Siberia. Radiocarbon dating has confirmed that the seed was 31,800 ±300 years old.

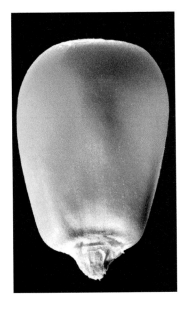

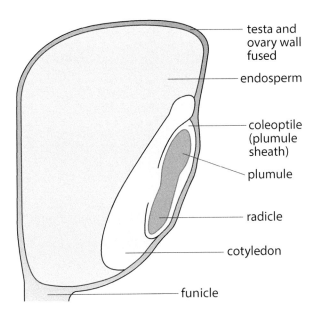

Maize (*Zea mais*) fruit and seed

Diagram of maize fruit and seed

Key Term

Dormant: Describes a seed when its active growth is suspended. Germination will only occur when specific conditions are met.

▼ **Study point**

The developing embryo's nutrition is provided by: (i) the nucellus then (ii) the endosperm then (iii) in some plants, one or two cotyledons.

▼ **Study point**

Flowering plants are divided into two main groups: monocots and dicots. The monocotyledons are important as they include cereals.

Monocots	Dicots
One cotyledon in seed	Two cotyledons in seed
Leaf veins are parallel	Leaf veins form a network
Sepals, petals and stamens in multiples of 3	Sepals, petals and stamens in multiples of 4 or 5
Vascular bundles scattered in stems	Vascular bundles in a ring in stems
Vascular bundles scattered in roots	Vascular bundles in the centre of roots

Seed dispersal

Seed dispersal is the movement of seeds away from the parent plant. If a seed were to germinate close to its parent, the parent plant would be more successful at obtaining water and minerals from the soil. It would be taller and would cast shade over the seedling, preventing it from photosynthesising adequately. The seedling would be out-competed. Seeds with features allowing them to be dispersed produce plants which avoid competition. These dispersal methods have been subject to natural selection:

YOU SHOULD KNOW ›››

››› How seeds are dispersed

››› The conditions required for germination

››› How food reserves are mobilised for use by the embryo

››› The role of gibberellin in seed germination

››› How seeds enhance survival

Dispersal method	Description	Example
Wind	Ash and sycamore fruits have sails that allow wind dispersal; dandelion fruits have a parachute of stiff hairs. The fruit of the poppy has pores, through which the seeds are shaken out when the stem is blown in the wind.	Dandelion seed
Transport	Birds eat seeds that pass through the digestive system and are dispersed in the faeces, e.g. cherries. Mammals, reptiles and fish can also disperse seeds this way. In a process called scarification, the digestive system weakens the testa by physical attack, by acid and by enzymes and the seeds of some species can only germinate when this has happened.	Song thrush eating berries
Rolling	When the fruit of a horse chestnut tree breaks open, the conker, which is the seed, falls to the ground and rolls away from the parent tree.	Horse chestnut fruit about to open and release conker
Bursting	When legume pods dry, they split and the seeds scatter, e.g. beans. In many species, the pods rotate as they burst open, sending the seeds in many different directions.	Pea pod splitting open
Water	Coconut palms grow by water. Coconuts are seeds and when they fall into the water, they float, because their air cavities make them buoyant, and are carried away.	Coconut palm
Carrying	Hooked seeds attach to animals coats and are carried away, e.g. burdock.	Burdock

Seeds and survival

The evolutionary development of seeds has contributed to the success of Angiosperms.

- Dormant seeds have a low metabolic rate and so they survive very cold weather.
- The testa is chemically resistant and so seeds survive adverse chemical conditions.
- The water content of a dormant seed is reduced below 10% and so seeds can survive very dry conditions.
- The testa can physically protect the embryo.
- The endosperm or cotyledons provide a supply of nutrients, which lasts until the emerging seedling can photosynthesise adequately.
- Seeds can be dispersed great distances from the parent plant and so do not compete with it.
- Dispersal allows the colonisation of new habitats.
- Inhibitors may only allow germination at a suitable time of year. They are broken down in very cold weather, in a process called 'vernalisation', so that the seed can germinate in the spring. In cabbages, the inhibitors are in the seeds and in tomatoes they are in the fruits.

30

Knowledge check

Link the appropriate terms 1–5 with the statements A–E:

1. Ovary wall.
2. Ovule.
3. Integument.
4. Triploid endosperm nucleus.
5. Embryo.

A. Becomes the seed.
B. Becomes the testa.
C. Develops into the fruit.
D. Develops from the zygote.
E. Becomes the food store.

Germination of the broad bean, *Vicia faba*

Conditions for germination

After a period of dormancy and when environmental factors become favourable, a seed will germinate. **Germination** is the process in which a plant grows from a seed. It begins with very vigorous biochemical and developmental activity. It lasts until the first photosynthesising leaves are produced, by which time, all the food stored in the endosperm or cotyledons will be used.

Key Term

Germination: The biochemical and physiological processes through which a seed becomes a photosynthesising plant.

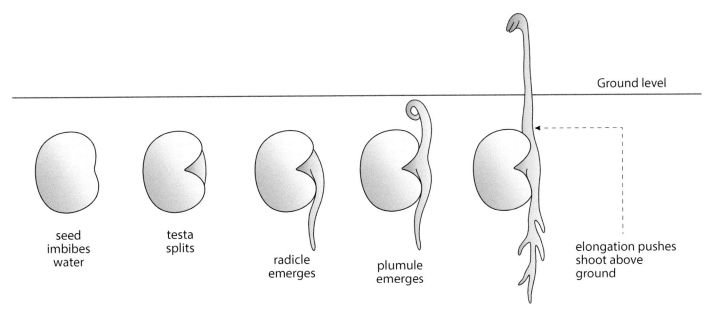

Ground level

seed imbibes water

testa splits

radicle emerges

plumule emerges

elongation pushes shoot above ground

Diagram of germination sequence in the broad bean

▼ **Study point**

Water, oxygen and a suitable temperature are needed for germination.

The three main requirements for successful germination are:

- A suitable temperature – the optimum temperature for germination is the optimum for the enzymes involved in the process. It varies from species to species, but is usually between 5°C and 30°C.
- Water – to mobilise enzymes for transport in the xylem and phloem and to vacuolate cells, making them turgid.
- Oxygen – aerobic respiration releases energy, which fuels metabolism and growth.

The effect of light varies between species. Daisy seeds need light to germinate but ivy seeds need darkness. Geranium seeds can germinate in the light or in the dark.

Mobilisation of food reserves and germination

When conditions are suitable, water is taken up, or 'imbibed' rapidly by the seed through the micropyle, the pore that was present in the ovary and that persisted in the seed. Water causes the tissues to swell and provides suitable conditions for enzyme activity.

Food reserves in seeds are insoluble in water and cannot, as such, be transported to the embryo. The reserves must be broken down into soluble molecules. Amylase hydrolyses starch into maltose and proteases hydrolyse proteins to amino acids. The soluble products are transported to the embryo and carried in the phloem to the apical meristems of the plumule and radicle, where rapid cell division occurs. Some of these sugars are converted to cellulose for cell wall synthesis. Aerobic respiration releases energy from sugars and amino acids are used to synthesise new proteins.

The swollen tissues rupture the testa and the radicle emerges from the seed. It is positively geotropic and negatively phototropic so it grows downwards. Then the plumule emerges. It is positively phototropic and negatively geotropic so it grows upwards.

Going further ▶

In some species, e.g. French beans, the cotyledons are pushed above ground and they photosynthesise until the first leaves emerge.

During germination the cotyledons of the broad bean remain below ground. The part of the plumule above the join between the embryo and the cotyledons elongates rapidly, pushing the plumule upwards. The plumule is bent over in the shape of a hook as it pushes its way up through the soil. This protects the tip from damage by soil abrasion.

If the seed has been planted at the correct depth in the soil, when the plumule emerges the hook straightens and the leaves unfurl and begin to photosynthesise. By now the food reserves in the cotyledons will have been depleted.

Germination sequence in *Vicia faba*

The graph shows that as a seed germinates, its dry mass has increased after about 10 days. The dry mass of the embryo as it develops into a seedling increases while the dry mass of the cotyledons, providing its food, decreases.

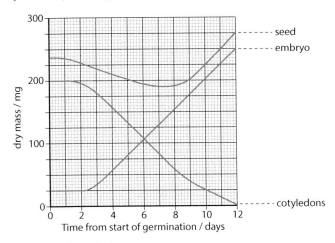

Changes in dry mass as a seed germinates

The effect of gibberellin

The brewing industry uses germinating barley seeds to make beer, so a lot of research has been done on barley germination. The terms 'malt' and 'malting' used in brewing refer to the maltose generated when the starch in barley is digested. It is likely that similar mechanisms occur in other species.

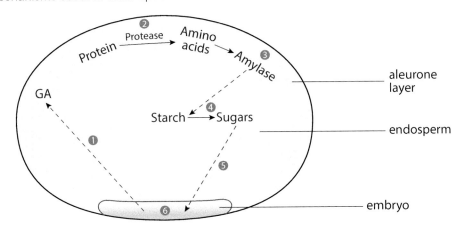

GA and germination

1. The barley embryo secretes a plant growth regulator, gibberellic acid, which diffuses through the endosperm to the aleurone layer. This is a layer of cells towards the outside of the seed, which has a high protein content.

2. The gibberellic acid switches on genes in the cells of the aleurone layer, resulting in transcription and translation, producing enzymes including protease and amylase.

3. The proteases hydrolyse protein in the aleurone layer to amino acids, which are used to make amylase.

4. The amylase diffuses out of the aleurone layer and hydrolyses the starch stored in the endosperm cells.

5. The maltose and glucose produced diffuse back through the endosperm to the plumule and radicle of the embryo.

6. They are respired for energy, which fuels biosynthesis and cell division and which brings the seed out of dormancy.

Specified practical exercises

To demonstrate the digestion of starch-agar using germinating seeds

Rationale

The food reserves in seeds are large, insoluble macromolecules. During germination enzymes are synthesised. The enzymes hydrolyse the macromolecules into small, soluble molecules that can move into the phloem of the embryo. They are translocated to the apical meristems where they are used as respiratory substrates or as metabolites in embryo development.

One of the enzymes that seeds synthesise is amylase, which digests the starch in the endosperm or cotyledon into maltose. The maltose moves into the embryo where it is hydrolysed into glucose, which is respired to generate energy. In this experiment, amylase digests the starch in the agar plate. Iodine solution fails to turn blue-black in areas that lack starch.

Design

	Name of variable
Independent variable	Live or dead seeds
Dependent variable	Width of halo around seed
Controlled variables	Starch concentration; temperature
Control	Boiled seeds
Reliability	3 seeds per plate
Hazard	Seeds are maintained at 20°C to reduce the risk of growing potentially harmful contaminants

Apparatus

- Broad bean seeds, soaked in water for two days at room temperature
- 10% bleach
- Sterile distilled water
- 250 cm³ beaker
- Starch-agar plates
- Sterile white tile
- Sterile fine scalpel
- Sterile fine forceps
- Iodine–potassium iodide solution
- Boiling tube
- Boiling water bath
- Timer

Method

1. Cover three soaked bean seeds with water in a boiling tube. Place the boiling tube in the boiling water bath for 10 minutes to kill the seeds. Place the boiled seeds on a white tile and allow them to cool to room temperature.

2. Surface sterilise three live beans in 10% bleach for 5 minutes, then rinse thoroughly in sterile distilled water.

3. Repeat the surface sterilisation with the boiled seeds.

4. Cut three live and the three boiled seeds in half, parallel with the long axis of the seeds, to separate the cotyledons.

5. Place the half seeds cut face down on the starch-agar.

6. Place the plates in an oven at 20°C for 48 hours.

7. Remove the seeds and cover the agar surface with 5 cm^3 iodine solution.

8. Pour the solution off after 2 minutes and observe the plates.

Observations

On the plate with boiled seeds, the agar stains uniformly blue–black. Boiling denatured the seeds' enzymes and no starch was digested.

In the plates with live seeds, some areas of the agar stain blue–black. But there is no staining immediately below the seeds, indicating that the starch there was digested. Surrounding each seed is a halo lacking stain, showing the extent to which amylase had diffused out of the seeds and digested the starch.

Further work

1. The width of the halo may be compared using different species of seed, e.g. broad bean, pea, runner bean.

2. The width of the halo may be measured at different times, e.g. every 6 hours after setting up.

3. The effect of different cereal seeds may be observed, e.g. barley, wheat and maize.

Dissection of wind- and insect-pollinated flowers

Rationale

Pollination is the transfer of pollen from the anther to the ripe stigma of a flower of the same species. In cross-pollinated species, this is usually done by insects or by the wind, and flowers have features to facilitate this pollination. Wind-pollinated flowers lack coloured petals and have exposed anthers and stigmas. Insect-pollinated flowers often have large, scented, coloured petals and anthers and stigmas contained within the ring of petals.

Apparatus

– Wind-pollinated flower, e.g. plantain (*Plantago* sp.)
– Insect-pollinated flowers, e.g. lily such as *Alstromeria*; herb robert (*Geranium robertianum*)
– White tile
– Fine forceps
– Fine scalpel
– Fine scissors
– Lens
– Microscope slides and cover slips
– Iodine–potassium iodide solution
– Dropping pipette

Method

Insect-pollinated flower

Herb robert (*Geranium robertianum*)

Alstromeria

1. Count the sepals and remove them from where they insert on the receptacle. If you are dissecting a monocot, such as a lily, the sepals may be coloured rather than green. They can be distinguished from petals because they insert into the receptacle further out than the petals.

2. Count the petals and remove them from where they insert on the receptacle.

3. Count the stamens and remove them from where they insert on the receptacle. Note the number of lobes on the anthers. If the anthers have dehisced, pollen grains may be exposed on the surface.

4. This leaves the carpel attached to the receptacle. Holding the carpel horizontally, slice it longitudinally, from the stigma down through the style and ovary into the receptacle and note the ovules in the ovary.

5. Microscopy

    (a) If the anthers have dehisced, brush a little pollen on to a slide and mount in two drops of iodine–potassium iodide solution. If the anthers have not dehisced, cut a section which is as thin as possible and crush it on the microscope slide, in two drops of iodine–potassium iodide solution. Add a cover slip. Find the pollen grains using the ×10 objective and then focus on well-separated grains with the ×40 objective. Focus up and down on an individual pollen grain to distinguish the surface markings.

    (b) Ovules should be visible in the ovary as small white spheres. Remove a small number of these and place them on a microscope slide in a few drops of iodine–potassium iodide solution. Macerate them so that the cover slip will be able to lie flat. Place the cover slip on the fragments and view under a ×4 and then a ×10 objective. Focus through the specimen so that you can distinguish individual cells.

Wind-pollinated flower

The inflorescence of plantain is a spike. This means that the flower head has many individual flowers arranged along the stem, with each flower attached directly to the stem, and not on the receptacle. Flowers ripen from the bottom to the top, with stamens then anthers being mature.

1. Remove a flower with an exposed stigma.
2. Using a lens and dissecting equipment, identify and separate the stigma, style and ovary.
3. Remove a flower with a mature stamen.
4. Using a lens and dissecting equipment, identify and separate the filaments and anthers.

flowers with exposed stigmas

flowers with mature stamens

pollinated flowers

Spike of the ribwort plantain, (*Plantago lanceolata*)

Scientific drawing of anther and calculations of size and magnification

To examine a TS anther, look at it on the slide with the naked eye to gauge its size and shape. Place the slide on the microscope stage, with the vascular strand at the top of the image. Focus using a ×4 objective and then a ×10 objective. Use a ×40 objective to examine the individual cells.

You should be able to distinguish the epidermis, the vascular strand, the parenchyma surrounding it, the pollen sac containing pollen cells undergoing meiosis, tetrads or mature pollen, the tapetum, sometimes called the inner wall, and the fibrous layer surrounding it, which is sometimes called the outer wall. In some sections, you may see the filament.

1. Draw a plan of an entire TS anther or, if it is too large to fit into one field of view, a representative portion. If only part of the anther is drawn, make a small sketch of the whole anther and indicate which part is shown in your tissue plan.
2. Align the eyepiece graticule with a clearly identifiable part of the specimen, e.g. the maximum diameter of a pollen sac, and count the number of eyepiece units that distance represents. Let us say that it measures 4 eyepiece units.
3. Having calibrated the microscope, the following calculation of the diameter may be made:

    With ×10 objective, maximum diameter of pollen sac = 4 eyepiece units

    From the calibration with a ×10 eyepiece, 1 eyepiece unit = 10 µm

    maximum diameter of pollen sac = (10 × 4) = 40 µm
4. The distance measured should be marked on the diagram with a straight line, with a small bar at each end, placed very precisely on the pollen sac.
5. The length marked on the diagram is measured in mm.
    Let us say it measures 168 mm.
6. Calculate the magnification of the diagram:

    Distance on drawing = 168 mm = (168 × 1000) µm = 168000 µm

    Actual distance = 40 µm

    $$\text{Magnification} = \frac{\text{image size}}{\text{actual size}} = \frac{168000}{40} = 4200$$

Inheritance

Variation between the individuals of a population is essential if a species is to survive in a constantly changing environment. Meiosis, as a prelude to sexual reproduction, is a major source of genetic variation. The way in which characteristics are passed from one generation to the next was demonstrated by the work of Gregor Mendel (1822–84). He was the first person to work out the ways in which characteristics are inherited. This was an extraordinary intellectual achievement because at that time, no one knew of DNA, genes or chromosomes. More recent work has shown exceptions to the rules Mendel described, including co-dominance, incomplete dominance, linkage, sex linkage, mutation and epigenetic modification.

Topic contents

By the end of this topic you will be able to:

- Define genetic terms.
- Predict the outcome of monohybrid crosses, including the test cross and incomplete and co-dominance.
- Predict the outcome of dihybrid crosses, including the test cross.
- Understand the concept of gene linkage and its consequences for genetic ratios.
- Carry out a chi^2 test.
- Explain mechanisms of sex determination.
- Describe the origins of mutations.
- Describe sex linkage using haemophilia as an example.
- Describe gene mutations using sickle cell anaemia as an example.
- Describe chromosome mutations using Down's syndrome as an example.
- Describe the relationship between carcinogens and genes.
- Explain the control of gene expression in epigenetics.

Genetic terms

A **gene** is the physical unit of heredity. It is a sequence of DNA that occupies a specific site, or **locus**, on a chromosome, and codes for a polypeptide. Polypeptides and proteins determine structures and functions in living organisms, and so the gene is also thought of as the sequence of DNA that codes for a characteristic.

There are several genes that control coat colour in dogs. In one, the nucleotide sequence codes for an enzyme that makes the fur black, but a small difference in the sequence gives a slightly altered pigment molecule that makes the fur brown. Small changes in nucleotide sequences of the same gene make small changes in the polypeptides they encode and produce different characteristics. The different sequences are called **alleles**. A single gene may have one, two or many alleles. Alleles always occupy the same locus because they are versions of the same gene.

One-year old female labradors

Organism	Gene	Alleles
Human	Number of fingers	5; 6
	Freckles	Present; absent
	Rhesus blood group	Positive; negative
Sweet pea	Height	Tall; dwarf
	Seed colour	Yellow; green
	Flower position	Terminal; axial

A diploid individual has one of each chromosome from each parent and, therefore, one copy of each gene from each parent. If the alleles of a particular gene are the same from both parents, the individual is **homozygous** for that gene. If they are different, the individual is **heterozygous** for that gene.

The **genotype** of an individual is all the alleles that they contain. Their **phenotype** can be described as their appearance, but it is much more than that, because it includes characteristics that cannot be seen, such as blood group. Both the genotype and the environment determine the characteristics of an individual so the phenotype is better thought of as the expression of the genotype in a specific environment.

There is a gene that codes for stem height in sweet pea and some other plants. An allele of this gene, given the symbol T, codes for the production of gibberellin, a growth regulator that elongates the stem, so plants with the T allele are tall. An alternative allele, given the symbol t, does not code for the production of gibberellin so the plants that do not have a T allele are dwarf. Each plant has two alleles, which gives three possible combinations:

Allele combination	Effect	Phenotype	Description of genotype
tt	No gibberellin produced	Dwarf	Homozygous recessive
TT	Gibberellin produced	Tall	Homozygous dominant
Tt	Gibberellin produced	Tall	Heterozygous

The T allele is always expressed when it is present, so it is described as **dominant**. The t allele is not expressed in the presence of the T allele, and is only expressed when it is homozygous. It is described as **recessive**.

YOU SHOULD KNOW ›››

››› The difference between genes and alleles

››› How to set out a genetic cross and predict ratios of offspring

››› Explain the derivation of the ratios 3:1, 1:1, 9:3:3:1 and 1:1:1:1

››› How to use the chi^2 test to compare observed and expected results of a genetic cross

▼ **Study point**

A gene controls a characteristic and alleles are different versions of the gene in the same way that a car ('gene') could be a BMW or a Porsche ('alleles').

Key Terms

Gene: A sequence of DNA that codes for a polypeptide and which occupies a specific locus on a chromosome.

Allele: A variant nucleotide sequence for a particular gene at a given locus, which codes for an altered phenotype.

Exam tip

Learn the definitions of genetic terms.

▼ **Study point**

For a heterozygous gene, the dominant allele is always written before the recessive allele e.g. Tt.

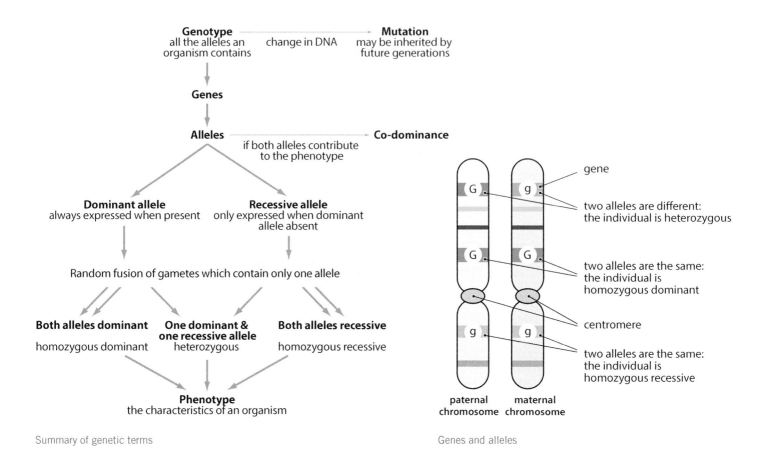

Summary of genetic terms

Genes and alleles

Monohybrid inheritance

Monohybrid inheritance is the inheritance of a single gene, such as that controlling plant height or seed colour.

Gregor Mendel's classic experiments investigated inheritance in plants of the garden pea, *Pisum sativum*. Because of their agricultural importance there were several varieties available that were true breeding, although nobody understood why. Peas were a useful choice for experiments because:

- They are easy to grow.
- Their flowers can self-fertilise and cross-fertilise.
- They make flowers and fruit in the same year.
- They make a large number of seeds from each cross. This means that when the phenotypes of the next generation are counted, their numbers make them statistically meaningful.

To study how characteristics were inherited, Mendel chose pairs of contrasting characteristics, such as tall or dwarf plants, plants with round or wrinkled seeds and plants with yellow or green seeds. He was fortunate, or perhaps skilled, in his choice because these characteristics are:

- Controlled by single genes
- Controlled by genes on different chromosomes
- Clear-cut and easy to tell apart.

These characteristics are examples of discontinuous variation. However, most characters show continuous variation. They have a range of values and are controlled by a number of genes, for instance height, in humans.

Going further ▶

Peas have been used in crop rotation in Europe since the 8th century, to increase nitrate levels in soil. By the time of Mendel, in the 19th century, they were very important in the 4-year crop rotation.

◀Link▶ Continuous and discontinuous variation are described on p218.

Genetic diagrams

A diagram of a genetic cross shows:

- The generations, i.e. parents, first generation (F_1), second generation (F_2) and, sometimes, further generations.
- The genotypes of parents and offspring.
- The phenotypes of parents and offspring.
- The alleles present in the gametes.
- The symbols for the alleles are defined.

F_1 stands for the first filial generation. It is the offspring of the parents of the cross. F_2 stands for the second filial generation. It is the offspring of a self-fertilised F_1 plant or of a cross between members of the F_1 generation. It is also the grandchildren of the original parents. Filial is a word that means 'relating to a son or daughter'.

Instructions for writing a diagram for a genetic cross

1. Choose suitable symbols for the alleles.
   - Use a single letter for each characteristic, for example, the first letter of one of the contrasting features.
   - If possible, use a letter in which the upper and lower cases differ in shape and size.
   - Use the upper case for the dominant characteristic and the lower case letter the recessive.
2. Write the parents' genotypes with the appropriate pairs of letters. Label 'Genotype of parents' and state their phenotypes.
3. Show the gametes produced by each parent. Circle and label them 'Gametes'.
4. Use a matrix, called a Punnett square, to show the results of the possible combinations that result from random crossing of all the gametes. Label 'Genotype of F_1'.
5. Show the phenotype of each F_1 genotype in the Punnett square.
6. Indicate the ratio of the phenotypes.

This genetic diagram shows the inheritance the of the gene for plant height:

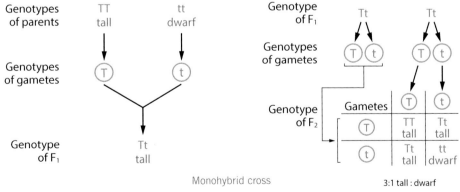

Monohybrid cross   3:1 tall : dwarf

- The gene has two alleles: T represents the allele for tall and t represents the alleles for dwarf.
- As the parents are homozygous, each only makes one type of gamete, so only one is shown.
- In the F_1, all offspring are heterozygous and their phenotype is tall.
- F_1 individuals each make two types of gamete.
- In the F_2, ¾ of the offspring are tall and ¼ are dwarf.

Test cross or back cross

If an organism has a dominant characteristic, it could be homozygous dominant or it could be heterozygous. The **test cross** shows if a dominant characteristic is determined by one or two dominant alleles.

A tall pea plant, for example, could be pure-breeding, TT, but it could also be Tt. It is not possible to tell from the appearance. To test its genotype, the tall plant is crossed with a dwarf plant. The dwarf phenotype has only one possible genotype: it is homozygous recessive, tt. If the tall plant were TT, all the F₁ would be tall, but if it were Tt, 50% would be tall and 50% would be dwarf:

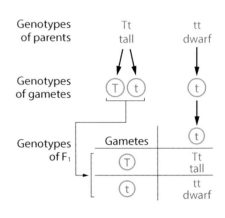

1:1 tall : dwarf

Monohybrid test cross

When neither allele is dominant

1. Co-dominance

When genes are co-dominant, both alleles in a heterozygote are expressed individually. As a result, the heterozygote has a combination of both of the homozygotes' characteristics, e.g. in the human ABO blood group system, the I gene has three alleles, I^A, I^B and I^O. Homozygous parents who have the genotype $I^A I^A$ have the A antigen on their red blood cells and are blood group A. Homozygous parents who have the genotype $I^B I^B$ have the B antigen on their red blood cells and are blood group B. Their offspring have the genotype $I^A I^B$. Both alleles are expressed: they have both A and B antigens on their red blood cells so they are blood group AB. This is an example of **co-dominance**.

Speckled hen showing co-dominance

Similarly, chickens with black feathers and white feathers can produce speckled offspring, as neither the allele for black feathers nor the allele for white feathers is dominant.

2. Incomplete dominance

For some genes, the phenotype of the heterozygote is intermediate between the two parental phenotypes, rather than their both being expressed, e.g. red flowered carnations crossed with white-flowered carnations have an F₁ with pink flowers. This is described as incomplete dominance. Neither the allele for red petals nor white petals is completely dominant and so the allele symbols are not given an upper or lower case letter. Instead, they have the symbols R or C^R for red and W or C^W for white.

Pink carnation showing partial dominance

In a cross between carnations with red and white flowers, all flowers of the F₁ generation are pink. They produce two different types of gamete so when they interbreed, the ratio of colours in the F₂ generation is 1 red: 2 pink: 1 white:

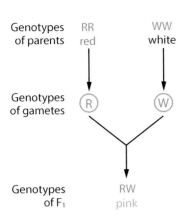

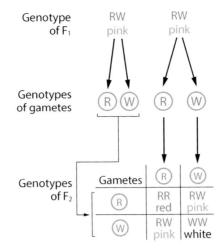

1 red : 2 pink : 1 white

Incomplete dominance

As a result of carrying out his experiments, especially those that produced the 1:2:1 ratio, Mendel formulated his first 'law of inheritance', the law of segregation, which states: The characteristics of an organism are determined by factors (alleles) which occur in pairs. Only one of a pair is present in each gamete.

Independent assortment

The diagrams show a cell at metaphase I of meiosis, with the maternal chromosomes coloured black and the paternal chromosomes coloured red. In relation to genes A/a and B/b, the mother's genotype is AB and the father's is ab. In diagram 1, the homologous chromosomes are arranged so that alleles A and B go to one pole and a and b to the other, so the gametes produced are written as (**AB**) and (**ab**).

Link Understanding genetics depends on understanding how chromosomes behave during meiosis. You learned the principles of meiosis in the first year of this course.

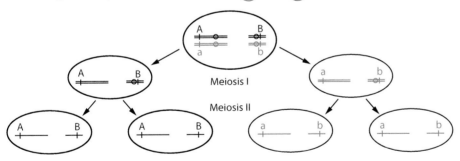

Independent assortment (1)

In diagram 2, the orientation of the shorter pair of chromosomes is reversed, so that the alleles A and b go to one pole and a and B to the other. The gametes produced are written as (**Ab**) and (**aB**).

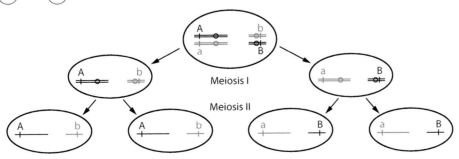

Independent assortment (2)

In this way, the mechanism of meiosis explains the observation that either of a pair of alleles can combine with either of another pair of alleles. The maternal and paternal chromosomes, and therefore the alleles they contain, mix with each other in any combination in the gametes. This is **independent assortment** of chromosomes. The genes A and B behave independently in relation to each other and so they are described as **unlinked**. A is equally likely to combine in a gamete with B or b. It accounts for the genetic ratios observed when the simultaneous inheritance of two genes is investigated in dihybrid inheritance.

Dihybrid inheritance

Dihybrid inheritance is the simultaneous inheritance of two unlinked genes, i.e. genes on different chromosomes. Mendel knew from his early experiments with monohybrid crosses that, for pea seeds, yellow colour was dominant to green and that round seed shape was dominant to wrinkled. He crossed plants with two pairs of contrasting characters:

- The allele for round seeds is R and the allele for wrinkled seeds is r.
- The allele for yellow seeds is Y and the allele for green seeds is y.

Plants with the genotype RRYY are homozygous for the two dominant characters, round and yellow seeds. They were crossed with plants with the genotype rryy, homozygous for the two recessive characters, wrinkled and green seeds. Each parent produced only one type of gamete, (RY) or (ry). All the F_1 plants had the genotype RrYy, and had the dominant characteristics, i.e. round, yellow seeds.

Each gamete of the F_1 plants contained an allele for shape and an allele for colour. In making the gametes, the R allele would combine with either the Y or y allele with equal probability, and therefore, in equal proportions. The Y allele would combine with either the R or r allele in equal proportions. Each parent, therefore, produced four types of gamete types in equal proportions: (RY), (Ry), (rY), and (ry).

When the F_1 plants self-fertilised, each gamete type in the pollen would combine with all four gamete types in the oospheres, with equal probability, and therefore, in equal proportions. When the F_1 were self-fertilised, they produced seeds that were the F_2 generation. The F_2 seeds were of four different combinations of shape and colour, as shown in the Punnett square:

> ▼ **Study point**
>
> In a dihybrid cross, both genes must be represented in the gamete, so they are written with two symbols, i.e. one allele for each gene.

> ▼ **Study point**
>
> In a dihybrid cross, two individuals heterozygous at both genes produce offspring with a phenotypic ratio of 9:3:3:1.

R represents round seed (dominant) **r** represents wrinkled (recessive)
Y represents yellow seed (dominant) **y** represents green seed (recessive)

parental phenotypes	pure-breeding round yellow	pure-breeding wrinkled green
parental genotypes (2n)	RRYY	rryy
gametes (n)	all (RY)	all (ry)
F_1 genotype (2n)	all **RrYy**	
F_2 parental genotype	RrYy	RrYy
gametes (n)	(RY)(Ry)(rY)(ry)	(RY)(Ry)(rY)(ry)

♀ \ ♂	(RY)	(Ry)	(rY)	(ry)
(RY)	**RRYY** round yellow	**RRYy** round yellow	**RrYY** round yellow	**RrYy** round yellow
(Ry)	**RRYy** round yellow	**RRyy** round green	**RrYy** round yellow	**Rryy** round green
(rY)	**RrYY** round yellow	**RrYy** round yellow	**rrYY** wrinkled yellow	**rrYy** wrinkled yellow
(ry)	**RrYy** round yellow	**Rryy** round green	**rrYy** wrinkled yellow	**rryy** wrinkled green

F_2 genotypes (2n) and phenotypes

9 round yellow : **3** wrinkled yellow : **3** round green : **1** wrinkled green

Dihybrid inheritance

The Punnett square shows that if the colour gene alone were considered, the ratio 12 yellow : 4 green, i.e. 3:1 would be obtained. Similarly, if the texture gene were considered, the ratio 12 smooth : 4 wrinkled, i.e. 3:1 would be obtained. In each case, the monohybrid ratio is obtained, because only one gene is being considered at a time.

To calculate the ratio of progeny, the total number is divided by the number of homozygous recessive individuals. In a cross, it is most unlikely that the ratio of progeny will be exactly as predicted. But if the ratio approximates to 9:3:3:1, it can be assumed that the progeny are derived from a dihybrid cross.

Phenotype	Round yellow	Round green	Wrinkled yellow	Wrinkled green
Total	315	108	101	32
Ratio	315/32 = 9.84	108/32 = 3.38	101/32 = 3.16	32/32 = 1

The formulation of the dihybrid ratio led to the 'second law of inheritance', also attributed to Mendel. It states: Either one of a pair of contrasted characters may combine with either of another pair. With our current understanding of genetics this statement can be rewritten as: Each member of a pair of alleles may combine randomly with either of another pair on a different chromosome.

The dihybrid test cross

A pea plant with round, yellow seeds could have any of the following genotypes: RRYY; RrYY; RRYy or RrYy. The monohybrid test cross tests a genotype by crossing it with a homozygous recessive, e.g. tt. The dihybrid test cross tests a genotype by crossing it with an individual that is homozygous recessive for both genes, i.e. rryy. The ratios of phenotypes in the progeny indicate the genotype of the parent. In each cross, the parental phenotypes are round, yellow and wrinkled, green.

Parents	RRYY rryy	(Rr)(YY) (rr)(yy)	(RR)(Yy) (rr)(yy)	(Rr) (Yy) (rr)(yy)
Gametes	(RY)(ry)	(RY)(rY) (ry)	(RY)(Ry) (ry)	(RY)(Ry)(rY)(ry) (ry)
F₁	RrYy	<table><tr><td></td><td>(ry)</td></tr><tr><td>(RY)</td><td>RrYy</td></tr><tr><td>(rY)</td><td>rrYy</td></tr></table>	<table><tr><td></td><td>(ry)</td></tr><tr><td>(RY)</td><td>RrYy</td></tr><tr><td>(Ry)</td><td>Rryy</td></tr></table>	<table><tr><td></td><td>(ry)</td></tr><tr><td>(RY)</td><td>RrYy</td></tr><tr><td>(Ry)</td><td>Rryy</td></tr><tr><td>(rY)</td><td>rrYy</td></tr><tr><td>(ry)</td><td>rryy</td></tr></table>
Ratio	all round, yellow	1:1 round yellow: wrinkled yellow	1:1 round yellow: round green	1:1:1:1 round yellow: wrinkled yellow : round green : wrinkled green

▼ **Study point**

In a genetic cross between two heterozygous parents with 96 offspring, the probability of the number of double recessive offspring, wrinkled and green

$$= \frac{96}{16} = 6.$$

Exam tip

Make sure you can quote Mendel's laws of inheritance and that you can apply them to genetic crosses.

Exam tip

In an exam question, the layout for your answer may be provided. Use the letters given to you for the allele symbols.

Key Term

Linked: Description of genes that are on the same chromosome and therefore do not segregate independently at meiosis.

▼ Study point

Linked genes are inherited together because they move together during meiosis and appear in the same gamete.

‹Link› You learned about crossing over in the first year of this course.

Going further ▶

Genes near the ends of chromosomes have so many cross-overs between them that they make almost equal proportions of different gamete types. Analysing the numbers of offspring phenotypes does not indicate linkage.

Going further ▶

A crossover value of 4.6% tells us that the genes are 4.6 map units apart on the chromosome. This is a relative value and is not related to an S.I. unit.

Linkage

The alleles of two genes that are on the same chromosome cannot segregate independently, i.e. they cannot move to opposite poles of the cell at meiosis. This is because they are on the same physical structure – the chromosome – so they must move together. They are described as **linked**. The cell in the diagram makes gametes (DE) and (de):

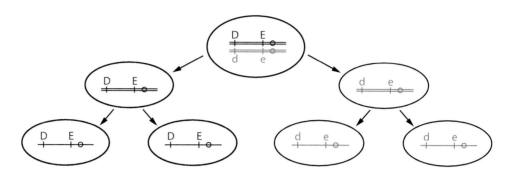

Inheritance of linked genes

In some cells undergoing meiosis, genetic crossing over would occur between genes D/d and E /e, producing the 4 gamete types (DE), (De), (dE) and (de). (DE) and (de) are the 'parental' genotypes and (De) and (dE) are 'recombinant' genotypes.

The combination of characteristics is described as parental when traits associated with D and E or with d and e occur together. The phenotypes are described as recombinant when traits associated with D and e or with d and E occur together.

Crossing over is a rare event and does not happen in most cells, so the majority of gametes would still be parental, (DE) and (de). Thus the number of gametes with the different genotypes is not equal, so Mendelian ratios are not produced in the next generation and the majority of phenotypes in the offspring would have parental phenotypes. The further apart two genes are on a chromosome, the more opportunity there is for a crossover to occur between them. This leads to more recombinant gametes and therefore more offspring with recombinant phenotypes.

Recognising linkage

If the numbers of progeny with different combinations of characteristics do not correspond with Mendelian ratios, it is likely that genes are linked. This is a common explanation for rejecting a null hypothesis in a genetics experiment.

In an experiment with *Drosophila*, the fruit fly, crossing flies that have red eyes and yellow bodies with flies that have brown eyes and dark bodies produced an F_2 generation with:

Red eyes, yellow body (parental) = 126

Brown eyes, dark body (parental) = 39

Red eyes, dark body (recombinant) = 5

Brown eyes, yellow body (recombinant) = 3

There are four phenotype classes, as expected but not in a 9:3:3:1 ratio, suggesting that the genes for eye colour and body colour in *Drosophila* are linked.

The 'crossover value', COV, can be calculated as $\dfrac{\text{number of recombinants}}{\text{number of progeny}} \times 100\%$.

For these data, $\text{COV} = \dfrac{5+3}{126+39+5+3} \times 100\% = 4.6\%$.

Probability

The presentation of a genetic cross is a prediction of the likely outcome. The actual results, though, are unlikely to agree precisely with the prediction. Imagine tossing a coin 100 times. It would be expected to land heads 50 times and tails 50 times. In practice, it would be surprising to obtain this result in the first 100 throws. If the coin landed 60 heads (H) and 40 tails (T), this could be due to a chance deviation from the prediction, or it could be because the coin is biased.

If two unbiased coins are tossed, there are four possible combinations: HH or HT or TH or TT.

Rule of addition: When a coin is tossed, it must give either H or T. The probability of getting either one result or the other is the sum of their independent probabilities. The probability of getting H is ½ and the probability of getting T is also ½. So the probability of getting one or the other is ½ + ½ = 1.

Rule of multiplication: Now consider both coins. The probability of throwing H on both is found by multiplying the two independent probabilities. In this case the probability of getting H on the first coin is ½, and the probability of getting H on the second coin is also ½. So the probability of HH is ½ × ½ = ¼.

The result of the Punnett square for the dihybrid cross shows a 3:1 ratio for each gene in the F_2 phenotypes. This can be used to calculate the probability of the cross when all four alleles, round, green, yellow and wrinkled are involved.

- The probability of the four alleles appearing in any of the F_2 offspring is:

  round (dominant) ¾

  wrinkled (recessive) ¼

  yellow (dominant) ¾

  green (recessive) ¼

- The probability of combinations of alleles appearing in the F_2 is as follows:

  round and yellow = ¾ × ¾ = 9/16

  round and green = ¾ × ¼ = 3/16

  wrinkled and yellow = ¼ × ¾ = 3/16

  wrinkled and green = ¼ × ¼ = 1/16

The chi² (χ^2) test

The expected ratio of phenotypes in the offspring of a monohybrid cross is 3:1 and of a dihybrid cross is 9:3:3:1. These ratios represent the probabilities of getting each of the phenotypes. It would be surprising if the results of a genetic cross gave exactly the expected ratio. The chi² test is used to test if the numbers of the different phenotypes are close enough to the predicted values to support the genetic explanation of how they arose. If the numbers are not close enough, this test tells us that they have arisen for another reason.

The null hypothesis

A statistical test is designed to test a null hypothesis, which, in genetics, is a statement that there is no difference between the observed and expected results of a cross. In genetics, a null hypothesis states that the observed results are due to Mendelian inheritance and that any deviation from the expected ratio is due to chance.

The reason that this is a null hypothesis is that Mendelian inheritance is based on there being no difference in:

- The number of the different types of gamete
- The probability of each gamete type fusing with another type of gamete
- The viability of the embryos, whatever their genotype
- And that genes are not linked.

Here are the numbers of each phenotype in the F_2 of a cross between a pea plants with round, yellow seeds and one with wrinkled, green seeds, as described on p200.

Characteristic	Round, yellow	Round, green	Wrinkled, yellow	Wrinkled, green
Totals	315	108	101	32

Carrying out a χ^2 test

- Calculate the expected values (E) based on an expected ratio of 9:3:3:1:
  - The total number of seeds = (315 + 108 + 101 + 32) = 556.
  - The class that is homozygous recessive for both genes, i.e. wrinkled green represents $\frac{1}{16}$ of the total = $\frac{556}{16}$ = 35 (0 dp).
  - The classes round, green and wrinkled, yellow are 3 times this value
  - = $(3 \times \frac{556}{16})$ = 104 (0 dp).
  - The round, yellow class is 9 times the value for the wrinkled, green
  - = $(9 \times \frac{556}{16})$ = 313 (0 dp).
- For each phenotype, calculate (O – E), i.e. the differences between the observed (O) and expected (E) results.
- For each phenotype, square the differences to calculate $(O – E)^2$.
- For each phenotype, calculate $\frac{(O – E)^2}{E}$ by dividing each value of $(O – E)^2$ by E, the expected value.
- Calculate $\Sigma \frac{(O – E)^2}{E}$ by adding all the values of $\frac{(O – E)^2}{E}$.

Phenotype	Observed (O)	Expected (E)	Difference (O – E)	$(O – E)^2$	$\frac{(O – E)^2}{E}$
Round, yellow	315	313	2	4	0.01
Round, green	108	104	4	16	0.15
Wrinkled, yellow	101	104	–3	9	0.09
Wrinkled, green	32	35	–3	9	0.26
				$\Sigma \frac{(O – E)^2}{E} = \chi^2$	0.51

▼ **Study point**

The figures have been rounded to whole numbers, for ease of calculation.

Degrees of freedom (df)

This is a measure of the number of values that can vary independently. In the analysis of monohybrid and dihybrid crosses, it is one less than the number of classes of data. For the dihybrid cross, df = 4 – 1 = 3.

Probability

Biologists make predictions based on their model of how a phenomenon works. Statisticians say that if the predicted results happen 5% or more of times the experiment is run, the biological reasoning is correct and any deviation from the exact prediction is not significant and is due to chance.

If the predicted results happen less frequently than 5% of times, the deviation from predicted results is significant; the assumptions on which the prediction was made are not correct and there must be a different explanation for the phenomenon.

The calculated value of χ^2 can be related to a probability that the observed and expected values are close enough that the prediction must have been correct. If that probability is greater than 5% (= 0.05), the null hypothesis can be accepted at the 5% level of significance. Any deviation from the predicted values is not significant and is due to chance. If it is less than 5%, the deviation from the predicted values is significant. The null hypothesis is rejected at the 5% level of significance and there must be another explanation for the results.

WORKING SCIENTIFICALLY

When drugs are tested for safety, a much lower level of significance is used, e.g. 0.0001%, implying that if 1 in 10^4 people suffers adverse reaction, the drug is not suitable for use.

Testing for significance

Degrees of freedom	Chi²			
	Probability that deviation is due to chance			
	0.99	0.50	0.10	0.05
	99%	50%	10%	5%
1	0.00	0.45	2.71	3.84
2	0.02	1.39	4.61	5.99
3	0.12	2.37	6.25	7.82
4	0.30	3.36	7.78	9.49

For three degrees of freedom, the calculated χ^2 value of 0.51 lies between 2.37 and 0.12. This is equivalent to a probability between 0.50 (50%) and 0.99 (99%). The probability value is greater than 5% and so we can accept the null hypothesis at the 5% level of significance and assume that inheritance is Mendelian and that any deviation from the 9:3:3:1 ratio is due to chance.

Constructing a conclusion

A complete statement of the conclusion must include certain points. In the example given here:

- The calculated value of χ^2 is less than the critical value of 7.82.
- This is equivalent to a probability of greater than 5%
- so the null hypothesis is accepted
- at the 5% level of significance.
- Inheritance is Mendelian and
- any deviation from the predicted ratio is due to chance.

If the calculated value of χ^2 is greater than the critical value, it is equivalent to a probability of less than 5%. It must be stated that the null hypothesis is rejected at the 5% level of significance. There must be some other explanation for the data.

▼ **Study point**

A null hypothesis is accepted or rejected at the 5% level of significance. If accepted, any deviation from the predicted ratio is due to chance. Always state the level of significance.

Sex determination

Most Angiosperm species are hermaphrodite and their flowers make both pollen and ovules, but there are two other main strategies:

- Monoecious plants have separate male and female flowers on the same plant, e.g. maize.
- Dioecious plants have separate male and female individuals, e.g. holly.

Among animals, there are hermaphrodites in the phylum Mollusca, e.g. the garden snail, and the phylum Annelida, e.g. the earthworm. Hermaphrodite vertebrates, however, are rare and most species have separate male and female individuals.

YOU SHOULD KNOW ›››

››› That, in humans, the sex of an individual is determined by X and Y chromosomes

››› Some alleles are carried on the sex chromosomes and are described as sex-linked

Crocodile hatching from its egg

Clownfish

Whether an individual animal is male or female can be controlled by many different factors, for example:

- Temperature: Lizard, crocodile and alligator eggs hatch as male when the temperature is above 32°C and female when it is below 32°C. Sea turtle eggs have the opposite response and hatch as females if laid in the full sun, but as males if laid in the shade.
- Sequential hermaphroditism: The common slipper limpet, a mollusc, makes stacks of individuals. Those at the top are male. As more males join the top of the stack, those below them become females.
- The male sewage sludge worm, *Capitella*, can become hermaphrodite and fertilise itself if females are not available.
- Clownfish live in hierarchies. When the dominant female dies, the dominant male changes sex and takes her place.
- Ploidy level: Bee, aphid and grasshopper eggs that are not fertilised are haploid and develop as males. Those that are fertilised are diploid and develop as females.
- Chromosome structure: In mammals, females have two X chromosomes. Males have an X and a Y chromosome. The reverse situation occurs in birds, moths and some fish, where the male sex chromosomes are ZZ but the females are ZW.

Human sex determination

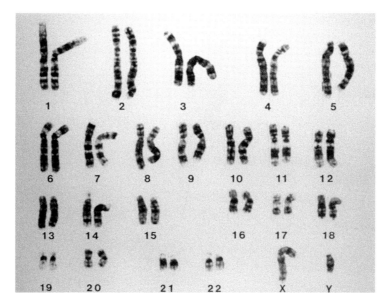

Karyotype of a normal human male – these chromosomes have been stained to show their banding patterns

Human cells contain 46 chromosomes, 23 from each parent. The chromosomes from the two parents can be arranged in homologous pairs, with each pair containing chromosomes of the same size and shape, with genes in the same order, coding for the same characteristics. Each homologous pair has one chromosome from each parent. The arrangement of homologous pairs in decreasing size order is called a karyotype.

Of the 23 pairs in the human karyotype, 22 pairs have identical genes, although they may have different alleles. These chromosomes are the **autosomes**. The 23rd pair constitutes the sex chromosomes. In females, they comprise two X chromosomes, which are identical. Males have an X chromosome and a Y chromosome, which is much shorter. These sex chromosomes are different sizes so they are called **heterosomes**.

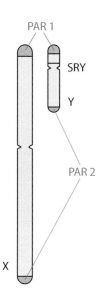

Regions of homology at the ends of the X and Y chromosomes

Two regions on the human X and Y chromosomes are homologous and can pair with each other at meiosis. They are called the 'pseudoautosomal regions', PAR1 and PAR2. They reflect the evolutionary loss of genes from one sex chromosome, leaving the larger X and the smaller Y chromosome that exist now.

Inheritance of sex

- All the female's secondary oocytes contain an X chromosome, so the female is the 'homogametic sex', i.e. the gametes are identical with respect to the sex chromosomes.

- In males, at meiosis I, an X chromosome passes into one secondary spermatocyte and a Y chromosome passes into the other. Consequently, half the male's sperm contain an X chromosome and the other half contain a Y chromosome. The male is the 'heterogametic sex', i.e. gametes are of different types with respect to the sex chromosomes.

- At fertilisation the oocyte may be fertilised by either an X-carrying sperm or a Y-carrying sperm, with equal probability. This gives an equal chance of the foetus being male or female.

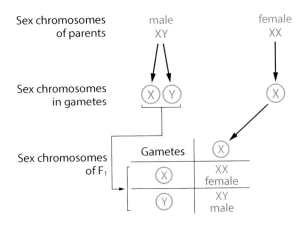

Sex determination in mammals

Maleness and the Y chromosome

The X and Y chromosomes probably have a common ancestor in our distant evolutionary past. But the Y chromosome, by now, has lost so many of its genes that it shares few regions of DNA with the X. One gene that the Y possesses that the X does not is the SRY gene. This is the 'sex-determining region on the Y chromosome' and its role appears to be to switch on the genes on other chromosomes, i.e. genes on the autosomes, which are responsible for the expression of male characteristics.

The commonest chromosomal abnormalities are those involving the number of sex chromosomes. An individual can have any number of X chromosomes and still be female but however many X chromosomes they have, all it takes is one Y chromosome, with its SRY gene, to make a male.

Sex linkage

Most of the lengths of the X and Y chromosomes are not homologous. In those regions, females have two copies of each gene as they have two X chromosomes, and males have only one copy, as they have only one X chromosome. If a female is heterozygous for one of those genes, the dominant allele will be expressed, as normal. Whichever allele the male carries is expressed, because even if he has the recessive allele, there is no second allele to be dominant over it.

There is a gene on the X chromosome that is associated with the blood disorder, haemophilia. Haemophilia is a potentially lethal condition occurring when an individual cannot produce enough of one of the 13 blood clotting proteins. Blood clots slowly, if at all, causing slow, persistent bleeding. This X-linked gene codes for the blood clotting protein, Factor VIII. The allele coding for the normal version has the symbol X^H and the allele coding for a mutant version has the symbol X^h. There are three possible genotypes for females:

- $X^H X^H$: blood clots normally.
- $X^H X^h$: a **carrier** female. She is heterozygous and carries a dominant allele so her phenotype is normal. She also carries a mutant allele, so she is described as a carrier.
- $X^h X^h$: a female with haemophilia.

There are two possible genotypes for males:

- $X^H Y$: blood clots normally.
- $X^h Y$: a male with haemophilia.

As a male only needs one mutant allele to have haemophilia, the condition is far more common in males than in females. The condition is therefore described as **sex-linked**. A gene on the X chromosome that gives rise to a sex-linked condition is a sex-linked gene.

Inheritance of sex-linked conditions

A male cannot pass the alleles on his X chromosomes to his sons, as they must receive his Y chromosome. His daughters, however, all receive an X chromosome from their father.

Duchenne muscular dystrophy (DMD) is caused by an X-linked recessive allele of the dystrophin gene. The gene codes for the protein dystrophin, which is a component of a glycoprotein that stabilises the cell membranes of muscle fibres. Symptoms of DMD begin at about the age of 2–3 years, and include loss of muscle mass and progressive muscle weakness.

The allele for the normal protein has the symbol X^D and the allele for the mutant protein has the symbol X^d.

A cross between a carrier female, $X^D X^d$, and a normal male, $X^D Y$, shows that statistically, carrier females have a 50% chance of having affected sons. Their daughters would be phenotypically normal but 50% of them would be carriers.

Gametes	X^D	X^d
X^D	$X^D X^D$ unaffected female	$X^D X^d$ carrier female
Y	$X^D Y$ unaffected male	$X^d Y$ affected male

The phenotypic ratio for the sons is 1:1 unaffected : affected and the ratio for daughters is 1:1 unaffected : carrier.

Affected males, $X^d Y$, cannot pass their mutant allele to their sons, who will all be unaffected. If the mother is unaffected, $X^D X^D$, their daughters will all be carriers.

Gametes	X^D
X^d	$X^D X^d$ carrier female
Y	$X^D Y$ unaffected male

To be affected, a female must receive the X^d allele from both her parents. Her father must be $X^d Y$ and her mother could be a carrier, $X^D X^d$ or be affected, herself, $X^d X^d$.

Pedigree diagrams

The inheritance of a medical condition marked on to a family tree allows a geneticist to infer an explanation of how it is inherited. The pedigree diagram can indicate whether the condition is sex-linked and whether it is the result of a dominant or recessive allele.

In these diagrams, females are symbolised by a circle and males by a square. Individuals with the condition are shaded and those unaffected are not.

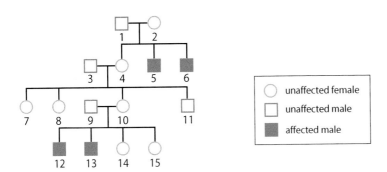

Family pedigree diagram showing the inheritance of DMD over 4 generations

The diagram shows:

- DMD occurs in males only, suggesting it is sex-linked.
- Within the family, DMD is inherited through the mother, suggesting that the gene responsible is on the X chromosome.
- DMD occurs in siblings, suggesting they received the DMD allele from the same parent.

To analyse a pedigree, it is helpful to assign genotypes when they are known. In this example:

- The affected males (individuals 5, 6, 12, 13) must be X^dY.
- The unaffected males (individuals 1, 3, 9, 11) must be X^DY.
- The mothers (2 and 10) and the grandmother (4) of affected males must carry the X^D allele because they are not affected. But they have passed on the mutant allele X^d. So individuals 2, 4 and 10 have the genotype X^DX^d.
- Females 7, 8, 14 and 15 are unaffected so they must carry the X^D allele, but their other allele cannot be deduced from this information.

When the genotypes are known, it is possible to predict the probability of members of a further generation being affected, e.g. if individual 12 and a homozygous, unaffected woman had a son and a daughter, their son would be unaffected, as the father's X^d allele would not be passed on to him. Their daughter would be a carrier.

Exam tip

If you answer an examination question with a genetic pedigree, write in as many genotypes and phenotypes as you can before you try to answer the question.

Going further ▶

After Mendel's work was rediscovered in the early 20th century, Hugo de Vries, the Dutch botanist, used the word 'mutation' for the sudden change he observed in the evening primrose, *Oenothera lamarckiana*.

▼ Study point

Mutations arising in somatic cells are not passed on to the next generation. Only mutations in gametes are inherited.

Going further ▶

In some bacteria, 'mutator' genes make other genes elsewhere on the bacterial chromosome more likely to be mutated.

Mutations

A mutation is a change in the amount, arrangement or structure in the hereditary material of an organism, either DNA or, in the case of some viruses, RNA. Mutations are described as:

- Spontaneous, as they may happen without an apparent cause.
- Random, as they appear to happen with equal probability anywhere in the genome of diploid organisms.

Mutations can occur in all cells, but only mutations that occur in gametes can be inherited. Most mutations are harmful, e.g. exposure to ultra-violet light is linked to mutations resulting in skin cancer. Beneficial mutations are rare but they give the individual a selective advantage.

Mutations may contribute to variation between individuals, which is the raw material for natural selection and therefore evolution. In haploid organisms, any mutation is expressed, i.e. can be seen in the phenotype, unless it is lethal. In diploid organisms, dominant mutations are expressed, but they are rare. Most mutations are recessive. They are likely to occur in a cell with a dominant allele and so are not expressed. Mutation in diploid organisms, therefore, has less impact on evolution than other sources of variation.

Mutation rate

Mutations can occur during DNA replication, prior to cell division, so, in general, organisms with short life cycles and frequent meiosis show a greater rate of mutation than others.

Mutation rates are normally very low but can be increased by:

- Ionising radiation, e.g. gamma rays, X-rays and ultra-violet light. Ultra-violet light of wavelength 260 nm is particularly mutagenic as it is the wavelength that DNA absorbs most efficiently. Radiation joins adjacent pyrimidine bases in a DNA strand so that at replication, DNA polymerase may insert an incorrect nucleotide.
- Mutagenic chemicals, e.g. the polycyclic hydrocarbons in cigarette smoke, methanal, mustard gas. Some chemicals, such as acridine, are mutagenic because they have flat molecules, which can slide in between base pairs in the double helix and prevent DNA polymerase inserting the correct nucleotide at replication.

Mutations can happen in various ways:

- Gene or point mutation: DNA is not copied accurately in S phase, before cell division. These errors involve one or a small number of bases.
- Chromosome mutation: chromosomes may get damaged and break. Broken chromosomes may repair themselves and the DNA and protein rejoin. But they may not repair themselves correctly, altering their structure and potentially affecting a large number of genes.
- Aneuploidy: a whole chromosome may be lost or added, in a phenomenon called non-disjunction, when chromosomes fail to separate to the poles of dividing cells at anaphase I or when chromatids fail to separate at anaphase II.
- Polyploidy: the number of chromosomes may double if the cell fails to divide following the first nuclear division after fertilisation.

Gene (point) mutations

If DNA polymerase changes the base sequence, a point or **gene mutation** occurs. These are:

- Addition – a base is added. If this happens in three places, an extra amino acid is added to the polypeptide chain at translation.

- Duplication – the same base is incorporated twice.

- Subtraction – a base is deleted. If this happens in three places, the polypeptide has one fewer amino acids when translated.

- Substitution – a different base is incorporated.

- Inversion – adjacent bases on the same DNA strand exchange position.

Point mutation changes the allele in which it occurs and changes the bases of the messenger RNA codon. The effect on the polypeptide produced at translation, and consequently the effect on the phenotype, depends on the nature of the mutation:

- The new codon may code for the same amino acid so there is no change to the polypeptide. This is a 'silent' mutation.

- If an amino acid with a similar chemical nature is substituted, the effect may be small, e.g. if valine replaces glycine or if glutamate replaces aspartate.

- If the mutation is at a significant site on the protein molecule, it may make a significant difference to the activity of the protein. If the protein were an enzyme, the structure of the active site could be destroyed.

Sickle cell anaemia

A substitution point mutation in the gene producing the β polypeptide of haemoglobin results in sickle cell anaemia. A DNA triplet on the coding strand, CTC, codes for the amino acid glutamate. Substituting A for T, at the second position, produces the triplet CAC, which codes for valine. The side chain of glutamate is large and hydrophilic whereas that of valine is small and hydrophobic. When the oxygen tension is low, the affected haemoglobin within the red blood cell aggregates. The cell membrane collapses on the precipitated haemoglobin and the red blood cell becomes sickle shaped. The cells become fragile and may break in the capillaries.

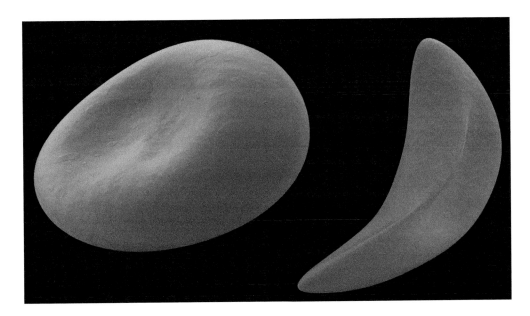

Normal and sickle cell red blood cells

Link You learned about SNPs (single nucleotide polymorphisms) when you studied DNA fingerprinting in the first year of this course.

WORKING SCIENTIFICALLY

The effect on the length of the polypeptide chain of adding or subtracting three bases provided evidence that the genetic code is a triplet code.

▼ **Study point**

If a polypeptide is modified by even one amino acid, it may not be able to perform its normal function in the cell.

▼ **Study point**

Bacteria are widely used in experiments because they have a short life cycle and a high rate of mutation.

▼ **Study point**

The mutation rate varies between species but is typically one mutation per 100,000 genes per generation.

Going further ▶

DNA polymerase's error rate is 10^{-10}, i.e. one incorrect nucleotide inserted in every 10^{10}. RNA polymerase's error rate is 10^{-5}. Natural selection can act on a variety of phenotypes, while the genotype is preserved.

Going further ▶

This mutation was historically described as being at position 6. The numbering system has been altered to include the start codon, methionine, so the mutation is now at position 7.

Going further ▶

The gene for the ß polypeptide of haemoglobin is on the short (p) arm of chromosome 11.

Key Term

Non-disjunction: A faulty cell division in meiosis following which one of the daughter cells receives two copies of a chromosome and the other receives none.

Going further ▶

Spermatogonia are constantly produced but oocytes are only produced in an embryo. An older mother has older oocytes, which have more chance of having accumulated genetic faults. So Down's syndrome is usually produced by a faulty oocyte, not faulty sperm.

Normal haemoglobin has the symbol HbA, produced by the allele Hb^A. The mutant haemoglobin has the symbol HbS and is produced by the allele Hb^S. Individuals who have the genotype Hb^SHb^S have sickle cell disease and may be severely affected, with joint pain and organ damage. The ability of their red blood cells to carry oxygen is reduced, resulting in anaemia and possible death. The alleles HbA and HbS are co-dominant and both HbA and HbS haemoglobins are produced. Heterozygous individuals, Hb^AHb^S, have sickle cell trait: at least 50% of their haemoglobin is HbA and their symptoms are less severe than people who have the genotype Hb^SHb^S.

Chromosome mutations

Chromosome mutations are changes in the structure or number of chromosomes in cells.

Changes in structure

During prophase I of meiosis, homologous chromosomes pair and exchange material at chiasmata. Mutation arises when a chromosome does not rejoin accurately at the corresponding position on its homologous partner. The homologous chromosomes, and therefore, the gametes they are in, end up with some different genes. Each gamete may still fuse with another and produce a new individual but further meiosis will be impossible as the mutant chromosomes will not be able to make homologous pairs at meiosis.

Changes in numbers of chromosomes

Changes in chromosome number are most likely to occur during meiosis, when homologous chromosomes separate at anaphase I, or when chromatids separate at anaphase II. A faulty spindle can result in the chromosomes not being shared equally between the daughter cells. This is **non-disjunction**. The faulty cell division means that one of the daughter cells receives two copies of a chromosome while the other gets none.

Down's syndrome

Down's syndrome occurs in about one in 1000 births. Chromosome number 21 is affected. If non-disjunction happens during oogenesis, a secondary oocyte has either no chromosome 21 or has two copies, instead of one. Those with no chromosome 21 cannot produce a viable embryo. A secondary oocyte with two copies of chromosome 21 that fuses with a normal sperm produces a viable embryo with cells containing three copies of chromosome 21, instead of two, with a total of 47 chromosomes. This condition is called trisomy 21 and produces Down's syndrome.

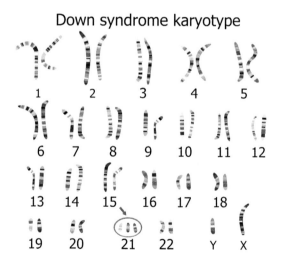

Down syndrome karyotype

Drawing of karyotype from a boy with Down's syndrome

Translocation Down's

In a translocation, a fragment of one chromosome has attached to another. Of people who have Down's syndrome, 5% have 46 chromosomes. During meiosis in a gamete that produced them, a fragment of chromosome 21 attaches itself to chromosome 14. When that abnormal gamete fuses with a normal one, it produces an embryo with two normal copies of chromosome 21 and an additional one attached to chromosome 14.

Changes in numbers of sets of chromosomes

You learned about ploidy levels when you studied cells and chromosomes in the first year of this course.

Cells with complete sets of chromosomes are described as euploid. If they have a small number of extra chromosomes or a small number too few, as may result from non-disjunction, they are aneuploid. If they have several sets of chromosomes they are **polyploid**. Polyploidy may arise in different ways:

- A defect in the spindle at meiosis may result in all the chromosomes at anaphase I, or all the chromatids at anaphase II, moving to the same pole of a cell. This makes gametes with two of each chromosome, instead of one. When a diploid gamete is fertilised by a normal haploid gamete, a triploid zygote with three sets of chromosomes is made. It may survive but it will not be able to make homologous pairs at meiosis. It will therefore not be able to make gametes so it will be infertile. Most plants can reproduce asexually and so triploidy does not prevent their reproduction.

- If two diploid gametes fuse, a tetraploid (4n) is produced.

- Endomitosis is the replication of chromosomes that is not followed by cytokinesis. If this happens in an early embryo, four sets of chromosomes are incorporated into the new nuclear envelope and successive rounds of mitosis continue to produce tetraploid cells. In very rare situations, infertile triploids (3n) undergo endomitosis, making hexaploids (6n). The hexaploids are fertile because they can make homologous pairs in meiosis.

Polyploidy is common in flowering plants and is associated with beneficial characteristics, such as vigour and disease resistance, e.g. 3n seedless water melons, 4n potatoes, 6n bread and 8n strawberries.

It is much more common in plants than in animals, possibly because many plants:

- Can reproduce asexually

- Are hermaphrodite and so do not use chromosomes to determine their sex.

Key Term

Polyploidy: Having more than two complete sets of chromosomes.

Going further ▶

Many plants produce mutagens in their anthers and stigmas, which encourage mutation, e.g. colchicine in the anthers of meadow saffron, *Colchicum autumnale*. An infertile triploid may acquire a mutation that enhances endomitosis, restoring fertility.

Going further ▶

Salmon and goldfish are among the few animals that have polyploid chromosomes.

▼ Study point

Mutation may affect a single gene, a whole chromosome or whole sets of chromosomes.

	Nicotiana sylvestris	*Nicotiana tabacum*
	2n = 24	4n = 48
Leaf		
Flower		
Chromosomes in root tip squash.		

Carcinogens and genes

An agent that causes cancer is called a **carcinogen** and is described as carcinogenic. Some mutagens are carcinogenic and cause mutations in DNA.

Tumour suppressor genes

Genes control cell division and division is halted when enough cells have been produced for growth and repair. Genes that regulate mitosis and prevent cells dividing too quickly are called tumour suppressor genes. A mutation may affect one of these genes and make it lose its regulatory function. The cell could then go through continual, repeated mitosis, which characterises cancer. If the cell escapes the attack of the immune system, it produces a collection of cells called a tumour. Tumours may be harmless, or benign. But sometimes the cells of a tumour are able to spread around the body and invade other tissues, making secondary tumours, or metastases. This type of tumour is described as malignant. Abnormalities in the tumour supressor gene *TP53*, which codes for the p53 protein, have been identified in more than half of all human cancers. The table shows how mutant p53 proteins can contribute to the development and spread of cancer:

Normal p53 protein	Mutant p53 protein
Activates repair of damaged DNA.	No DNA repair.
Prevents the cell from entering S phase, holding it in G_1 while damaged DNA is repaired.	Cell with damaged DNA enters S phase and DNA is replicated.
Initiates apoptosis if damaged DNA cannot be repaired.	Mutant cells survive and undergo mitosis.

Oncogenes

A proto-oncogene codes for a protein that contributes to cell division. Mutation may switch on such a gene permanently, so excessive amounts of the protein are made, causing rapid, repeated mitosis, i.e. cancer. When a proto-oncogene is mutated such that it causes cancer, it is called an **oncogene**. This may happen if:

- A mutation causes chromosomes to rearrange, and places the proto-oncogene next to a DNA sequence that permanently activates it.

- There is an extra copy of the proto-oncogene, resulting in too much of its product being made, causing excessive mitosis.

Tobacco smoke contains over 4000 chemicals, including tar, nicotine and carbon monoxide. Over 40 are known to be carcinogenic and over 400 others known to be toxic. Tar collects in the lungs as the tobacco smoke cools. It is a mixture of many chemicals. Some, e.g. polycyclic hydrocarbons, can enter the nuclei of alveolar cells and slide between the base pairs in their DNA, causing mutation by preventing accurate replication. This is one way in which tobacco smoke is carcinogenic.

Key Term

Oncogene: A proto-oncogene with a mutation that results in cancer.

▼ Study point

Proto-oncogenes cause cancer when they are permanently switched on. Tumour suppressor genes cause cancer when they are switched off.

Going further ▶

About 25% of all cancer deaths in developed countries are due to carcinogens in the tar of tobacco smoke.

34

Knowledge check

Link the appropriate terms 1–4 with the statements A–D.

1. Carcinogen
2. Sickle-cell anaemia
3. Non-disjunction.
4. Down's syndrome.

A. An example of a point (gene) mutation.
B. Substance that causes cancer.
C. An example of a chromosome mutation.
D. Chromosomes shared unequally between daughter cells.

The control of gene expression

Variation is the differences between members of a species. It has traditionally been related to:

- Differences in DNA nucleotide sequence, i.e. different alleles.

- Physiological effects of the environmental, e.g. higher light intensity increasing plant growth.

Evidence has accumulated that, in addition, the environment can alter the expression of genes by affecting how they are transcribed, without changing their nucleotide sequence. The changes are **epigenetic**, i.e. they affect the genes but not their nucleotide sequences.

Key Term

Epigenetics: The control of gene expression by modifying DNA or histones, but not by affecting the DNA nucleotide sequence.

Epigenetic modifications

- DNA methylation: cytosine can have a methyl or hydroxymethyl group added. Methylated cytosine can be read as cytosine, pairing with guanine at transcription. But if regions of DNA are heavily methylated, they are less likely to be transcribed.

Cytosine (C)	5-Methylcytosine (mC)	5-Hydroxymethylcytosine (hmC)

Methylation of cytosine

- Histone modification following translation, e.g. by attaching an acetyl group to the amino acid lysine, a methyl group to lysine and arginine or a phosphate group to serine and threonine. These changes to histone proteins alter their interaction with DNA. It changes the arrangement of the nucleosomes. When unmodified, the nucleosomes pack more tightly. The DNA is less accessible to enzymes and so transcription is reduced. When histones are modified, the coiling is more relaxed and transcription factors and RNA polymerase have access to the DNA, so transcription is increased.

Link You learned about post-translational modification in the first year of this course.

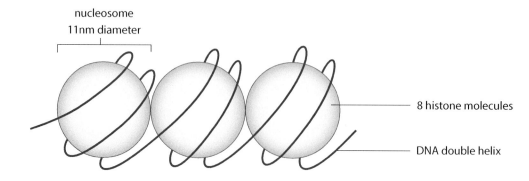

nucleosome
11nm diameter

8 histone molecules

DNA double helix

Histones not modified; nucleosomes tightly packed; gene not accessible

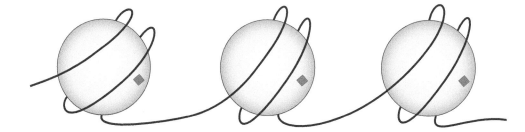

Histones modified; nucleosomes separated; gene accessible

Monozygotic twins have epigenetic differences

The stem cells of the embryo progressively differentiate, switching off genes coding for enzymes that are not needed. Such changes mean that differentiated cells only express the genes that are necessary for their own activity. Thus, for example, skin cells can produce melanin but retinal cells produce rhodopsin. Different epigenetic changes can happen to cells within the same tissue and to cells in different tissues, which allows the vast difference in gene expression in the different cells of an organism.

DNA damage happens about 60,000 times a day in each cell of the human body. Most is repaired, but at the site of repair, epigenetic changes might remain. Monozygotic twins provide evidence: they have identical nucleotide sequences as they come from the same fertilised oocyte. Their DNA methylation and histone acetylation is very similar initially but the differences between them increase as they get older, and they increase the longer the twins live apart.

Consequences of epigenetic changes

- Genomic imprinting: if genes are inactivated in gametes, the inactivation may be transferred to the next generation. This is the basis of genomic imprinting, in which a gene may be permanently switched off by DNA methylation on the chromosome derived from one parent. If this switching is damaged, a medical condition may ensue. An example of imprinting involves the NOEY2 gene. Only the father's copy is expressed. If, like the mother's, it is not, the risk of breast and ovarian cancer is increased.

- X inactivation: epigenetic changes can switch off a whole chromosome. Cells of female mammals use only one X chromosome. The other is inactivated and becomes a mass of densely staining chromatin called the Barr body. The patchwork of tortoiseshell cats reflects the random inactivation of either X chromosome, and therefore the fur colour gene they carry, with alternative X chromosomes inactivated in adjacent groups of cells.

Epigenetic changes have been implicated in autoimmune conditions, mental illness, diabetes and in many cancers. Their study is likely to become very important in disease detection, treatment and prevention.

WORKING SCIENTIFICALLY

A classic Swedish study shows that epigenetic changes may be inherited. For couples who experienced famine as children: their son's sons have less cardio-vascular disease but their son's daughters die earlier. If food is available, all the grandchildren have a higher risk of diabetes.

Tortoiseshell cat

Variation and evolution

Some characteristics are controlled by one gene, with no environmental influence on its expression. Most characteristics have a range of values and are controlled by many genes and the environment. For a species to survive in a constantly changing environment, variation is essential. Population genetics describes how allele frequencies change and it allows calculations of allele and genotype frequencies in a population.

Darwin's observations of variation led him to propose the theory of natural selection. Selection pressures result in some individuals reproducing more efficiently than others and transmitting their alleles in higher proportion to the next generation. Isolation of breeding groups leads to gradual changes of phenotype and the formation of new species.

Content

By the end of this topic you will be able to:

- Explain that variation is the result of a combination of genetic and environmental factors.
- Describe continuous and discontinuous variation.
- Describe competition for breeding success and survival.
- Describe selection pressures.
- Use the Hardy–Weinberg equation to calculate allele and genotype frequencies.
- Explain the founder effect and genetic drift.
- Describe the mechanisms of natural selection.
- Explain how isolation can result in speciation.
- Describe Darwin's theory of evolution by natural selection.

Key Term

Variation: The differences between organisms of the same species.

Variation

Organisms differ in their phenotypes for several reasons:

- They have a different genotype.
- They have the same genotype but have different epigenetic modifications.
- They have different environments.

The differences in phenotype within a species are referred to as **variation**. If they result from different DNA nucleotide sequences or, in some cases, different epigenetic modification, they can be inherited and so the variation they produce is **heritable**. In asexually reproducing organisms, heritable variation can only be increased by mutation. Sexual reproduction, on the other hand, has several mechanisms that generate heritable variation:

- The crossing over between homologous chromosomes during prophase I of meiosis.
- The random distribution of chromosomes during metaphase I of meiosis.
- The random distribution of chromatids at metaphase II of meiosis.
- The mixing of two different parental genotypes at fertilisation.

Variation imposed by the environment that is not a result of epigenetic change cannot be transmitted to offspring and so it is **non-heritable.**

These processes establish a new combination of alleles in each generation, but it is the rare occurrence of mutation that generates long-lasting variation of a novel kind. Variation can be described in two ways.

Discontinuous variation

For some characteristics, there is only a small number of possibilities, e.g. you can have five or six fingers; the peppered moth, *Biston betularia*, may be speckled (light) or dark (melanic); a plant may be tall or dwarf. There are no intermediate types and the characteristics are discrete, i.e. clear-cut and easy to tell apart.

Speckled peppered moth

Melanic peppered moth

Each of these characteristics is controlled by single genes, and is described as **monogenic**. The environment has no influence on the gene's expression. The distribution of phenotypes is best shown as a bar chart:

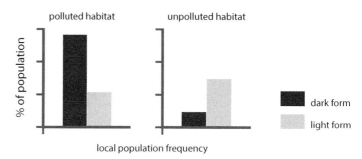

Distribution of peppered moth varieties

Continuous variation

For some characteristics, there are many possible values, e.g. babies born at 9 months tend to weigh between about 2 kg and 5 kg; plants may have any number of leaves. The values show a gradation from one extreme to the other within a range. Such characteristics are controlled by many genes and are described as **polygenic**. The environment has considerable influence on the genes' expression and therefore has a role in determining phenotypic variation.

- Human height: an individual inherits a number of alleles from their parents contributing to height. If the individual has the potential to grow tall, their alleles determine a potential maximum size, but whether or not this is reached depends on environmental factors, such as nutrition and exercise.

- If organisms of identical genotype are subject to different environmental influences, despite their identical genotypes, they show considerable variation, e.g. plants produced by asexual reproduction, such as strawberries from runners, may show much variation in their height, in response to light intensity, or in the direction of growth, in response to light direction.

- Plants need light for normal growth and development. Without it, auxin synthesis elongates their stem cells and they are unable to synthesise chlorophyll. If a seed germinates in the dark, the seedling is etiolated: it shows chlorosis and grows taller than in the presence of light.

A bar chart could be plotted showing the distribution of a characteristic showing continuous variation, e.g. height in a human population. Most members of the population have a certain height with fewer who are taller or shorter. If the tops of the bars are joined, a smooth, symmetrical curve is produced, showing a normal distribution around the mode, which is also the mean and the median.

▼ **Study point**

Both the genotype and the environment contribute to continuous variation. Discontinuous variation is caused by differences in genotype, and the environment has no influence on it.

Cress seeds from the same batch of seeds grown in the presence (left) and absence (right) of light

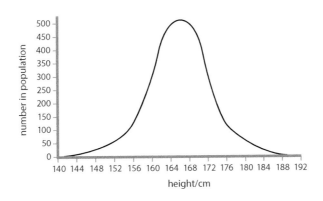

Normal distribution of height in a human population

Many characteristics are polygenic and they are largely responsible for continuous variation in a population.

	Type of variation	
	Discontinuous	Continuous
Heritable	✓	✓
Non-heritable	✗	✓
Number of genes	1	Many
Description	Monogenic	Polygenic
Environmental influence	None	Some
Number of values	Few	Many
Human example	Number of fingers	Height
Plant example	Height	Number of leaves

35

Knowledge check

Complete the paragraph by filling in the spaces:

Variation is the differences between members of the same .... Monogenic characters generate .... variation and their expression is not influenced by the .... Continuous variation is ....

▼ **Study point**

In any population, the number of young produced is far greater than the number surviving to become adults. They compete for limited resources and those with particular alleles are more successful.

 ‹Link› Competition is a factor in the formation of new species by natural selection, described on p232.

 **Key Terms**

Selection pressure: An environmental factor that can alter the frequency of alleles in a population, when it is limiting.

Natural selection: The increased chance of survival and reproduction of organisms with phenotypes suited to their environment, enhancing the transfer of favourable alleles from one generation to the next.

▼ **Study point**

Competition, environmental factors and human influence exert selection pressures, increasing the chances of some phenotypes and therefore some alleles being passed on to the next generation, and decreasing the chance of others.

Competition for breeding success and survival

When there is no environmental resistance, organisms over-produce. This means that two parents usually have more than two offspring, so populations tend to get bigger from one generation to the next. But in most situations there are limits to population growth: some resources may be limited so individuals have to compete for them, e.g. plants compete for light, soil, water and mineral ions; animals compete for food and shelter. As a result, fewer offspring are produced, or offspring may die before maturity, and so not reproduce.

Competition can be looked at in two ways:

- Inter-specific competition occurs between individuals of different species competing for the same resources, e.g. plants compete for space; desert animals compete for water.
- Intra-specific competition occurs between individuals of the same species, e.g. giant pandas compete for bamboo; animals compete for mates or nesting sites.

Selection pressure

Variation within a population of organisms means that some individuals have characteristics that give them an advantage. Charles Darwin wrote about competing organisms being in a 'struggle for existence'. Those with an advantage can gain access to the limited resources more easily and as a result, they reproduce more than those without.

Consider a population of rabbits: the female may produce several litters each year, with 2–10 rabbits in each. If all the rabbits survived to become adults and they, in turn, reproduced successfully, the rabbit population would increase rapidly. Eventually, as the increasing number of rabbits eat more and more of the vegetation, food would become in short supply. When this happens, rabbits that are better able to find the food, and eat it before the others can, have an advantage. They reproduce more successfully, having more offspring and the alleles that gave them the advantage are therefore represented more in the next generation. These alleles have been 'selected for' and the food supply was the **selective agent** or **selection pressure**. It was the environment that determined which rabbits would reproduce more successfully and so this is an example of **natural selection**.

Many environmental factors can act as selection pressures and they affect an individual's capacity to reproduce. Phenotypes that allow an organism to be more adaptable enhance the likelihood of survival and reproduction when conditions are not optimal, e.g. if an individual can reproduce at slightly higher temperature or use a slightly different nesting site.

Selective agencies include:

- Availability of nesting sites: some animals raise young in very specific situations, e.g. crows build their nests at between 10 and 20 m above ground, in the angle between tree branches.
- Day length affects reproductive behaviour: most pregnancies in female rabbits are in May, following the enlargement of the males' testes, which begins in November, in response to the short days.
- Overcrowding allows diseases to spread. Among humans, tuberculosis is much more common in densely populated places. It can be fatal and, throughout history, has been a selection pressure. People with alleles that give them a more effective immune system are more likely to survive and pass those alleles on.
- Predation: when a population grows, predators have more food and their numbers increase. Some individuals in the prey population are more likely to survive and reproduce, passing the alleles on that have made them successful, e.g.:

- Some are better camouflaged, e.g. melanic peppered moths camouflaged by soot on trees and speckled forms are camouflaged by lichens,
- Mimic species are harmless but have characteristics of other species that are toxic to the predator, e.g. the harmless Mexican milk snake mimics the highly toxic Texas coral snake, with characteristic red, black and yellow markings.

- Temperature affects survival: mosquito larvae survive best in water that is between 24°C and 28°C; wolverines can only reproduce if there is very heavy snowfall. They dig their dens under the snow, which must be deep enough to last until May, when the cubs are well-enough developed to emerge.
- Human impact: habitat loss has destroyed breeding grounds, e.g. cyanide fishing in which sodium cyanide is squirted into water to make it easier to catch fish in coral reefs, kills coral and algae; sea turtles hatch on land but need a dark night sky to orient themselves so that they move towards the sea, but artificial lights on the beaches attract them and they move away from the sea and do not survive.

Wolverine

The value of an allele depends on the environment

It is the phenotype that is suited, or not, to a particular environment. The phenotype is determined in part by the genotype. If the phenotype provides an advantage, the alleles that produced it are transmitted to the next generation more successfully than other alleles. The same allele may produce a phenotype that is suited to one environment and not to another, e.g. the coat colour of rabbits and hares may vary. Most have alleles that produce a brown coat. A small number may be homozygous for a recessive allele that gives a white coat. A white animal is more likely than a brown one to be killed by a predator, so it is unlikely to survive to become a mature adult. The chances of it reproducing and passing on its allele for white fur are very small and the allele remains rare in the population. However, in the Arctic, white fur would be camouflaged against the snow, and therefore, the animal would be more likely to survive and pass on its alleles. Consequently, in such an environment, the allele producing a white coat is more common.

White snowshoe hare

Selection pressure is an abstract factor that shapes populations over time. A selection pressure acts in a consistent fashion over long timeframes and affects the reproductive and survival rates of the individuals in a species. The term 'fit' in Biology describes how likely an organism is to survive, reproduce and pass on alleles to the next generation. Selection pressures allow the fittest to survive and increase the overall fitness of a population in a particular environment. The environment exerts a selection pressure and this determines the spread of any allele within the gene pool. But many selection pressures act at the same time, and these, in combination with mutation and genetic drift, determine the phenotypic outcome.

36

Knowledge check

Link the appropriate terms 1–4 with the statements A–D.

1. Selective advantage.
2. Discontinuous.
3. Interspecific.
4. Selection pressure.

A. Variation due to characters controlled by a single gene.
B. Competition between individuals of different species.
C. An environmental factor altering the frequency of alleles in a population.
D. Characteristic that enables an organism to survive and reproduce better than other organisms in a population in a given environment.

Key Terms

Gene pool: All the alleles present in a population at a given time.

Allele frequency: The frequency of an allele is its proportion, fraction or percentage of all the alleles of that gene in a gene pool.

Genetic drift: Chance variations in allele frequencies in a population.

Exam tip

Memorise the nine conditions under which the Hardy–Weinberg principle operates.

Population genetics

The gene pool and allele frequencies

Up to this point, the transmission of genes and alleles between individuals in a population has been described. In contrast to considering individuals, population genetics describes the behaviour of the genes and alleles of an entire population. A population of sexually reproducing organisms displays variation. All the alleles of all the genes of all the individuals in a population at any one time constitute the **gene pool**. The genotypes of the individuals are a selection of the allele combinations that are possible, at that time. Population genetics is not concerned with the genotypes of individuals, but describes the proportions of the different alleles in the whole gene pool, i.e. the **allele frequencies**.

The proportions of alleles in the gene pool remain stable if the environment is stable. However, environments normally change. Some phenotypes will be advantageous and will be selected for, so the alleles that produced them will be transmitted to the next generation. Other phenotypes will be selected against, and the alleles that produced them will not be passed on. The gene pool is constantly changing, some alleles becoming more frequent and others less frequent. In some circumstances, alleles may be totally lost from the gene pool.

Genetic drift

If a sexually reproducing population is in a stable environment and there is no mutation and no immigration or emigration, the frequencies of all the alleles stays constant.

Consider an allele that occurs in 1% of the population:

- If there are 1 000 000 individuals, then 10 000 individuals possess the allele. If mating is random with respect to this allele, in the next generation, the frequency will be the same.

- If the population is much smaller, e.g. 1000 individuals, only ten will carry the allele. If, by chance, one of the ten fails to reproduce and pass on the allele, its frequency in the next generation is reduced by 10%.

- In a very small population, it may be that, by chance, none of those possessing the allele mate. Then the allele will be lost from the population altogether.

This chance variation in allele frequency in a population is **genetic drift**. It is most significant in small or isolated populations when a small number of alleles form a large proportion of the total, and, in those situations, it may be an important evolutionary mechanism.

The Hardy–Weinberg principle

The Hardy–Weinberg principle states that in ideal conditions, allele and genotype frequencies in a population are constant from generation to generation. The ideal conditions are:

The organisms	Organisms are diploid
	The allele frequencies are equal in both sexes
	They reproduce sexually
	Mating is random
	Generations do not overlap
The population	The population size is very large
	There is no migration
	There is no mutation
	There is no selection

The Hardy–Weinberg equation

The Hardy–Weinberg principle is expressed in an equation which allows allele frequencies to be calculated from genotype frequencies, and vice versa:

Consider dominant and recessive alleles, A and a respectively, for a single gene. If the frequency of A is p, i.e. f(A) = p, and the frequency of a is q, i.e. f(a) = q, then because A and a are the only alleles, p + q = 1.

If a population containing only these alleles mates at random, with all possible allele combinations, the genotypes of the population can be deduced:

Males		Females	
		A	a
	A	AA	Aa
	a	Aa	aa

The symbols A and a are not put in circles in this table, as they would be in a Punnett square for a genetic cross, because they are not gametes, but the alleles in the gene pool.

The population comprises genotypes in the ratio 1 AA : 2 Aa : 1 aa.

Using p and q as the frequencies for A and a, we can describe the frequencies of the genotypes:

$f(AA) = p^2$

$f(Aa) = 2pq$

$f(aa) = q^2$

Because these three genotypes comprise the whole population, $p^2 + 2pq + q^2 = 1$ or $(p + q)^2 = 1$. This is the Hardy–Weinberg equation. Under the conditions in which the Hardy–Weinberg principle operates, the allele frequencies, p and q, remain constant over the generations and the population is said to be in **Hardy–Weinberg equilibrium.** The equation shows that a large proportion of recessive alleles exist in the heterozygotes. Heterozygotes are, therefore, a reservoir of genetic variability.

An analysis of the population genetics of sickle cell disease

Sickle cell disease (SCD) affects 9 people in 100 in some parts of Africa. Knowing that the condition is related to an autosomal recessive allele, it is possible to calculate the allele and genotype frequencies:

Sickle cell disease is caused by an autosomal recessive allele
$\therefore q^2 = 0.09 \therefore q = \sqrt{0.09} = 0.3$.

$p + q = 1$

$p = 1 - 0.3 = 0.7$.

If the dominant allele has the symbol Hb^A and the recessive allele has the symbol Hb^S:

$f(Hb^AHb^A) = 0.7^2 = 0.49$, i.e. 49% of the population is phenotypically normal.

$f(Hb^AHb^S) = 2pq = 2 \times 0.7 \times 0.3 = 0.42$ (2 dp), i.e. 42% of the population has sickle cell trait.

▼ **Study point**

An allele's frequency in a population remains constant from one generation to the next if the population is in Hardy–Weinberg equilibrium.

▼ **Study point**

$(p + q) = 1$ and $p^2 + 2pq + q^2 = 1$, so if you know p, you can find q and then you can calculate the frequencies of all the genotypes.

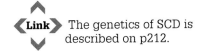 ◆**Link**▶ The genetics of SCD is described on p212.

▼ **Study point**

The Hardy–Weinberg equation is used to calculate the frequency of alleles in the human gene pool associated with genetic diseases, such as cystic fibrosis.

Evolution and selection

Under the conditions in which the Hardy–Weinberg equilibrium is maintained, the average phenotype of the population does not alter. But in general, those ideal conditions do not exist. Allele frequencies change and phenotypes alter over time. This is what is meant by the term **evolution**. It is a property of a population not of an individual and it is a change in the average phenotype.

If the change in phenotype is profound enough, organisms with the altered phenotype are unable to reproduce successfully with the initial population. Then **speciation** has occurred, i.e. a new species has formed.

The founder effect and genetic drift

When a small number of individuals become isolated and start a new population, e.g. colonising a new habitat such as an island, the founder members of the new population are a small sample of the population from which they originated. By chance they may have a very different allele frequency from the original population. This is the **founder effect.** While the founder population remains small, it may undergo genetic drift and become even more different from the large parental population. In a small population, chance variation in allele frequency from one generation to the next can represent a large change in a phenotype for a large proportion of the population. As many genes are subject to such fluctuation, the average phenotype can change rapidly. The smaller the population, the more significant this effect is.

The founder effect and genetic drift contributed to the evolutionary divergence of Darwin's finches after strays from the South American mainland reached the remote Galapagos Islands. It accounts, in large part, for the many unique species in small islands adjacent to large landmasses, such as Madagascar.

Coquerel's sifaka, one of the many species unique to Madagascar

Natural selection

There are three types of natural selection.

Stabilising selection

A characteristic showing continuous variation has a range of values. In a certain environment, the average phenotype may provide a greater advantage than either extreme. In that case, the extreme values will be selected against. The normal curve displaying the range of phenotypes in the next generation will have a smaller standard deviation but a higher peak. In other words, the average has stayed the same, but more individuals have that value. If the environment remains unchanged, this continues until most members of the population share the same value.

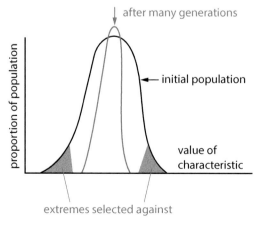

Stabilising selection

A classic example is in birth weight in humans. Very small, light babies lose heat rapidly and they are susceptible to infection, so they survive less well than those of normal weight. Very large, heavy babies do not pass through the birth canal as easily as normal weight babies and they also survive less well. Most babies are born at around 3.5 kg. Extremes of birth weight survive less well, so stabilising selection is operating.

Directional selection

In a changing environment, an extreme phenotype may become advantageous. Then, other values are selected against and over time, the average phenotype changes.

Lake Victoria in East Africa has many species of cichlid fish. The lake has changed in extent and depth many times in the past. Imagine a species of small cichlids with no predators. Change to a wetter climate may have united the lake with nearby bodies of water containing bigger fish that are potential predators. When the fish mixed, the predators acted as a selection pressure on the smaller fish. Those with larger fins swam faster than those with smaller fins so they had a selective advantage. They escaped predation and survived. When they reproduced, they transmitted the alleles for larger fins to the next generation. Over time, the allele frequency for larger fins increased and the average fin size increased, so evolution occurred. Events such as this have contributed to the enormous number of cichlid species in Lake Victoria.

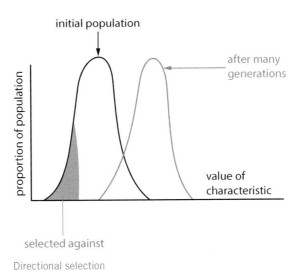

Directional selection

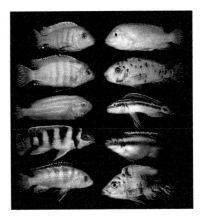

Ten of the estimated 3000 species of cichlid species

> ▼ **Study point**
>
> The allele leading to smaller fins remained in the gene pool as a recessive allele in heterozygous fish.

Disruptive selection

In some situations, the average phenotype does not provide an advantage and is selected against. Over several generations, a lower and a higher value are selected and so a curve displaying the proportions of the population with values of the characteristic is bimodal, i.e. it has two peaks.

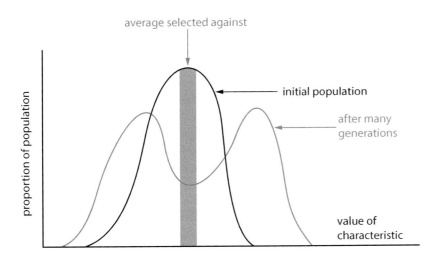

Disruptive selection

A population of coho salmon, for example, has large and small males. Large males gain access to females by fighting, whereas the small males gain access by 'sneaking'. Medium sized males cannot use either strategy and are selected against.

Both directional and disruptive selection change the average phenotype. When the phenotype is so different that mating between the new and original types cannot produce fertile offspring, then speciation has occurred.

> ▼ **Study point**
>
> Natural selection provides an explanation for both the lack of phenotypic change in a stable environment and for phenotypic change in a changing environment.

Key Terms

Species: A group of phenotypically similar organisms that can interbreed to produce fertile offspring.

Reproductive isolation: The prevention of reproduction and, therefore, gene flow between breeding groups within a species.

▼ Study point

The major mechanisms of speciation are the founder effect, coupled with genetic drift, and natural selection.

Isolation and speciation

A **species** is a group of organisms that can interbreed to form fertile offspring. In some situations, it is not possible to tell whether organisms may interbreed, for example, if the organisms:

- Are few in number and widely separated
- Are only known as fossils
- Have a very long life cycle
- Only reproduce asexually.

In these cases, if the physical characteristics are very similar, organisms are assumed to be in the same species.

New species may arise:

- Abruptly, by polyploidy when their chromosome number doubles by endomitosis. This is rare and more common in plants than in animals; the example of bread wheat is described on p230.
- Gradually, by isolating groups of individuals.

A population is an inter-breeding group of one species. Sub-groups within the population may breed more often with each other than with the rest of the population – these are **demes**. If a deme becomes isolated, it cannot breed with members of other demes and so the gene flow in and out of this deme is prevented. The mechanism preventing gene flow is an isolating mechanism and the groups are **reproductively isolated**.

If the demes are isolated for many generations, they undergo changes in allele frequency and accumulate so many different mutations that they are no longer able to inter-breed successfully with the members of the initial population. Speciation has occurred and the separate species each have their own gene pool. In practice, demes can be isolated and phenotypic differences can be identified while interbreeding remains possible. But after many generations, reproductively isolated demes diverge to such an extent that they become two separate species.

Reproductive isolation can be:

- Pre-zygotic – gametes are prevented from fusing and so a zygote is never formed.
- Post-zygotic – gametes fuse and a zygote forms. Even if the organism develops and grows, it is sterile and so the genes of the parent species are kept separate and the species do not merge.

Pre-zygotic isolation

Geographical isolation

This occurs when the population becomes split by a physical barrier into separate demes. An example of geographical isolation may be described in this way:

The birds of a population feed and breed only in the cool conditions of a valley. The mountain peaks are too cold for them to survive. The climate becomes warmer so the birds move up the mountains, to where it is cooler. They are split into separate demes on the different mountains, each with its own gene pool. Over time the birds are subjected to different selection pressures so the allele frequencies of the demes change and the populations accumulate different mutations. The populations become very different, e.g. birds isolated on one mountain adapt to feed on insects in crevices. Birds with long beaks feed successfully and survive to reproduce and pass on the alleles that produce a long beak. On another mountain, the birds adapt to feed on fruit and so a different beak type provides an advantage. If the temperature reverts to its original value, the birds return to the valley and they come into contact again. Their appearances have altered and the different-shaped beaks produce different mating calls. The birds from the two groups are no longer attracted to each other. The two populations have established different gene pools and can no longer interbreed, so two distinct species have evolved.

The physical barrier may be a mountain, a river, desert or other feature that prevents interbreeding. Where the demes are in different areas, speciation is **allopatric**, e.g. the sticky cinquefoil, *Potentilla glandulosa*, grows in California. Those growing in the lowlands are bushier and taller than those growing at high altitude, showing a difference in phenotype, although, for the time being, they can still interbreed and are still in the same species.

Short, bushy cinquefoils growing at high altitude

Key Term

Allopatric speciation: The evolution of new species from demes isolated in different geographical locations.

The African elephant was considered to be one species but it is now clear that the elephants of West Africa, the savanna elephants of Central, Eastern and Southern Africa and the forest elephants of Central Africa have diverged so much following their geographical isolation, that they are now three separate species.

Behavioural isolation

Many animals have mating rituals and courtship displays that are recognised by other members of the species as a prelude to mating. The male grasshopper, *Chorthippus*, vibrates its forewings against its hind legs and makes a stridulation, or song, characteristic of its species. If a female does not recognise the song, she does not mate with the male that makes it. If, however, a male of species A is prevented from singing while the song of species B is played to a species B female, she will mate successfully with the species A male. The two groups have diverged so much that they will only mate under these abnormal circumstances. In the wild, they form two separate mating groups. In time, after many generations, they are likely to accumulate other mutations that will prevent these successful matings, by which time they will be two species.

The individuals share a habitat and so this type of speciation is **sympatric speciation**.

It has been suggested that *Homo sapiens* and *Homo neanderthalis* remained as two separate species as they did not recognise each others' mating behaviour. That modern humans have 1–4% *H. neanderthalis* genes shows that the mechanism of reproductive isolation did not always operate.

Chorthippus

Key Term

Sympatric speciation: The evolution of new species from demes sharing a geographical location.

Morphological isolation

The exoskeleton of an insect is rigid, so the genitalia of males and females must be complementary for sperm to be transferred to the female and for successful mating to take place. This prevents genes mixing, unless individuals carry the alleles that produce the appropriately shaped body parts. The reproductive barrier is related to the morphology and it is sometimes called mechanical isolation. The individuals that are reproductively isolated in this way share a habitat, so speciation resulting from morphological isolation is sympatric.

Morphological isolation also occurs in plants that have a pollinator that is unable to pollinate other species, e.g. the star orchid, *Angraecum sesquipedale*, which has a spur 27–43 cm long, with a nectary at the base. The nectary can only be reached by the Madagascan sphinx moth, *Xanthopan morganii praedicta,* which has an extremely long mouthpart. Other insects do not visit this orchid. It has only one pollinator so its genes are isolated from those of other orchids.

Going further ▶

The moth species has the suffix '*praedicta*' because Darwin predicted its existence. This example shows that evolution is well-enough established to be a predictive science.

Anagraecum sesquipedale   Madagascan sphinx moth

Gametic isolation

Gametes in open environments often meet gametes of other species, e.g. in marine coral reefs, corals of many species release their gametes in the water at the same time. In about a third of cases, they make viable inter-species hybrids, an important feature in coral evolution but two-thirds are incompatible. Barriers prevent the fusion of gametes of different species, so the populations are isolated:

- In many cases, there are molecules in gamete membranes that prevent inter-species fusion.

- In some cases, chemo-attractants secreted by female gametes are only recognised by male gametes of the same species.

- The pollen of many Angiosperms can germinate on the stigma of another species but the pollen tube only penetrates a small distance down the style and does not reach the ovule.

- The sperm of many animals does not survive in the oviduct of another species.

The isolated populations may accumulate mutations and undergo speciation. Because the gametes mix in the same environment, gametic isolation leads to sympatric speciation.

▼ Study point

Speciation can occur as a result of the founder effect, genetic drift and natural selection. It is enhanced by reproductive isolation.

Seasonal isolation

Seasonal isolation is sometimes called temporal isolation. Reproductive organs of different demes are mature at different times of the year. They are therefore unable to hybridise and are genetically isolated. They grow in the same locality, so seasonal isolation leads to sympatric speciation. The creeping buttercup, *Ranunculus repens*, grows in similar habitats to the lesser celandine, *Ranunculus ficaria*. If the two interbred frequently, the two species would lose their distinctive characteristics and become one. But this is prevented because *R. ficaria* flowers in April and *R. repens* flowers in June.

Flower of *R. ficaria*, the lesser celandine   Flower of *R. repens*, the creeping buttercup

Knowledge check

Link the appropriate terms 1–4 with the statements A–D.

1. Hybrid.
2. Species.
3. Speciation.
4. Gene pool.

A. All the alleles in a population at any one time.

B. The evolution of new species from existing species.

C. A group of similar organisms that can interbreed to produce fertile offspring.

D. The offspring resulting from cross breeding between different species.

Post-zygotic isolation

Hybrid inviability

Fertilisation occurs but incompatibility between the genes of the parents prevents the development of an embryo, e.g. **hybrid** embryos of the northern leopard frog, *Rana pipiens* and the wood frog, *Rana sylvatica* survive for less than a day.

Hybrid sterility

In some cases, an embryo formed from the gametes of two species can develop. But if the chromosomes are not sufficiently similar, they are unable to pair at prophase I of meiosis and so gametes cannot form. This means that the hybrid is sterile. This sterility prevents gene flow between the parental species and so the species remain distinct. A classic example is that of the mule. It is a hybrid between a female donkey and a male horse. The converse is the cross between a male donkey and a female horse, which produces a hinny. Mules and hinnies are sterile. A horse has 64 chromosomes, with 32 in the gametes, and a donkey has 62, with 31 in the gametes. Mules and hinnies have 63 chromosomes and, being unable to make homologous pairs at meiosis, are sterile.

Infertile F$_1$ hybrids in plants are well known, e.g. *Spartina*, the cordgrass that grows along the south cast of Britain. On rare occasions, endomitosis occurs and nuclear division in the early embryo is not followed by cytokinesis. The chromosome number doubles and so there are two of each chromosome. Pairing at meiosis can happen and the hybrid is fertile.

Key Term

Hybrid: The offspring of a cross between members of different species.

A hinny

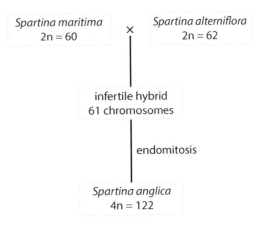

Spartina maritima
2n = 60

×

Spartina alterniflora
2n = 62

infertile hybrid
61 chromosomes

endomitosis

Spartina anglica
4n = 122

Spartina chromosomes

Attempts have been made to reconstruct the history of bread wheat, by analysing the chromosome complements of wheat and its relatives, including wild grasses in the Middle East, where wheat was domesticated. Repeated hybridisation and chromosome doubling have produced the hexaploid wheat, now one of the most widely grown crops in the world. The scheme below shows one suggested history for domesticated wheat.

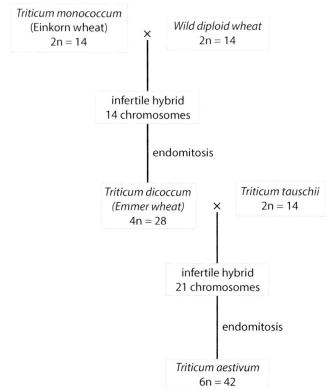

Wheat chromosomes

38

Knowledge check

Match the methods of reproductive isolation A–C with their examples 1–3.

A. Geographical.

B. Seasonal.

C. Hybrid sterility.

1. A liger is the infertile offspring of a female lion and a male tiger.
2. The lungfish (*Dipnoi*) that live off the west African coast cannot interbreed with those that live off eastern South America.
3. Adults of one species of cicada emerge every 13 years and of another every 17 years, so they can only hybridise every 13 × 17 = 221 years.

Hybrid breakdown

Some F_1 hybrids are fertile, but their F_2 is sterile. This is more common in plants than in animals and is best known in cotton, legumes and rice. Incompatibility between nuclear genes and genes in the mitochondria and chloroplasts has explained some cases of this.

A summary of reproductive isolation methods

	Type of isolation	Sympatric or allopatric	Example
Pre-zygotic	Geographical	Allopatric	Sticky cinquefoil
	Behavioural	Sympatric	Grasshopper
	Morphological	Sympatric	Insects
	Gametic	Sympatric	Sea urchin
	Seasonal	Sympatric	Buttercup and lesser celandine
Post-zygotic	Hybrid inviability	Sympatric	Northern leopard frog and wood frog
	Hybrid sterility	Sympatric	Mule
	Hybrid breakdown	Sympatric	Cotton

Darwinian evolution

WORKING SCIENTIFICALLY

Evolution is a property of a population, describing a change in the average phenotype, produced by changes in allele frequency in the population. These changes are brought about by:

- Mutation
- Gene flow
- Genetic drift
- Natural selection.

Charles Darwin's name is traditionally associated with the theory of evolution, although Alfred Russel Wallace's contribution has become more widely acknowledged. In 1859, Darwin published 'On the origin of species by means of natural selection' in which he described natural selection as the mechanism that causes gradual changes from one generation to the next over long periods of time. Each generation was very slightly different from the preceding generation, which Darwin called 'descent with modification'. Our current understanding of stabilising selection shows that natural selection can also maintain characteristics in a population.

After failing to become a doctor or clergyman, following enormous efforts to persuade his father of his ambitions to be a naturalist, Charles Darwin (1809–82) joined the crew of HMS Beagle, a scientific survey ship. He was employed as a naturalist and companion to the ship's captain in a voyage around the world, between 1831 and 1836. The visit to the Galapagos Islands, a group of volcanic islands about 600 miles off the coast of Ecuador, showed Darwin the variability of life there in comparison with the mainland. It was formative in the intellectual strides he made in the development of his view of the natural world.

Galapagos giant tortoise

Cuban stamp from 2009 commemorating the 150th anniversary of the publication of *On The origin of Species* showing the route of HMS Beagle

Darwin's observations of variation in a population and the tendency for the adult population to be stable in number led him to propose that those organisms that are better adapted to their environment are more likely to survive and reproduce, producing offspring that are similarly successful. Darwin observed:

- Members of a population show variation.

- Individuals within a population have the potential to produce large numbers of offspring, yet the number of adults remains constant from one generation to the next.

- Resources are limited.

From these observations, Darwin deduced:

- There is a 'struggle for existence' with only the 'fittest' surviving. Herbert Spencer introduced the term 'survival of the fittest' in 1864.

- Those that survive reproduce and their offspring have the characteristics that enable them to succeed. In time, a group of individuals undergoes many changes and becomes sufficiently distinct to belong to a new species.

- If the environmental conditions change, the features needed to survive in it will change, so natural selection is a continuous process.

In more modern terms, this could be expressed as:

There is competition for limited resources and those individuals that have alleles producing a better-adapted phenotype reproduce more successfully. They pass the alleles that produced this phenotype to their offspring, and so, as this continues over the generations, those alleles are represented in increasing proportion in the population. The average phenotype therefore changes. This change is evolution. If the changes are significant enough, an organism may no longer be able to interbreed successfully with a representative of the original population and so a new species has formed. Natural selection maintains phenotypes in a stable environment and changes them in a changing environment.

Evolutionary biologists distinguish between two types of evolutionary product:

- Adaptations are features used in an original form, such as the larynx of a bat making echolocation signals.

- Exaptations are structures that appear to have had their original use modified, such as the co-option of feathers maintaining body heat into feathers used for flight.

Natural selection, the founder effect and genetic drift can only modify the phenotype by acting on the alleles already present. Evolution, therefore, can only modify what is there, and it does so in response to environmental conditions. If the environment changes, different structures evolve. This means that:

Caudipteryx was a feathered dinosaur about the size of a peacock

- Evolution is not a direct process going from simple to complex or small to large. It is entirely contingent upon the environment, e.g. ancestors of the tapeworm, *Taenia*, were free-living, having evolved complex nervous, locomotory and excretory systems. These have been greatly reduced in *Taenia*. This shows that evolution does not necessarily follow a linear pathway.

- Some structures do not appear to be ideally suited to their current function, e.g. most humans at some time have back problems, suggesting that our upright stance may still be in the process of evolutionary modification.

Going further ▶

Many aspects of biology and geology provide evidence for evolution but evidence must be interpreted with care:

Sometimes an organism is identified as being from a taxon thought to be extinct. The taxon is called a **Lazarus taxon**, because it is as if it were dead and has come back to life, e.g. the Coelacanth is a fish that was thought to have become extinct 65 million years ago. But it was rediscovered in 1938, off the east coast of South Africa.

Research may show that a fossil was identified in error, because it has shown convergent evolution with an extinct taxon. It is in an **Elvis taxon** – organisms resemble others no longer alive, but they are not the real thing, e.g. external features of fossils of *Rhaetina gregaria* suggested it was a brachiopod from the late Triassic (210 million years ago). But the internal structure showed many differences from other late Triassic brachiopods and actually, it was a member of the genus, *Lobothyris*, so it is now called *Lobothyris gregaria*. *L. gregaria* is an Elvis species.

A fossil that has been freed from its rock, but re-fossilised in a younger rock is a **zombie taxon**, e.g. trilobites from Cambrian limestone, from 470 million years ago, have been found re-fossilised in Miocene siltstone, from only 15 million years ago.

Going further ▶

The fossil record suggests that one species gradually changing into another is much rarer than adaptive radiation, where many species form from a single ancestral species.

CH 13

Application of reproduction and genetics

Powerful techniques to sequence and use DNA, therapeutic stem cell cloning and tissue engineering have been developed with tremendous potential for treating genetic diseases, but ethical issues must be resolved. The polymerase chain reaction enlarges DNA samples for testing. Its use, coupled with genetic fingerprinting, has raised concern over the misuse of stored personal data. Recombinant DNA technology has important applications, including the production of medically important molecules. Gene insertion into crop plants may improve yield and introduce tolerance to environmental threats.

Topic contents

By the end of this topic you will be able to:

- Describe the aims and achievements of the Human Genome and 100K Project.
- Describe the ethical concerns surrounding the use of this knowledge.
- Describe the potential of sequencing DNA from non-human organisms.
- Describe the processes and limitations of genetic fingerprinting and the polymerase chain reaction.
- Describe the role and impact of genetic profiling in society.
- Describe the use of recombinant DNA technology in bacteria.
- Describe examples of genetically modified crops and issues surrounding their use.
- Discuss the implications of genetic screening and gene therapy.
- Discuss the ethical issues surrounding gene therapy.
- Discuss the use of genomics in healthcare, and implications for its use in the future.
- Describe the use of stem cells and their use in tissue engineering.
- Describe the ethical issues surrounding the use of stem cells.

The Human Genome Project and the 100K Genome Project

The Human Genome Project

Sequencing the nucleotides in the human genome was proposed in 1985 and the Human Genome Project accordingly began in 1990. It was planned to take 15 years but rapid advances in DNA sequencing and in computing allowed a working draft to be published in 2000, followed by a more complete draft in 2003. Though the project is complete, analysis continues and it will take many years to study all the data. The Human Genome Project was designed to improve knowledge and understanding of genetic disorders and consequently, improve their diagnosis and treatment.

Its aims were to:

- Identify all the genes in the human genome and identify which chromosome each is on.
- Determine the sequence of the 3 billion base pairs in human DNA and store this information in databases.
- Improve tools for data analysis.
- Transfer related technologies to the private sector, to develop medical innovation.
- Address the ethical, legal and social issues that may arise from the project.

The main findings of the Human Genome Project were:

- Humans have about 20 500 genes, far fewer than anticipated.
- There are more repeated segments of DNA than had previously been suspected.
- Fewer than 7% of the families of proteins were specific to vertebrates, emphasising the close relationships between all living organisms.

Sanger sequencing

The project used a method of sequencing DNA called 'Sanger sequencing' or the 'chain termination' method. Sanger won his second Nobel Prize for this invention:

- DNA was broken into single-stranded fragments of different lengths, up to about 800 bases long.
- Complementary strands were synthesised but these were incomplete because the four nucleotide triphosphates (NTP) were altered. For each NTP, the 3'OH of the deoxyribose was removed, to make a dideoxynucleotide, i.e. lacking both the 2'OH and the 3'OH. So when it was incorporated into the newly synthesised strand, DNA polymerase would not be able to bind the next nucleotide and the chain could not lengthen, i.e. the chain was terminated. This last nucleotide was marked with a radioactive isotope, an antigen or fluorescent marker, with a different one for each of the four nucleotides.
- From each original DNA fragment, a large number of complementary DNA strands of increasing length were produced. They were separated by gel electrophoresis, according to their size, and the terminal, marked nucleotide was identified. As all the terminal nucleotides were known, in fragments of increasing size, the base sequence of the original DNA fragment was known.

Using this method, it took a year to sequence a million base pairs. There are many faster techniques in use now, including passing DNA through nanopores in protein molecules. New rapid techniques are collectively known as Next Generation Sequencing (NGS) and they can sequence an entire genome within a few hours.

Going further ▶

The Human Genome Project investigated about 90% of the genome, but did not include centromeres or telomeres.

Exam tip

Remember the aims of the Human Genome Project.

Key Term

Electrophoresis: A lab technique that separates molecules on the basis of size, by their rate of migration under an applied voltage.

Exam tip

You will not be asked to recall the stages of the Sanger sequencing method.

Samples being loaded automatically on to a gel for electrophoresis

▼ Study point

The current approach to health is to encourage prevention rather than cure. It is cheaper and beneficial to the population as a whole.

The 100K Genome Project

Following the success of the Human Genome Project, The 100K Genome Project was launched in 2012, to use NGS to sequence 100 000 genomes from NHS patients with cancer or a rare disease, and from members of their families. It is run by Genomics England, under the Department of Health. The project aims to:

- Create an ethical, transparent programme based on consent.
- Set up a genomic service for the NHS to benefit patients.
- Enable medical and scientific discovery.
- Develop a UK genomics industry.

It is anticipated that by combining sequence data with medical records, the causes of diseases will be recognised and understood and their diagnosis and treatment improved.

Morals and ethical concerns

A vast quantity of data has been produced by the Human Genome Project and the 100K Genome Project and its potential is profound. We do not know how this information might be used in the future. Society has yet to decide how it should be treated and where legal and moral responsibilities for it lie. Ethical issues cover many areas, including:

- Ownership of genetic information: once base sequences are known, it must be clear who owns the information. If it is the property of the individual, then safeguards must be put in place that it is not misused, e.g.
    - If a person is identified as having a DNA sequence that might predispose them to heart disease, this information should not be used to set their insurance premiums or to deny them life and health insurance.
    - If a DNA sequence suggests a particular ancestry, this should not be a pretext for social discrimination.
    - No company should make financial profit from using a DNA sequence without permission.
- The identification of allele sequences: a patient's DNA can be scanned for mutated sequences that may be correlated with future health problems. Some people do not wish to have this knowledge about themselves. But if the same health problems could affect close relatives, it must be clear whether or not the relatives have a right to the information.
- Genetic screening can be useful in association with genetic counselling. If a family has a history of a genetic defect, family members can consult a genetic counsellor for advice on the risk for themselves or, potentially, for their children. Advice may be based on: which members of the family have the condition, whether parents are closely related and the frequency of the relevant mutated gene in the general population. If a DNA sequence is available, the genetic counsellor's advice has more significance.
- Embryos made during the process of in vitro fertilisation can be screened for the presence of alleles leading to conditions including cystic fibrosis, Huntington's disease and thalassaemia. A choice can then made to only implant a healthy embryo. A legal framework already exists for the use of 'spare embryos' for research.
- The popular press has made much of embryo screening and the potential for choosing alleles to ensure specific characteristics. Many of the characteristics that are considered desirable, such as sporting ability or appearance, are the result of many interacting genes, their epigenetic modifications and the environment. Many of these genes are, as yet, unknown. It is not possible that they can be selected or that the environment can be so finely controlled that the desired outcome is produced. If, in the fullness of time, techniques are developed that will allow this, society will have to decide whether,

▼ Study point

The incidence of some genetic diseases, e.g. thalassaemia, an inherited blood disease common in some Mediterranean countries, is falling as a result of genetic testing.

how and by whom such choices may be made. A prudent society would begin the discussion far in advance of decisions being needed.

- Parents may wish to have their children screened, to know if they carry sequences that might pre-dispose them to adult diseases, such as Alzheimer's disease, breast or ovarian cancer. A decision must be made as to whether or when to tell the child the results of such a test. Society must decide if a parent has the right to such knowledge about a child, as it suggests that the child's DNA is the property of the parent. This has yet to be resolved.

- Storage and security of genomic data is a concern because of the potential for computer storage to be hacked.

▼ **Study point**

Genetic research will continue. It is essential to ensure that ethical guidelines are developed and followed, enabling safe and responsible use of information.

DNA sequencing of non-human organisms

Genomes have been sequenced from organisms in all domains and kingdoms of life and from a large number of phyla. The table shows some landmark organisms for which the genome has been sequenced. Species are chosen on account of their scientific or medical significance, or for their economic or cultural importance.

Organism	Date sequenced	Reason	Number of bases
MS2 an RNA bacteriophage	1976	1st organism	3569
ØX174 a DNA bacteriophage	1977	1st DNA sequence	5386
Haemophilus influenzae	1995	1st bacterium	1.8×10^6
Saccharomyces cerevisae baker's yeast	1996	1st eukaryote	12.1×10^6
Caenorhabditis elegans a nematode worm	1998	1st animal	10×10^6
Arabidopsis thaliana a small Angiosperm	2000	1st plant	119×10^6
Drosophila melanogaster fruit fly	2000	1st insect	165×10^6
Homo sapiens human	2001	Human	3200×10^6

The examination of genomes of closely related organisms allows inferences to be drawn concerning evolutionary relationships. These provide a true phylogenetic classification and are compared with schemes based on phenotypic characteristics. In some cases, unexpected relationships have been discovered. Primate sequences, such as that of the chimpanzee, have been of particular interest as they contribute to the understanding of human origins. Comparisons also indicate to conservation scientists which species need particular protection.

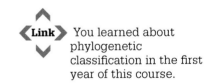

Link You learned about phylogenetic classification in the first year of this course.

The challenge of malaria

Malaria is common in the tropical and subtropical regions around the equator, in sub-Saharan Africa, Asia, and Latin America. In 2013, the World Health Organisation reported 198 million cases with over half a million deaths, of which 90% were in Africa. It is of enormous importance that research into malaria and its treatment continue. Chemicals have been used to attack both the vector, *Anopheles gambiae*, and the parasite, *Plasmodium falciparum*.

Going further ▶

12 insecticides in 4 classes are used against *Anopheles*. They are the organochlorines, organophosphates, pyrethroids and carbamates.

Going further ▶

The CRISPR–Cas9 system can cut DNA at any desired location and insert other DNA sequences. It could treat genetic diseases, fight infections and increase food crop yields, but its application generates ethical concerns.

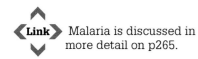

Link Malaria is discussed in more detail on p265.

Going further ▶

Substandard or fake drugs and unregulated and poorly supervised drug use are the main reasons that *Plasmodium* has become drug resistant.

Killing the vector, *Anopheles gambiae*

Insecticides are used in indoor sprays to kill mosquitos in buildings, where they rest on walls after feeding on blood. But the mosquitos are increasingly resistant to these insecticides, particularly in Africa. If a mosquito becomes resistant to an insecticide, it becomes resistant to all insecticides in that particular class. Pyrethroid resistance is a particular problem, as that is the only insecticide recommended for use with the nets under which people sleep, to protect them at dawn and dusk, when the mosquitos feed.

The DNA sequencing of the *Anopheles gambiae* genome was performed in 2002. The sequence is used to try to develop chemicals that can prevent the mosquito from transmitting malaria, by making it susceptible to insecticides once more.

In 2015, a genetically modified mosquito was produced using a gene-editing technology called CRISPR–Cas9, that allows genes to be 'written into' a genome. Mosquito eggs were modified with the addition of a gene that would allow them to synthesise an antibody against *Plasmodium*. Then if the mosquito acquired *Plasmodium* when taking blood from an infected person, the *Plasmodium* would not survive in the mosquito. The mosquito would not then spread infection when biting people in future. In the lab, 99.5% of the offspring of modified mosquitos carried the gene. This mosquito will not be released into the wild, but it provides a model system for future progress in malaria control.

Killing the parasite, *Plasmodium falciparum*

Drugs to kill *P. falciparum* have been used since the early 17th century, when extracts from *Cinchona* bark were first used. They contained quinine, which disrupts *Plasmodium*'s digestion of haemoglobin in the red blood cells. A toxic derivative of haemoglobin accumulates and kills the *Plasmodium*.

Spontaneous mutations in *Plasmodium* have caused resistance to quinine and subsequently, to other drugs. For some drugs, a single point mutation is all that is needed to generate resistance; for others, several mutations are needed. Over time, resistance has become established in populations and it can be very stable, persisting long after a particular drug is no longer used.

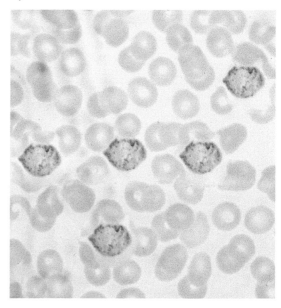

Normal red blood cells and red blood cells infected with *Plasmodium*

- The drug chloroquine, like quinine, disrupts the digestion of haemoglobin in *Plasmodium*'s food vacuole. But mutant *Plasmodium* expels chloroquine from its food vacuole 50 times faster than normal *Plasmodium*, so there is not enough time for the drug to have its effect. The mutant is consequently resistant to the drug.

- Atovaquone kills *Plasmodium* by acting on the electron transport chain in its mitochondria. Resistance to it develops very rapidly, caused by single point mutation in the gene for cytochrome b.

- Artimesinin, from sweet wormwood, *Artimesia*, is used in combination with other drugs. It also acts on the *Plasmodium* in the red blood cells, although the mechanism is not well understood. Resistance has now been detected to artimesinin as well.

The genome of *Plasmodium falciparum* was sequenced in 2002. It is hoped that knowledge based on a better understanding of the genetic control of *Plasmodium* infection will allow the development of more effective drugs.

Genetic fingerprinting

A person's DNA profile is their 'genetic fingerprint'. About 99.9% of the human genome is the same in every person, but apart from monozygotic (identical) twins, the remaining 0.1% makes an individual's genetic fingerprint unique. A genetic fingerprint is not the same as a DNA sequence, because it represents only non-coding portions of DNA. It relies on two techniques:

- The polymerase chain reaction to make large numbers of copies of DNA fragments.
- Gel electrophoresis, to separate the DNA fragments based on their size.

The DNA of a genetic profile

Less than 2% of human DNA codes for proteins, in sections of the genome called exons. They have within and between them base sequences called introns that do not code for proteins. The introns have sequences of nucleotides in which up to 13 bases repeat up to several hundred times. These are called 'short tandem repeats' or STRs.

The number of repeats within a particular STR is different in different individuals and this is what makes a genetic fingerprint unique, e.g. chromosome 7 in humans has an STR called D7S280, in which the bases GATA repeat between 6 and 15 times in different alleles. The number of repeats is inherited and so can be used for identifying people and for tracing family relationships.

Here is part of the DNA sequence of chromosome 7 set out in the standard fashion, with bases written in lower case letters, numbered and written in groups of 10. The repeated sequence is shown in red:

61-120	tattttaagg	ttaatatata	taaagggtat	gatagaacac	ttgtcatagt	ttagaacgaa
121-180	ctaacgatag	atagatagat	agatagatag	atagatagat	agatagatag	atagacagat
181-240	tgatagtttt	tttttatctc	actaaatagt	ctatagtaaa	catttaatta	ccaatatttg

The polymerase chain reaction (PCR)

PCR is semi-conservative replication of DNA in a test-tube. It greatly amplifies the DNA, i.e. it makes millions of copies, and it works rapidly. This makes PCR useful with very small or degraded samples.

The DNA sample is dissolved in a buffer and mixed with:

- Taq polymerase. This is DNA polymerase from the bacterium *Thermus aquaticus*, which lives in hot springs and hydrothermal vents. Taq polymerase has an optimum temperature of about 80°C but even at 97.5°C, it remains active for 9 minutes before denaturing.
- Nucleotides containing the four DNA bases.
- Short single-stranded pieces of DNA between 6 and 25 bases long, called **primers**. They are complementary to the start of the DNA strand and bind to it, signalling taq polymerase to start replication.

The technique depends on rapid temperature change, which happens in a device called a 'thermocycler'.

YOU SHOULD KNOW ›››

››› How the technique of genetic fingerprinting is carried out

››› The uses of genetic fingerprinting

››› How DNA is copied using the polymerase chain reaction

››› Ethical considerations of these techniques

Going further ▶

A variable number of STRs occur in the gene for huntingtin, the protein implicated in Huntington's disease. People with more repeats have earlier onset of the disease and more severe symptoms.

Going further ▶

An STR is sometimes called a hypervariable region (HVR). Two mitochondrial HVRs are used extensively in research into human family relationships and human origins.

◢ Key Term

Primer: A strand of DNA about 10 nucleotides long that base pairs with the end of another longer strand, making a double-stranded section, to which DNA polymerase may attach prior to replication.

Going further ▶

Kary Mullis won a Nobel Prize in Chemistry for his development of PCR. This achievement is unusual for someone who believes in astrology and extra-terrestrial visitation and who denies climate change and that HIV causes AIDS.

▼ **Study point**

Forensic scientists often use PCR when producing a genetic fingerprint, to increase the quantity of DNA because the sample obtained at a crime scene may be very small. After 40 cycles, the number of copies made from each original fragment $= 2^{40} = 1.1 \times 10^{12}$ (1 dp).

▼ **Study point**

It is important that the fragments of DNA used in PCR are not contaminated with any other biological material as the contaminants may contain DNA, which would also be copied.

▼ **Study point**

The limitations to using PCR include the error rate and its limit on suitable fragment size, the presence of inhibitors and contaminants and the biochemical limits to the process.

The stages of PCR are:

- The original 'target' DNA is heated to 95°C, separating it into two single strands.

- The solution is cooled to 55°C, which is cool enough for the primers to anneal to the complementary base sequences on each of the single strands of DNA.

- The solution is heated to 70°C and taq polymerase catalyses the synthesis of a complementary strand by adding complementary nucleotides and catalysing the formation of phosphodiester bonds in the sugar-phosphate backbone. This is the elongation or extension phase. For each initial fragment of double-stranded DNA, two identical double strands are produced.

- The sequence is repeated many times.

Placing samples in a thermocycler

Limitations of PCR

Before PCR was invented, copies of genes were made by inserting them into replicating micro-organisms. PCR is an increasingly valuable tool but the traditional method will still be used because PCR has limitations:

- Contamination: any DNA that enters the system by accident can be amplified. The contaminating DNA may be air-borne, come from the experimenter or from contaminated reagents. But most contamination comes from previous PCR reactions using the same apparatus.

- Error rate: all DNA polymerases sometimes insert a nucleotide containing the wrong base. They usually proofread and correct their errors, but taq polymerase cannot do this. It makes an error about once in every 9000 nucleotides. After 30 cycles of PCR, the error rate becomes 1 per 300 nucleotides, because each cycle copies and multiplies the previous errors, so they accumulate.

- DNA fragment size: PCR is most efficient for making DNA about 1000–3000 base pairs long because taq polymerase cannot correct its errors. If a lower temperature, higher pH and a proofreading polymerase in addition to taq polymerase are used, a length of 40 000 base pairs can be generated. But many genes, including human genes, are much longer than this.

- Sensitivity to inhibitors: molecules in the sample may act as inhibitors and PCR is very sensitive to them, e.g.:

  – Phenolics, especially in plant material

  – Humic acids in archaeological specimens

  – Haem breakdown products, which bind with the Mg^{2+} needed for DNA polymerase to function

  – The traditional blue dye used on denim.

- Limits on amplification: at the start of PCR, the number of DNA molecules made increases exponentially. After about 20 cycles, it slows down: the increase becomes linear and then plateaus because:

  – Reagent concentrations become limiting

  – The enzyme denatures after repeated heating

  – DNA in high concentrations causes the single-stranded molecules to base pair with each other rather than with the primers.

Gel electrophoresis

- DNA is extracted from biological material and cut into thousands of fragments of varying lengths, using restriction endonucleases.

- The DNA fragments are separated by length with gel electrophoresis, on an agarose gel. Agarose is a polysaccharide extracted from seaweed and makes a gel with pores, through which small molecules can move:
  - DNA samples are loaded into wells at one end of the gel.
  - A voltage is applied across the gel. The phosphate groups of the DNA backbone have a negative charge so the fragments are attracted to the anode. Smaller fragments move more easily through the pores and so they migrate through the gel faster than the larger fragments.
  - If fragments of known length are separated on the same gel at the same time, making a 'DNA ladder', the lengths of the fragments under test can be estimated.

- The electrophoresis trough is covered with a nylon membrane, which touches the gel and picks up the DNA fragments. This process is called Southern blotting.

- Radioactive or, more commonly, luminescent DNA **probes** that contain sequences complementary to the STRs attach by base pairing to specific parts of the fragments. Any unbound probes are washed off.

- A film that is sensitive to X-rays or the wavelengths emitted by the luminescent probe is placed over the Southern blot overnight.

- The film is exposed and the autoradiograph reveals a banding pattern in which the dark bands show the position of the probe, and therefore the repeated sequences. This pattern is the genetic fingerprint.

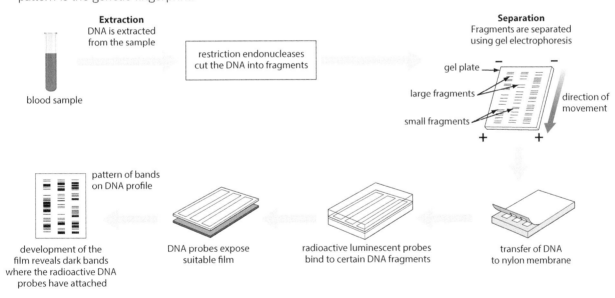

Extraction
DNA is extracted from the sample

blood sample

restriction endonucleases cut the DNA into fragments

Separation
Fragments are separated using gel electrophoresis

gel plate

large fragments

small fragments

direction of movement

pattern of bands on DNA profile

development of the film reveals dark bands where the radioactive DNA probes have attached

DNA probes expose suitable film

radioactive luminescent probes bind to certain DNA fragments

transfer of DNA to nylon membrane

Genetic fingerprinting

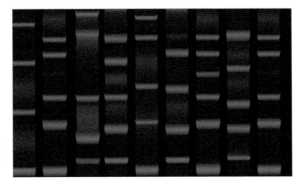

Nine DNA profiles produced by gel electrophoresis

DNA profiling in society

The technique of DNA fingerprinting was invented by Professor Sir Alec Jeffreys. It has been a vital tool in ensuring justice. Its first use, in 1985, was to confirm the identity of a British boy originally from Ghana, showing his very close relationship with other members of his family. The first forensic use was to identify the killer of two girls in the same Leicestershire village who were raped then murdered three years apart, in 1983 and 1986. Not only did it verify the identity of the murderer, but it exonerated someone who had initially been a prime suspect and would, wrongly, have been given a most severe prison sentence.

DNA profiling has been used in many situations, including testing for:

- Paternity – the DNA from white blood cells is used to construct DNA profiles. The bands in a child's profile are compared with the mother's. Any bands that they share were inherited from her. The remaining bands in the child's profile must have been inherited from the father. If they do not match the alleged father, he is not the child's biological parent. If they do match, there is a high probability that he is the father, but this technique cannot provide absolute proof.

- Twins – monozygotic (identical) have identical banding patterns in their DNA profiles whereas dizygotic (fraternal) twins do not. At birth, twins may look very similar so this is a useful tool for distinguishing which type of twin they are.

- Siblings – people who have been adopted may wish to confirm that alleged biological siblings are in fact their blood relatives. If they are, their DNA profiles will show many similarities, as, on average, half their genes will be identical.

- Immigration – some visa applications depend on proof of relatedness.

- Forensic use to identify and rule out suspects in criminal cases.

- Phylogenetic studies – profiles of members of different taxa can be compared to determine whether they have been classified suitably and to determine how closely they are genetically related.

DNA profiling has been of increasing significance since its instigation in the 1980s. Arguments for and against its use are related to moral and ethical aspects of possession of personal knowledge. The arguments resemble those used in the context of DNA sequencing.

The pros of DNA profiling

- It does not require an invasive method to obtain a biological sample, as mouth swabs, urine or hair can be used to obtain DNA, rather than blood samples.

- The technique can be used on samples that would be too small for blood testing.

- It has reversed wrongful convictions when used with other forensic tools and evidence.

- DNA profiling can rule out non-matches of DNA samples, to exonerate people who have been falsely accused.

- Efforts are being made to store genetic material from people around the world, before isolated groups are intermixed and lost.

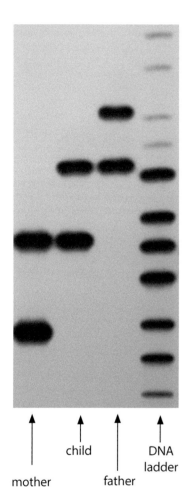

child

mother   father   DNA ladder

DNA profile indicating paternity

▼ Study point

Remember the difference between DNA profiling and DNA sequencing: in sequencing, the nucleotide sequence is determined but in profiling only the short repeated sequences are considered.

The cons of DNA profiling

- Some people consider that any request for a DNA sample is a violation of an individual's right to privacy and of their civil liberties.

- DNA profiles are held in computer databases, which are vulnerable to misuse and hacking. Individuals may suffer loss of privacy; information taken ostensibly for one purpose may be used for another.

- Profiles offer probabilities, not absolutes. With a given number of samples on a database, there may be a one in a million chance that an apparent match does correctly identify someone. If more samples are available, the probability of an apparent match being correct may rise to one in one hundred million.

- Access to and use of data must be carefully regulated, as with DNA sequencing. Health insurers could conceivably use it to deny coverage or claims; prospective employers could avoid hiring those who have certain genetic traits or risks for certain diseases; private information, such as gender reassignment, could be made public without consent.

- DNA profiling may produce wrongful convictions if:
  - It is used inappropriately to influence juries and judges, especially if they lack an understanding of the significance of the results.
  - Errors may occur in the procedure.
  - People conducting the tests may not be trustworthy.
  - DNA evidence is planted at a crime scene.

Genetic engineering

Genetic engineering allows genes to be manipulated, altered and transferred from one organism or species to another, making a genetically modified (GM) organism. Applications of genetic engineering include the transfer of genes or gene fragments into:

- Bacteria, so that they can make useful products, such as insulin.

- Plants and animals, so that they acquire new characteristics, for example resistance to disease.

- Humans, to reduce the effects of genetic diseases, such as Duchenne muscular dystrophy.

When genetic material from two species is combined, the result is **recombinant DNA**, so genetic engineering is sometimes referred to as recombinant DNA technology. Organisms that have DNA from another species introduced into their cells are described as **transgenic** organisms. The introduced DNA is the **donor DNA** and the organism into which it has been introduced is the **host**. When a cell has incorporated a plasmid containing a foreign gene, it is described as **transformed**. Organisms which have had genes altered or deleted are genetically modified, but they are not transgenic because they do not contain any foreign genes.

The process of producing a protein using genetic engineering technology involves:

- Isolation of the DNA fragments.
- Insertion of the DNA fragment into a **vector**.
- The transfer of the DNA into a suitable host cell.
- Identification of host cell that have taken up the gene, using gene markers.
- Cloning the transformed host cells.

Knowledge check

Link the appropriate temperatures 1–3 with the descriptions A–C of PCR.

1. 95°C.
2. 55°C.
3. 70°C.

A. Taq polymerase catalyses the synthesis of a complementary strand.
B. DNA separates into two single strands.
C. Primers anneal to the DNA.

YOU SHOULD KNOW ›››

››› Two methods of isolating DNA fragments

››› How recombinant DNA is introduced into host cells

››› How vectors, such as bacterial plasmids, are used to transfer DNA fragments

››› That markers are used to identify successful gene uptake

Key Terms

Recombinant DNA: DNA produced by combining DNA from two different species.

Transgenic: An organism that has been genetically modified by the addition of a gene or genes from another species.

▼ **Study point**

With this method of insulin production, human, rather than animal, insulin is used. It reduces the use of animals for medical processes and carries less risk of an adverse immune response.

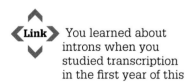

Key Terms

Restriction enzyme: Bacterial enzyme that cuts the sugar-phosphate backbone of DNA molecules at a specific nucleotide sequence.

Sticky end: A sequence of unpaired bases on a double-stranded DNA molecule that readily base pairs with a complementary strand.

Link You learned about introns when you studied transcription in the first year of this course.

DNA identification and isolation

To explain the principles of using gene technology to produce useful molecules on a large scale, the production of human insulin in the bacterium *Escherichia coli (E. coli)* is described. Insulin for human use has been made this way since 1978.

Locating the gene

A donor molecule of human DNA contains the gene that codes for insulin. Locating the correct piece of DNA is not easy: the insulin gene is one of about 20 500 genes in the human genome and the cell contains only two copies of it. The gene can, however, be identified with a gene probe, i.e. a specific segment of single-strand DNA that is complementary to a section of the gene.

Isolating the gene

The identified and located gene can be isolated with either of two enzymes, restriction endonuclease or reverse transcriptase.

1 – Using restriction endonuclease

Restriction endonucleases are bacterial enzymes that cut DNA at specific nucleotide sequences. As the sequences occur in many places, the DNA is cut into many small fragments and individual genes can be isolated. Some **restriction enzymes** cut straight across a DNA double helix, making a blunt cut. But many make a staggered cut, which leaves unpaired bases on both strands. These bases pair with complementary sequences very readily, so they are called **sticky ends**.

The bacterium *E.coli* produces a restriction endonuclease called EcoR1. It was called this because it was the first to be isolated and it came from *E. coli* strain RY13. EcoR1 catalyses the formation of breaks in the DNA backbone in a specific sequence of nucleotides, where a nucleotide containing guanine is next to one containing adenine. The line of cut is staggered and leaves sticky ends when the cut strands are separated. The four unpaired bases at each end of the two strands are in the reverse order, so they form a palindrome. If EcoR1 made a cut either side of the insulin gene, the gene would be isolated from the rest of the genome.

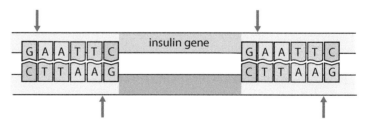

arrows indicate points where the enzyme breaks the strand

Use of restriction enzymes to cut DNA

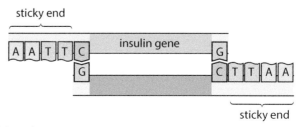

Production of sticky ends

There are two main drawbacks with using restriction endonuclease to excise a gene:

- If the recognition sequence occurs within the gene of interest, the gene will be broken into fragments that have no function.

- Eukaryotic genes contain **introns**, i.e. nucleotide sequences that do not contribute to the mRNA that codes for the polypeptide. In normal situations, introns are removed from RNA transcripts read from the DNA in the nucleus, before the mature mRNA moves to the cytoplasm. But using the whole gene means that the introns are present and so will be incorporated into the plasmids. Bacteria do not have introns in their genome and may not have the appropriate enzymes to process the RNA to remove them after transcription. Any protein translated will, therefore, contain extra amino acids representing the intron sequences, and will not be functional.

2 – Using reverse transcriptase

Although each cell contains only two copies of the insulin gene, there may be many molecules of mRNA that have been transcribed from it. This is especially true in cells that synthesise and secrete the gene product: the cytoplasm of the ß-cells of the pancreas has large quantities of the mRNA transcribed from the gene coding for insulin. This mRNA can be extracted.

Reverse transcriptase is an enzyme that produces DNA from an RNA template. It made by a group of viruses called retroviruses. The enzyme synthesises DNA, called copy DNA or cDNA, which is complementary to the RNA. With this method, many copies of cDNA that is complementary to the mRNA for insulin can be made.

This method does not have the problem of introns because the cDNA is made from mRNA from the cytoplasm. The RNA in the nucleus that has been transcribed from the DNA has been processed to remove introns so they do not appear in the mature mRNA. DNA polymerase then catalyses the synthesis of DNA that is complementary to the single-stranded cDNA, making a double-stranded molecule containing the gene for insulin.

cell of human pancreas

mRNA coding for insulin

| reverse transcriptase |

single-stranded copy DNA formed

| DNA polymerase |

double-stranded DNA (copy of insulin gene)

Reverse transcriptase and gene isolation

Making a recombinant plasmid

A bacterial cell is unlikely to take in a gene spontaneously, so the gene must be carried into the cell by a **vector**. Viruses have been used as vectors, but in this instance, the vector is a **plasmid**, i.e. a small, double-stranded circle of DNA found in bacteria. A plasmid is much smaller than the bacterial chromosome and contains only a few genes. Plasmids can move in and out of cells, which makes them useful for introducing genes into bacteria.

To isolate plasmids, the bacteria containing them are treated with:

- EDTA to destabilise the cell walls

- Detergent to dissolve the phospholipid cell membrane

- Sodium hydroxide to make an alkaline environment that denatures the membrane proteins.

The plasmids can then be separated from the cell debris.

The circular DNA molecule making up the plasmid is cut open using the same restriction endonuclease as was used to isolate the gene, which means it has the same nucleotide sequence in its sticky ends. The vector and gene are mixed and their complementary base sequences base pair with each other. The gene is now loosely bound to the plasmid

Key Term

Reverse transcriptase: Enzyme, derived from a retrovirus, that catalyses the synthesis of cDNA from an RNA template.

Going further　▶

The retrovirus group includes HIV. The retrovirus genome is RNA, not DNA. When they infect cells, retroviruses use their genome RNA as a template to synthesise cDNA, using reverse transcriptase.

▼ Study point

Using reverse transcriptase rather than restriction enzymes overcomes the problems of locating the gene, cutting it into non-functional fragments, the presence of introns and the need to process RNA to make it functional.

Key Terms

Vector: A virus or plasmid used as a vehicle for carrying foreign genetic material into a cell.

Plasmid: Small circular loop of self-replicating double-stranded DNA in bacteria.

▼ Study point

The same restriction enzyme is used to cut open the donor DNA and the plasmid, so that their sticky ends are complementary.

Key Terms

DNA ligase: Enzyme that joins together portions of DNA by catalysing the formation of phosphodiester bonds between their sugar-phosphate backbones.

and the join is made permanent with the enzyme **DNA ligase**. A ligase binds molecules together, in this case, the sugar-phosphate backbones of the gene and the plasmid. The gene has been 'spliced' into the vector, so the plasmid is now recombinant DNA.

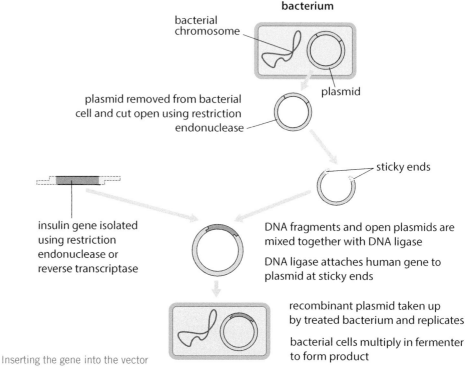

bacterium

bacterial chromosome

plasmid

plasmid removed from bacterial cell and cut open using restriction endonuclease

sticky ends

insulin gene isolated using restriction endonuclease or reverse transcriptase

DNA fragments and open plasmids are mixed together with DNA ligase

DNA ligase attaches human gene to plasmid at sticky ends

recombinant plasmid taken up by treated bacterium and replicates

bacterial cells multiply in fermenter to form product

Inserting the gene into the vector

To make a good vector, a structure should:

- Be self-replicating
- Be small
- Not be broken down by host cell enzymes
- Not stimulate an immune response in the recipient
- Be able to be screened to confirm that the gene that was actually inserted into the plasmid
- Have markers to allow host cells that have successfully taken up the vector to be identified.

Transfer of DNA into the host cell

When plasmids are mixed with bacterial cells, as few as 1% of the bacterial cells take up the plasmid and become transformed. This can be increased with calcium chloride: the positive charge on the calcium ions binds the negatively charged DNA backbone of the plasmid and the membrane lipopolysaccharides. The plasmid DNA passes into the cells with a heat shock, in which cells chilled to 4°C are briefly heated to 42°C.

The use of genetic markers

To obtain bacteria that contain plasmids with the gene in them, the plasmid must successfully incorporate the gene and the bacteria must successfully take up plasmids. Both of these events can be confirmed.

- A vector that has not taken up the gene of interest is described as empty. To ensure that the vector has taken up the gene, its DNA can be sequenced.
- To identify which cells have been transformed, plasmids with antibiotic-resistant genes are used. These confer resistance to one or more antibiotics, such as ampicillin, tetracycline and chloramphenicol. The cells are cultured in a growth medium containing

Link You learned about plasmids when you studied cell structure in the first year of this course.

the antibiotic and if they have incorporated the plasmid, they also contain a gene for antibiotic resistance. They break down the antibiotic and can grow. If they do not contain the plasmid, they do not have the resistance gene and the antibiotic kills them. Surviving cells must, therefore, contain the antibiotic resistance gene and therefore must contain the plasmid. An antibiotic resistance gene is a **marker gene** because it marks the presence of the plasmid.

- To distinguish which transformed bacterial cells have taken up empty plasmids, i.e. lacking the gene of interest, 'blue-white screening' is used: bacterial cells are grown on medium containing a lactose analogue, X-gal. They turn white if they contain a plasmid with the gene, but blue if the plasmid is empty.

Bacterial cells with recombinant plasmids, i.e. containing the insulin gene, are cultured in large volumes in fermenters. Each culture forms a clone. Cloning of the recombinant bacteria produces multiple copies of the recombinant plasmid. Each bacterial cell contains about 40 plasmids and when the cells replicate, the plasmids do also. The bacterial enzymes transcribe the insulin gene in the plasmid and translate the mRNA they produce. Insulin is made in large quantities and is purified for medical use.

The pros of genetic engineering in bacteria

Incorporating novel genes into bacteria to produce specific molecules or to allow bacteria to perform specific metabolic functions has been hugely beneficial, e.g.:

- Medical products: large amounts of pure human proteins for use in medicine have been made, e.g. insulin, clotting factors to treat haemophilia, human growth hormone to treat some kinds of dwarfism. These are safer than the older products, which were purified from cadavers and could transmit disease, e.g. clotting factors to treat haemophilia caused HIV-AIDS and hepatitis C; preparations of human growth hormone transmitted Creutzfeldt-Jakob disease.

- Tooth decay: the oral bacterium *Streptococcus mutans* makes lactic acid and is a major contributor to tooth decay. Modified strains that do not make lactic acid are available. They out-compete lactate-producing *S. mutans* in the mouth and lead to a reduction in cavity formation.

- Prevention and treatment of disease: bacteria have been modified to produce vaccines and treat disease in other ways, e.g. to hinder tumours and fight Crohn's disease.

- Enhancing crop growth, e.g. modified bacteria can secrete molecules that are toxic to plant pests.

- Environmental use: detecting and removing environmental hazards is an increasingly important field. Modified bacteria have cleaned up mercury pollution and detected arsenic in drinking water.

The cons of genetic engineering in bacteria

- Plasmids are easily transferred. They may exchange genes with other bacteria and there is potential for the antibiotic resistance marker genes to be transferred. If they are taken into potentially pathogenic species, the infections they cause will not be treatable with antibiotics. This compounds the problem of the overuse of antibiotics selecting for resistant mutants in medical, agricultural and veterinary situations.

- Fragments of human DNA used to make gene samples and cytoplasmic mRNA used to make cDNA may contain oncogenes or gene switches that activate proto-oncogenes in recipient cells. The public must have confidence that the protocols in place ensure these do not contaminate the final product.

- A micro-organism with a new gene may become a threat if released into the environment.

- A newly introduced gene may disrupt the normal function of other genes in ways not yet understood.

Going further

Other methods used to confirm the presence of a plasmid in the host cell include fluorescence and the detection of specific enzymes.

Key Term

Clone: A population of genetically identical cells or organisms formed from a single cell or parent, respectively.

Going further

Over 87 days in 2010, in the largest spill ever, oil flowed into the Gulf of Mexico after the BP oilrig Deepwater Horizon exploded. Bacteria occurring naturally metabolised some of the oil. Modified bacteria may be used following other such disasters in the future.

40

Knowledge check

Link the terms 1–4 with the statements A–D.

1. DNA ligase.
2. Restriction endonuclease.
3. Reverse transcriptase.
4. Marker gene.

A. Enzyme which joins together portions of DNA.
B. Enzyme used to synthesise DNA from an RNA template in specific cells.
C. Enzyme which cuts DNA molecules at specific base sequences.
D. Gene for antibiotic resistance, indicating the presence of the plasmid.

Genetically modified crops

10 000 ybp (years before the present) the global human population was about 5 million. It took most of human history for the population to reach 1 billion, in the year 1800. Each successive billion appeared more quickly than the previous billion, and by the end of 2015, the global human population was over 7.3 billion. These people all need food and throughout human history, the challenge of providing enough food for all has remained. Increasing affluence has been correlated with the increased consumption of meat, despite this being a very inefficient use of land, water and food. The lower the trophic level at which people consume, the more food is available for all. Making this worse is that it is estimated that up to 70% of all crops grown are lost between harvesting and supplying the home. Transformed plants have been suggested as a way of increasing the food supply, enhancing its nutritional value and making crop plants disease resistant and drought tolerant. Genetically modified food crops have been used since the 1980s.

There are many ways of introducing a novel gene into plant cells:

- The 'gene gun' fires small spheres, often of gold or tungsten, coated with a preparation of the gene at plant cells. Some penetrate the cell wall and are taken up through the cell membrane.

- Electroporation – an electric field increases the permeability of cell membranes, enhancing gene uptake.

- Microinjection, in which a membrane is pierced with an ultra-fine needle and the gene is injected into the cytoplasm or even the nucleus. This technique is much more developed for use with animal cells than plant cells.

- Using the bacterial vector *Agrobacterium tumefaciens* is the most common method for making transgenic plant cells.

Going further ▶

Following recent changes to classification and naming, *A. tumifaciens* is now officially *Rhizobium radiobacter*, but its traditional name is more commonly used.

Transforming plants with *Agrobacterium tumifaciens*

1. Plasmid extracted from the *A. tumifaciens*.

2. Restriction enzyme is used to cut the plasmid and remove the tumour-forming gene.

3. A section of DNA containing a gene for disease resistance is located and isolated using the same restriction endonuclease.

4. The gene is inserted into the plasmid, replacing the tumour-forming gene. DNA ligase is used to join the donor and vector DNA together.

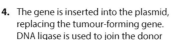

gene for disease resistance ➝

The plasmid is inserted back into the bacterium.

5. The bacterial cell is introduced into plant cell. The bacterial cell divides and gene is inserted into plant chromosome.

6. Transgenic plant cells are grown in tissue culture and transformed plants are regenerated.

Making disease-resistant transformed plants

A. tumifaciens is a soil bacterium. It infects plants and T-DNA, a section of the bacterium's plasmid, can integrate into the plant's chromosomes. Plasmid genes are transcribed and translated and the auxins they produce cause a tumour, or gall, to form, giving the plant crown gall disease. The plasmid is called the Ti (tumour-inducing) plasmid. By splicing chosen genes into the plasmid, the genes can be transferred into plant cells, which are subsequently grown in tissue culture and regenerated into plants which express these introduced genes.

Soya beans

In many countries soya beans are very important as a source of food. Products from soya are used as ingredients in a wide range of foods such as flour, protein and oil. In the UK, about 60% of manufactured foods, including bread, biscuits, baby foods, and soya milk contain soya. Crop plants are routinely treated with herbicides but they can damage the plants they are treating, which means the crops cannot be sprayed while they are growing. 'Roundup Ready' soybeans, grown commercially since 1996, are genetically modified to contain a gene that makes them herbicide-resistant. The crops can be sprayed to remove weeds, without inhibiting their growth.

Genetically modified soya bean plants

Going further ▶

Roundup is a very effective herbicide because it contains glyphosate, which inhibits the production of amino acids needed for protein synthesis, and it breaks down in the soil into harmless compounds.

Tomatoes

Bt tomatoes

Bacillus thuringiensis is a bacterium that lives in the soil and contains a plasmid with a gene that codes for a protein that acts as an insecticide. The gene has been incorporated into the cells of Bt tomatoes. The insecticidal proteins are not made in the fruits, which humans eat, but in the leaves, which insects eat. As the plants make their own insecticide, the farmer does not need to spray the crop, which protects all the unintended targets of applied insecticide.

Bt tomatoes have been given resistance to tobacco hornworm (*Manduca sexta*), tomato fruitworm (*Heliothis zea*), the tomato pinworm (*Keiferia lycopersicella*) and the tomato fruit borer (*Helicoverpa armigera*). A 91-day feeding trial in rats showed no adverse effects from eating them, but the Bt tomato has never been commercialised. Tomato farming has a significant ecological footprint and the use of Bt technology has the potential to reduce this, as well as to increase farm income.

Antisense tomatoes

Tomatoes ripen naturally when they produce the enzyme polygalacturonase, which breaks down the pectin in their cell walls. But if they are transported long distances from their supplier, they may over-ripen and not be suitable to sell. The 'Flavr Savr' tomato was grown commercially between 1994 and 1997 to overcome this problem. *Agrobacterium tumifaciens* was used to introduce a second copy of the polygalacturonase gene into the tomato plant, but this copy had a base sequence complementary to that of the normal gene, i.e. it was an 'antisense' gene. The mRNA transcribed from the antisense gene is complementary to the mRNA strand of the original gene. The two types of RNA base pair in the cytoplasm to form a double-stranded molecule. This prevents the mRNA of the original gene from being translated and blocks the production of the enzyme. Tomato ripening is delayed and Flavr Savr tomatoes have a longer shelf life. Traditional plant-breeding techniques of crossing and hybridising improved their flavour, but Flavr Savr tomatoes were not a commercial success.

Going further ▶

The *B. thuringiensis* plasmid has a gene for an insecticidal protein, called 'cry' protein because it crystallises readily. It is described as a δ-endotoxin.

Going further ▶

When an insect ingests the Bt bacterium, the cry protein destroys its gut wall. The bacterium enters the blood and the insect dies from blood poisoning.

Going further ▶

Antisense tomatoes have been used successfully in making tomato purée.

▼ **Study point**

Bt tomatoes are genetically modified and transgenic. Flavr Savr tomatoes are genetically modified but they are not transgenic because they do not contain a gene from another species.

Benefits and concerns about genetically modified crops

Since the introduction of genetically modified food plants, questions have been asked about potential risks, labelling, nutritional properties and effects on the environment. The techniques of genetic modification and their applications have been challenged, as is appropriate with any new technology.

Arguments in favour of GM crops

- Higher crop yield: ever more crops are lost to disease and as the world's climate changes, to droughts and floods. Incorporating genes for insect, fungus and worm resistance or for drought or salt tolerance are likely to increase crop yield.

- Pesticide reduction: genes for pathogen resistance have the additional advantage of reducing the quantities of pesticides applied to farmland.

- Improved food: nutritional quality can be enhanced, as with Golden Rice, which contains an added gene to increase the content of vitamin A precursors and, were it to be licensed, prevent blindness in children in some parts of the world; foods can also be enhanced with improved flavour and better keeping qualities.

- Introducing genes that confer resistance to herbicide decreases plant loss in the field.

- 'Pharming' refers to the production of pharmaceutical molecules in genetically modified crop plants. Plants have been modified to make antibodies, blood products, hormones, recombinant enzymes and human and veterinary vaccines.

Most people's diets have a small number of staples, i.e. a food that is the dominant part of a diet and eaten routinely. The top three, in terms of world production, are all cereals: maize, then rice, then wheat, and their cultivation requires huge quantities of nitrogenous fertilisers. These have disrupted the global nitrogen cycle and had severe environmental effects. Attempts continue to try to make transgenic cereals that contain the nif (nitrogen-fixing) genes from nitrogen-fixing bacteria. The plants would then make their own fertiliser and less would be added artificially. This would make a valuable contribution to restoring damaged habitats.

 The planetary boundary for nitrogen is discussed on p113.

Arguments against the use of GM crops

- Pollen from GM plants might transfer genes to wild relatives. In this way, it is feared that herbicide resistance might spread to wild plants, and produce what the popular press has called 'superweeds'. If GM crops do not have wild relatives in the UK, e.g. potatoes, this fear may be unfounded.

- Pest resistance: plants with introduced genes that enable them to resist insect attack may lead to a population of resistant insect or fungal pests. Long-term field trials will establish whether these concerns are well founded. However, if crops are modified to synthesise more than one pesticide, it is unlikely that resistance to two would simultaneously develop.

- Marker genes: genetically modified organisms contain marker genes, some of which confer antibiotic resistance. There is concern that these genes may be transferred to the bacteria in the intestine of the consumer.

- Plant breeding may fall into the hands of a few commercial companies, limiting the number of crop varieties available to the farmer. This could lead to the elimination of old varieties. Reduction in biodiversity decreases the range of potentially useful genes.

- New proteins: it is claimed that there may be adverse health effects from eating a crop that is expressing a new gene, making a new protein. Extensive trials in the decades since genetically modified foods have been produced have yet to provide any evidence for this claim in any organism.

- 'Organic' farming: it is claimed that pollen from genetically modified crops could compromise organic crops.

- Economic concerns: genetically modified organisms are subject to intellectual property law and it is feared that the associated expense will be borne by the farmer.

These controversies have led to litigation, international trade disputes, protests and to legislation severely restricting the use of genetic modification in food production in some countries. Many manufacturers and retailers have banned ingredients from genetically modified plants. GM crops are grown widely throughout the world, although none are grown in the UK. The UK does, however, import GM animal feed and GM products, largely maize and soya, for incorporation into processed food. There is no doubt that they have significant economic and nutritional value. Public confidence in their use in the UK, however, does not reflect this fact.

Genetic screening and gene therapy

Genetic diseases are caused by abnormalities in the genome. They may be:

- Singe gene conditions, e.g. cystic fibrosis, Tay–Sachs disease
- Chromosomal, e.g. Down syndrome
- Multifactorial, e.g. Alzheimer's disease, some cancers.

The World Health Organisation reports that there are over 10 000 monogenic diseases, i.e. diseases associated with one gene, and that there is a genetic component to:

- Disease in 1–3% newborns
- Most miscarriages in developed nations
- 30% postnatal infant mortality in developed countries
- 30% of paediatric and 10% of adult hospital admissions
- Medical conditions in over 10% of all adults.

There is, therefore, an overwhelming need for understanding genetic diseases and developing treatments and cures.

Genetic screening

A patient and, perhaps, their family, may undergo genetic screening to determine the nature and inheritance of a genetic condition. It provides useful information in individual cases, and has potential to:

- Confirm a diagnosis
- Indicate an appropriate treatment
- Allow families to avoid having children with devastating diseases
- Identify people at high risk for conditions that may be preventable.

There are concerns, however, about widespread screening:

- Many believe it is an invasion of privacy.
- Defective alleles identified in prenatal tests may increase the number of abortions. Observations show that these are disproportionately of female foetuses.
- Individuals with defects may be placed in a high-risk group for insurance purposes to cover the cost of treatment, so their insurance cover would be very expensive or even impossible to obtain.

> **YOU SHOULD KNOW** ›››
>
> ››› The meaning of the term gene therapy
>
> ››› The difference between somatic and germ-line therapy
>
> ››› The cause of Duchenne muscular dystrophy
>
> ››› The treatment of Duchenne muscular dystrophy by gene therapy

> ▼ **Study point**
>
> Knowing the nucleotide sequence of a normal, functioning gene makes it possible to devise techniques to eliminate the risk of a genetic disease, by correcting or replacing the faulty allele.

The main uses of genetic testing

Hospitals provide genetic screening and interviews so that the patient can be fully informed of the results and their implications, e.g.:

- Carrier screening to identify if an unaffected person carries a recessive allele associated with a genetic disease. People who are carriers may decide not to have children or to have an antenatal genetic test to check if their child will be born with the disease.

- Pre-implantation genetic diagnosis to screen embryos generated from in vitro fertilisation.

- Pre-natal diagnostic testing.

- Newborn baby screening.

- Pre-symptomatic testing for predicting adult-onset disorders such as Huntington's disease, adult-onset cancers and Alzheimer's disease. People at greatest risk can have regular screening and make decisions concerning their lifestyle.

- Confirmation that an individual has a suspected disease.

- Forensic and identity testing.

Commercialised gene tests

Commercially available gene tests are targeted at healthy people. They give a probability of developing a few conditions, including heart disease, colon cancer and Alzheimer's disease, and some may indicate the body's ability to metabolise alcohol and certain drugs. Limitations on their use include:

- Commercial products are not regulated or independently verified.

- They test only a small number of the approximately 20 500 genes in the human genome.

- It is difficult to interpret a positive result. Some people who carry a disease-associated mutation never develop the disease. These mutations may work with other genetic and environmental factors to cause disease, and their effects cannot be predicted.

- Laboratory errors may occur, e.g. misidentification, contamination.

- There may be no available medical options for treating these diseases.

- The tests may provoke anxiety.

- There are risks for discrimination and social stigmatisation for people who have taken tests, whatever the results.

Gene therapy

Some, but not all, genetic diseases can be treated by either replicating the function of genes using drugs or by gene therapy. Gene therapy is a technique in which a defective allele is replaced with one cloned from a healthy individual, providing a treatment or cure. The main challenge is in developing a gene delivery system, so that it is inserted correctly into the genome and functions correctly when there.

To introduce the DNA into the target cells, gene therapy uses:

- A virus as a vector or

- A plasmid as a vector or

- Injection of naked plasmid DNA.

Going further ▶

Genetic testing during pregnancy can be carried out on a foetus at 8–10 weeks, using chorionic villus sampling, or at 15–20 weeks, with amniocentesis.

WORKING SCIENTIFICALLY

Gene therapy raises ethical issues, particularly the possible long-term consequences of germ-line therapy. Ethics committees consider the advantages and disadvantages of research proposals for gene therapy and their potential outcomes.

There are two main approaches:

- **Somatic cell** therapy targets body cells in the affected tissues. This method may be therapeutic, but the genetic changes are not inherited in daughter cells of the treated cells, and do not appear in future generations.

- **Germ-line therapy** introduces the corrective genes into germ-line cells, in this case, the oocyte, so the genetic correction is inherited. But germ-line gene therapy is controversial. Genes interact with each other, e.g. some are switches that control other genes. Potentially influencing such genes in the oocyte has unpredictable effects in future generations.

Duchenne muscular dystrophy (DMD)

Duchenne muscular dystrophy (DMD) is a recessive, sex-linked form of muscular dystrophy, which affects about one in 3500 live male births.

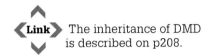

Link The inheritance of DMD is described on p208.

Most cases of DMD are caused by one or more deletions in the dystrophin gene. This gene has 79 exons and deletions in any of these alter the reading frame of the dystrophin mRNA. The ribosome meets a stop codon too soon and the dystrophin protein is not produced. Dystrophin is a structural protein in muscle, so people with DMD have severe muscle loss and many become wheelchair-bound by the time they are teenagers. Life expectancy is only 27 although survival into the 30s, and even into the 40s and 50s, is becoming more common.

Going further ▶

Drisapersen is designed to skip over exon number 51, and it could help 13% of boys with DMD.

The drug drisapersen is an antisense oligonucleotide, i.e. a 50-nucleotide sequence that is complementary to the mutated sequence. It treats DMD by acting as a 'molecular patch': it binds to the mRNA over the exon with the deletion. That portion of RNA therefore becomes double-stranded, so the ribosome cannot translate it. This restores the reading frame, so that a shorter, partially functional dystrophin can be synthesised. This type of treatment is called exon skipping.

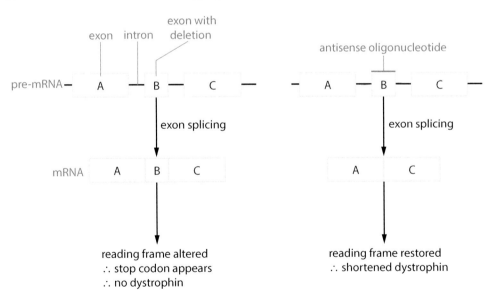

Exon skipping

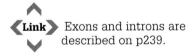

Link Exons and introns are described on p239.

Drisapersen is delivered to the patient in subcutaneous injections. Clinical trials have not yet provided clear evidence as to the best age for treatment or for its length. Meanwhile, other approaches to DMD treatment include:

- Gene therapy – a shortened version of the healthy gene has been designed because the normal gene is too big to put into a virus.

- Research into stem cell treatment.

Exam tip

When discussing applications of reproduction and genetics, take care with your use of words. Avoid clichés such as 'slippery slope' and 'designer babies'. Use appropriate scientific terminology to convey exactly what you mean.

Going further ▶

In the commonest mutation, the deletion of three nucleotides results in the deletion of the amino acid phenylalanine in the 508th position.

Link The inheritance of autosomal recessives is described on p197.

Key Term

Liposome: Hollow phospholipid sphere used as a vehicle to carry molecules into a cell.

WORKING SCIENTIFICALLY

Liposomes can be charged with cations or anions. DNA is negatively charged. Plasmid DNA containing the therapeutic gene binds with cationic (positive) liposomes and the plasmids are absorbed, forming 'lipoplexes'.

41

Knowledge check

Identify the missing words.

To treat genetic diseases with gene therapy, a healthy gene can be introduced by a virus or a ...., acting as a vector, or by injection. Duchenne muscular dystrophy is caused by .... in one or more exons of the dystrophin gene causing a change in the .... frame of the mRNA. An .... oligonucleotide is used to bind to the pre-mRNA and allow a shortened, functional protein to be made.

Cystic fibrosis (CF)

People with cystic fibrosis are homozygous for an autosomal recessive allele. Carriers can be identified with a blood test. The normal allele codes for the cystic fibrosis trans-membrane regulator (CFTR), a cell membrane protein that transports chloride ions out of cells. Sodium ions follow and water leaves by osmosis so the extra-cellular mucus is watery. Mutant CFTR cannot transport ions and so the water does not move through the membrane and the mucus remains thick and sticky:

- The bronchioles and alveoli become clogged, causing congestion and difficulty in breathing. The mucus is difficult to remove and leads to recurrent infections. Chest physiotherapy is needed to keep the airways open.

- The pancreatic duct becomes blocked and pancreatic enzymes cannot reach the duodenum so food digestion is incomplete and absorption is limited. Children with CF have large appetites to compensate.

- The vas deferens in males may become blocked, reducing fertility.

The gene coding for the CFTR protein has been isolated and cloned. In early attempts, an inactivated virus was used to deliver the gene but there was more success using **liposomes**, i.e. hollow phospholipid spheres containing the gene preparation. They are inhaled with an aerosol and fuse with the phospholipid bilayer of the lung epithelial cell membranes. The DNA enters the cells, which transcribe the inserted gene and make the CFTR protein. The gene remains functional providing relief from CF symptoms, but as soon as the epithelial cells that have taken up the gene are replaced, the treatment has to be repeated. This is a treatment, not a cure.

A more recent approach, appropriate in those cases where a specific CF mutation produces a misfolded CFTR protein, is to use a drug, ivacaftor, that corrects the protein folding.

The effectiveness of gene therapy

Gene therapy has had some success, and the advantages to those who receive it far outweigh the disadvantages. However:

- Only a small proportion of the introduced genes are expressed.

- There may be an immune response in the patient.

Seeking to provide opportunities for children with genetic disease with the chance of a healthier life is a moral obligation. The fear that commercial companies will abuse the technique of gene therapy, such as in providing a chance to choose or modify the characteristics of a child, must be addressed by society as a whole.

Genomics and healthcare

Genomics is concerned with analysing the structure and functioning of genomes and has applications in many fields, including medicine, biotechnology, anthropology and social sciences. It combines recombinant DNA techniques, DNA sequencing, fine-scale genetic mapping and bioinformatics, i.e. the development of software tools to analyse biological data.

DNA is 'annotated', which means that using base sequences, predictions are made about whether sequences code for RNA, for proteins or have a regulatory function. This can be used to infer what metabolic pathways are controlled and genomes can be compared. Even the most complex biological systems, such as the brain or the genetic basis of drug response and disease, can be analysed.

The Human Genome Project and the 100K Project provide examples of how genomics is used. It is anticipated that there will be improvements in healthcare:

- More accurate diagnosis, e.g. two brothers suffered from inherited nerve damage causing the muscle loss and weakness called peripheral neuropathy. The 100K Project showed they had a new mutation. They now have the opportunity to join a treatment trial, which could benefit other members of their family with the same mutation.

- Better prediction of the effect of drugs, e.g. codeine is converted to morphine, which is a painkiller. Morphine is detoxified and excreted. If too much morphine is produced or if its excretion is impaired, normal doses of codeine may be toxic. Individual genetic differences affect metabolic pathways, so genomic information about a patient can inform a decision about drug dose.

- New and improved treatments for disease, e.g. testing a tumour may identify genetic changes for which a specific drug is beneficial.

- NGS technology sequences genomes very quickly and it may allow patients to have individual therapies based on their DNA sequence, e.g. Warfarin is one of the commonest anticoagulants used to prevent stroke and blood clots. Calculating the dose is very complex because of drug-drug interactions, dietary interactions, age, body surface area and genetic factors, especially two particular gene variants, CYP2C9 and VKORC1. Genomics has made it possible to recommend Warfarin doses depending on which genotype a patient has.

42

Knowledge check

Match the terms A–D with their descriptions 1–4.

A. Genetic disease.
B. Gene therapy.
C. Liposome.
D. Antisense technology.

1. A lipid vesicle used for gene delivery into cells.
2. Binding complementary nucleic acid sequences, to prevent transcription or translation.
3. A disorder resulting from a faulty genome, at the genetic or chromosomal level.
4. Replacing a faulty gene with a healthy copy.

Tissue engineering

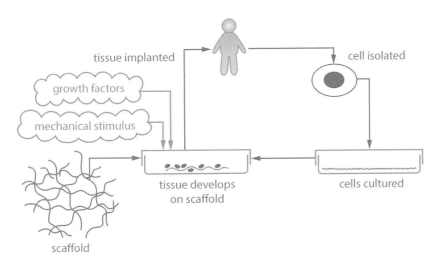

Tissue engineering

YOU SHOULD KNOW ›››

››› What is meant by the term tissue engineering

››› The properties and sources of stem cells

››› The advantages and disadvantages of using stem cells

››› The ethical concerns about cloning human tissues and organs

››› The ethical concerns about using human stem cells

Tissue engineering uses the methods of biochemistry, cell biology, engineering and materials science to repair, improve or replace biological functions. Its goal is to produce 'off the shelf' bio-artificial organs and to regenerate injured tissue in the body.

Tissue engineering has allowed the replacement of many tissues and organs, e.g. trachea, bone, bladder and skin. An artificial skin called 'Apligraf', which in 1998 became the first engineered tissue to be licensed, is widely used for burns patients in place of skin grafts. Artificial support systems that mimic liver or pancreas function have been constructed and the techniques are also used to make artificial meat. In the future, it is hoped to treat diabetes, traumatic spinal cord injury, Duchenne muscular dystrophy, heart disease, and vision and hearing loss.

Cultured human skin to be used in a skin graft

▼ **Study point**

In tissue engineering, cells are induced to grow on a framework of synthetic material to produce a tissue such as skin tissue.

Cells for tissue engineering

Telomeres are the repeated nucleotide sequences on the ends of chromosomes, which shorten at each division and limit cell mortality. In 1998 the technique for elongating them was developed, extending the life of differentiated cells. Cells can be separated from tissues with enzymes, e.g. trypsin and collagenase, and following elongation of their telomeres, used to grow tissue replacements. Examples include fibroblasts for making skin replacement or chondrocytes for replacing cartilage. Stem cells, however, are the preferred engineering materials. They can be extracted from blood and from solid tissues.

Cells for tissue engineering are classified by their source:

- Autologous cells are from the same individual. These have fewest problems with rejection and pathogen transmission, but are not always available, e.g. if the patient has a genetic disease or severe burns, or is very ill or old.

- Allogeneic cells come from a donor of the same species.

- Xenogenic cells are from another species, e.g. cells from pigs and from the Chinese hamster have been used to develop cardiovascular implants. Research with pig cells has shown that one of the dangers comes from the viral sequences in the pig DNA, which are harmless to pigs, but dangerous for humans. It has been claimed that pig cells with these sequences removed have been developed, increasing the chance of xenogenic cells for therapy.

- Syngenic or isogenic cells are from genetically identical organisms.

Scaffolds

Cells are 'seeded' on to a scaffold, i.e. an artificial structure that can support a 3D tissue. It must:

- Allow cells to attach and move

- Deliver and retain cells and biological molecules

- Be porous to allow diffusion of nutrients and waste products

- Be biodegradable and be absorbed by the surrounding tissues. The rate it degrades should match the rate of tissue formation so eventually it will break down leaving the 'neotissue'.

Tissue culture

Cells that are grown in tissue culture form cell lines that are clones, as all the cells derived from a single parent cell are genetically identical. The cell lines are used to produce cloned tissue samples and, in some, cases, to generate organs. This production of cloned material is referred to as therapeutic cloning, to distinguish it from cloning whole organisms, i.e. reproductive cloning. It has a major advantage in that a patient is unlikely to reject tissue or organs cloned form their own cells. The law permits therapeutic cloning in the UK, but not reproductive cloning.

During the period of tissue culture, cells must be given oxygen, nutrients, growth factors and the correct pH, humidity, temperature and water potential. Diffusion is normally adequate in tissue culture but as structures get larger, they need other mechanisms, e.g. capillary networks.

Sometimes, special physical or chemical stimuli are needed for the correct structures to differentiate, e.g. chondrocytes need low oxygen concentration, mimicking their development in skeletal tissue; endothelial cells need shear stress to mimic blood flow in blood vessels; cardiovascular tissue, such as heart valves need mechanical stimuli, such as pressure pulses to stimulate their development.

Going further ▶

The facility to freeze and store stem cells taken from the umbilical cord at birth may allow autologous ESCs to be used to grow tissues and organs in the future.

Going further ▶

Kidneys continue to develop even after birth and the urine of premature babies contains a significant number of kidney stem cells. They can be isolated and used for engineering kidney tissue.

▼ **Study point**

Cell cultures have been used since the 1880s for medical and research purposes. New uses for tissue culture techniques, such as cell replacement therapy and tissue engineering, are being developed.

Stem cells

Stem cells are unspecialised cells and can develop into many different cell types. When a stem cell divides by mitosis, each daughter cell can either remain a stem cell or become another type of cell with a more specialised function, e.g. muscle fibre or red blood cell.

Key Term

Stem cell: An undifferentiated cell capable of dividing to give rise to daughter cells, which can develop into different types of specialised cell or remain as undifferentiated stem cells.

Types of stem cell

- Embryonic stem cells (ESC) are found in 3–5-day-old embryos. The blastocyst has ESCs, which can form every cell type in the body. They are described as totipotent. ESCs were first isolated in 1981, from mouse embryos, and then from human embryos in 1998.

- Some adult tissues, e.g. bone marrow, muscle, brain contain adult stem cells, residing in a specialised area within the tissue, called the 'stem cell niche'. They can replace cells that are lost through normal wear and tear, injury, or disease but cannot they form all cell types. In some organs, e.g. the pancreas and the heart, stem cells only divide under special conditions.

- In 2006, adult cells were genetically 'reprogrammed' to become induced pluripotent stem cells (iPSCs), which behave as ESCs and can differentiate into any cell type.

Stem cells have huge potential:

- For tissue engineering to regenerate tissues and organs. For example, regenerating bone, using cells derived from bone marrow stem cells, developing insulin-producing cells to cure type I diabetes and repairing damaged heart muscle with cardiac muscle cells.

- For cell-based therapies to treat disease. The need for transplantable tissues and organs far outweighs the available supply. Stem cells stimulated to differentiate into specific cell types are a renewable source of replacement cells and tissues to treat diseases, e.g. macular degeneration, spinal cord injury, stroke, burns, heart disease, diabetes, osteoarthritis and rheumatoid arthritis.

- To screen new drugs. Cancer cell lines have long been used to screen potential anti-tumour drugs, but stem cells allow drug testing in many more different cell types.

- To develop model systems to study normal growth and identify the causes of birth defects.

- To investigate the events that occur during human development, and how gene switches turn undifferentiated stem cells into differentiated cells and form tissues and organs.

Study point

Stem cells are undifferentiated dividing cells in adult animal tissues. They can develop into any other types of cell, given the correct conditions.

Advantages of using stem cells

- Embryonic stem cells can become any cell type but adult stem cells are more limited.

- A blastocyst may contain about 100 ESCs so they can be isolated in useful numbers. Adult stem cells are rare, so isolating them from an adult tissue is challenging.

- Embryonic stem cells grow easily in culture and large quantities can be readily produced. Techniques for growing adult stem cells are less developed.

- The use of stem cells will make the acute problem of a shortage of organs for transplantation less significant.

- A patient receiving tissue derived from ESCs is likely to need immunosuppressive drugs, and the drugs themselves may cause side effects. A patient's own adult stem cells, or tissues derived from them, are less likely to provoke immune rejection after being transplanted.

Knowledge check

Fill in the spaces.

Stem cells are unspecialised cells that can ....into many different cell types. They are found in embryos and in adult tissues such as bone .... They are increasingly important in medicine as they have to read many applications, including .... engineering and .... testing.

Going further ▶

The 14-day limit was chosen as this is the earliest time that any cell that might eventually become part of the nervous system can be distinguished.

Disadvantages of using stem cells

- Techniques for extracting, culturing and manipulating stem cells are still under development and the behaviour of cell cultures is not always predicable. As a result, currently, products of stem cell technology, such as the artificial trachea, are expensive and rare.

- The use of stem cells is very new and so long-term studies have not yet been possible. There are concerns related to the premature aging of cells and other as yet unpredictable events.

Ethics and the use of embryonic stem cells in therapeutic cloning

The UK has more stringent regulations than many countries concerning the use of embryos and the cells derived from them. At the time of writing, the Human Fertilisation and Embryology Authority (HFEA) grants licences for such research on embryos up to 14 days after fertilisation.

Their requirements include:

- That any stem cells or cell lines created are maintained indefinitely and may be used in many different research projects or clinical therapy.

- That stem cells are deposited in the UK Stem Cell Bank so that they are available to other research groups, nationally and internationally.

- That there is no financial reward for any development or discovery made using them, although they can be patented.

- Donors must give specific consent to embryos created with their gametes being used in stem cell research.

Adult stem cells have been used in research and medicine for many years, for example, bone marrow transplants for disorders such as leukaemia use the stem cells of the bone marrow. There do not appear to be ethical problems using adult stem cells. The issues surrounding the use of embryonic stem cells include:

- The source of embryonic stem cells: recent legislation allows researchers to create embryos for research but prior to this, they used 'spare embryos' from in vitro fertilisation. Some people argued that creating embryos specifically for research contravenes the principle that human life should never be created as a means to an end. These embryos, though, cannot legally be transferred to a uterus so a new individual could never be born from one of them.

- The moral status of the embryo. Under the Human Fertilisation and Embryology Act (1990), an embryo does have moral rights, but not to the same degree as a living person. Some groups, including the Catholic Church, claim that new life begins at conception, so a foetus at any stage should have full human rights. Others, including some other religious traditions, say that as a foetus develops, for example as it acquires a nervous system, its rights increase.

- The potential rights of a foetus are balanced against the potentially large benefits that other people may gain from the research and treatments it produces. Some say the use of ESCs is never justified as adult stem cells or iPSCs could be used.

People involved in research, on the other hand, say that ESCs are still important because:

- They will clarify fundamental biological mechanisms.
- They will indicate which types of stem cell will be most useful in cell-based treatments.
- A pre-14-day embryo is a ball of cells with no possibility of independent existence, so its use is justified.

- Some people fear that stem cells may lead to humans being cloned, an act that fundamentally devalues human life. Reproductive cloning of humans is illegal in the UK but it is feared that any research knowledge gained could be used to clone humans elsewhere.

▼ **Study point**

Ethical concerns about the use of ESCs include their source and the moral status and rights of the embryo from which they were obtained, and the danger of developing human reproductive cloning.

Going further ▶

There are about 200 different types of cells in the human body. They all differentiate from embryonic stem cells, which themselves are derived from a single cell, the zygote. Understanding the process of differentiation is important in developing the treatment and prevention of disease, and stem cells will be of major use in this. Other uses include drug development and in producing tissues and organs for transplant. But researchers must first learn how generate enough stem cells, how to induce them to differentiate into different cell types, how to ensure that newly differentiated cells survive if transplanted into a body and how to make sure they integrate into and function with tissues that are already present. Techniques must be developed to prevent such transplanted cells being attacked by the immune system and to ensure that they have no way of harming the recipient.

Much research needs to be done. Throughout the process, society must continually ask questions and re-evaluate answers to be sure that research is carried out within a moral and ethical framework. Laws are framed by parliament and our MPs have an obligation to represent us in this. You can find out your MP's voting record, to determine if you wish them to continue to represent you. Can your MP answer these questions? Can you?

- Should spare embryos from in vitro fertilisation be used at all?
- Should spare embryos from in vitro fertilisation be used only with specific permission from the parents?
- What should happen if the parents disagree whether or not their embryos should be used?
- Should embryos be created specifically in order to extract embryonic stem cells?
- If not, are there any other circumstances in which it is acceptable to deliberately create an embryo?
- Would it be acceptable to create an embryo if it were never to be implanted into a uterus?
- Does an embryo have rights?
- If so, are they the same as those of a foetus?
- Are they the same as those of a newborn?
- Do the rights of an embryo depend upon its stage of development?
- If it could be guaranteed that great benefit would accrue by using embryos for research, does that make using them a moral obligation?
- Do we know enough about induced pluripotent stem cells or adult stem cells to say that their use is equivalent to that of embryonic stem cells, and to use them instead?

CH 14

Option A: Immunology and disease

The cause of disease may be external to the body, e.g. an infection, or internal, e.g. an autoimmune condition. The use of antibiotics is of critical importance in treating bacterial disease. Overuse, especially in farming, has led to the selection of antibiotic-resistant bacteria, making some infections hard to treat.

Immunology concerns the immune system and how it protects the body. It studies an organism's response to invasion by foreign proteins and by microbes and their products. The key organs of the immune system are the bone marrow and the thymus gland.

Topic contents

By the end of this topic you will be able to:

- Explain that the body is a host to many other organisms.
- Describe characteristics and treatment of cholera, tuberculosis, smallpox, influenza and malaria.
- Describe the relationship between the pathogenic action of viruses and their mode of reproduction.
- Describe the features of antibiotics, including the mechanism of action of penicillin and tetracycline.
- Understand how the overuse of antibiotics has resulted in the spread of antibiotic-resistant strains of bacteria.
- Understand how natural barriers reduce the risk of infection.
- Distinguish the innate from the adaptive immune system.
- Describe the primary and secondary immune responses.
- Distinguish active from passive immunity.
- Describe the differential effectiveness of vaccines.
- Consider the ethical and moral implication of vaccination programmes.

Disease

The body as a host

It has been said that we should not think of ourselves as individuals, but as colonies. That is because although we have 10^{13} cells in our bodies, we have at least that number of other individual organisms living in or on the body. They may be internal, living in our cells, tissue fluid and our guts, or external, on our skin and hair and at body openings. They include microbes, such as fungi, protoctista and over 1000 bacterial species in the gut flora. We may carry other larger parasites, including insects, such as fleas, or worms, such as tapeworms.

Many of these organisms are parasites and have the potential to cause disease if, for example, they secrete toxins, if their populations increase too much or if they are transferred to an inappropriate part of the body. But many contribute to our general health and with those we have mutualistic relationships.

- The bacteria *E. coli* in the large intestine synthesise vitamin K, but in the stomach and small intestine, they can cause gastro-intestinal disease.
- The mites in the hair follicles of our eyelashes eat dead cells. They stick to eye make-up, so if you don't remove your mascara, you risk their population building up too much and causing inflammation.
- *Entamoeba* is the protoctistan that grazes on the dead cells of our gums. If you don't brush your gums enough when you clean your teeth, they will reproduce in large numbers and cause gingivitis.

The bacteria and fungi that colonise our skin prevent the build-up of harmful microbes, as do many of the gut flora. If these useful organisms are removed, for example when antibiotics are used to kill bacteria, then other organisms, such as yeast, increase in number and cause disease. There is an ecological balance in a healthy body, in which the many different organisms keep each other in check.

Important diseases

All groups of organisms suffer from **infection** and disease. Even bacteria can be infected by viruses. Susceptibility to pathogens is a property of life, as one organism provides a habitat for another. Some microbial diseases with significant social and medical consequences are described here.

Cholera

- Cholera is caused by the Gram-negative, comma-shaped bacterium *Vibrio cholerae*. It can only reproduce when it is inside its human host.
- Cholera is **endemic** in parts of the world. People become infected through contaminated food or water. They become **carriers** and, acting as **reservoirs** of disease, can contaminate other water supplies and spread the disease.

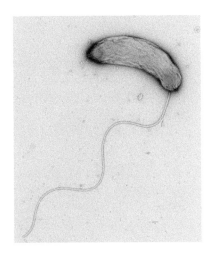

Vibrio cholerae

▼ **Study point**

Many organisms live in or on the body in a mutualistic or parasitic relationship with us. They may cause disease or defend us from disease.

▼ **Study point**

The word 'mutualism' refers to organisms of two species that live together to their mutual advantage. Sometimes the word 'symbiosis' is used.

Key Terms

Infection: A transmissible disease often acquired by inhalation, ingestion or physical contact.

Endemic: Disease occurring frequently, at a predictable rate, in a specific location or population.

Carrier: An infected person, or other organism, showing no symptoms but able to infect others.

Disease reservoir: The long-term host of a pathogen, with few or no symptoms, always a potential source of disease outbreak.

Key Terms

Toxin: A small molecule, e.g. a peptide made in cells or organisms, that causes disease following contact or absorption. Toxins often affect macromolecules, e.g. enzymes, cell surface receptors.

Link Faulty CFTR channel proteins are associated with cystic fibrosis, as described on p254.

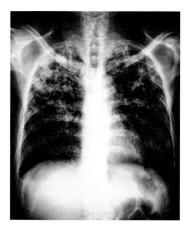

Lung X-ray showing cavities often seen with TB

Going further

A person who has been exposed to TB may carry anti-TB antibodies, which are detected by a skin test. A negative skin test means they have no antibodies, so they are offered the vaccine.

- A **toxin** is produced by *V. cholerae* in the small intestine that affects the chloride channel proteins, called CFTR. Water and many ions, including Cl^-, Na^+, K^+ and HCO_3^- are not absorbed into the blood and the patient has severe, watery diarrhoea. This causes dehydration so the blood pressure falls dramatically and the patient may die within hours.

- Cholera can be prevented with good hygiene and sanitation. Better sewage treatment, water purification, safe food handling and regular hand washing have all reduced the incidence of cholera. A vaccine is available, providing temporary protection but it is only given to people at very high risk.

- Treatment has two strands:
  - Water and ions are replaced by giving patients electrolytes, either orally or, in severe cases, intravenously.
  - The bacteria are treated with antibiotics.

There is a famous example of epidemiology, i.e. the study of the spread of disease, which dates back to 1854. At the time, it was thought that cholera was caused by 'bad air'. John Snow, a physician, suspected that it was transmitted by water so he mapped the cases during a cholera outbreak in Soho, London. He identified the water from a pump in Broad (now Broadwick) Street as the source. When its handle was removed, the number of new cases dropped immediately, supporting his hypothesis.

Tuberculosis (TB)

- Tuberculosis is caused by the bacillus bacterium *Mycobacterium tuberculosis*. It is named for the tubercles, or nodules of dead and damaged cells in the lungs of people who are infected. Tubercles may contain gas-filled cavities, which are easily seen in X-rays.

- The infection spreads rapidly by aerosol transmission, i.e. the inhalation of bacteria-laden droplets from the coughs and sneezes of infected people. This is why in many countries you see signs saying 'No spitting'. TB spreads very rapidly

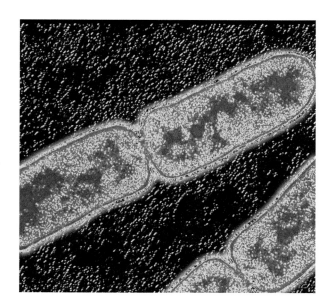

Section through *Mycobacterium tuberculosis*

in crowded conditions; in densely populated cities, it is once again a public health concern. The decreased efficiency of immune systems of HIV-AIDS patients is partly correlated with the recent increase in TB cases in some areas.

- The bacteria mainly infect the lungs so patients develop chest pain and they cough up phlegm (sputum), which often contains blood. The bacteria may infect lymph nodes in the neck, which swell. People lose their appetite and develop a fever.

- TB is treated with a long course of antibiotics but *M. tuberculosis* does show some antibiotic resistance.

- To prevent TB, the BCG vaccine is given to babies and, if a skin test proves negative, to people up to the age of 16. It provides about 75% protection, but only for 15 years. It is less effective in adults, and only given to those at risk. The vaccine is made from an attenuated (weakened) strain of a related bacterium, *M. bovis*. BCG stands for 'bacillus of Calmette and Guérin', the French scientists who developed it.

Smallpox

- Smallpox is caused by a DNA-containing virus, *Variola major*.

- The virus is inhaled or transmitted in saliva or from other bodies if there is close contact with an infected person.

- It enters small blood vessels in the skin, mouth and throat and is dispersed around the body. It causes a rash and then fluid-filled blisters, which leave scars in survivors. Some survivors also suffer blindness and limb deformities.

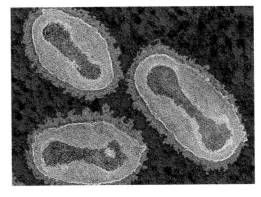

Section through *Variola major*

- Infected people are given fluids, drugs to control the fever and pain and antibiotics to control bacterial infections, but up to 60% die.

- The smallpox vaccine produces a strong immune response. It is made using live *Vaccinia* virus, a close relative of the smallpox virus, and has been very effective at preventing the disease. Before it was available, infected people were isolated, to prevent the virus spreading.

- Throughout history, millions have died from smallpox, including up to 500 million between 1900 and 1979, when it was declared eradicated, following vaccination campaigns that began in the 19th century. It is the only species that humans have deliberately made extinct. The only remaining virus is kept in research labs with very high levels of biosecurity. Ethical debates concerning the final total extinction of the species are ongoing.

Going further ▶

The earliest evidence of smallpox is on the mummy of Pharaoh Ramses V, who died in 1145 BCE, but DNA analysis suggests it emerged as a new disease in humans about 12 000 years ago.

Influenza

- There are three subgroups of the influenza virus, called 'flu A, B and C. 'Flu A is the best known and will be described here. Influenza viruses infect many species; avian (bird) and swine 'flu have provided recurrent sources of new influenza viruses that infect humans. When a new strain of 'flu appears, with new proteins on the virus surface, the human immune system is not able to provide adequate protection. As a result of this lack of immunity, **pandemics** occur, e.g. in 1918–1920, Spanish 'flu infected 500 million people and killed over 50 million, about 4% of the world's population, in one of the deadliest natural disasters in human history.

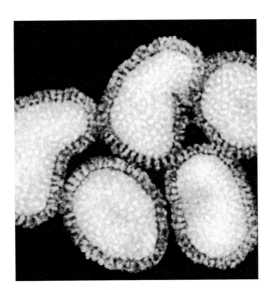

Influenza viruses

⬐ Key Terms

Pandemic: An epidemic over a very wide area, crossing international boundaries, affecting a very large number of people.

Antigen: A molecule that causes the immune system to produce antibodies against it. Antigens include individual molecules and those on viruses, bacteria, spores or pollen grains. They may be formed inside the body, e.g. bacterial toxins.

- The influenza virus contains RNA as its genetic material, but it is unusual for a virus in that its RNA is in 8 single strands, rather than one. The virion is surrounded by a phospholipid envelope, derived from the host's cell surface membrane. The envelope contains two important proteins which are **antigens** and are the spikes on the surface of the virus particle:

  – Haemagglutinin (H) has a role in the virus entering a host cell.

  – Neuraminidase (N) has a role in the virus leaving the host cell.

▼ Study point

Gastric 'flu is not caused by an influenza virus. It is caused by bacteria or viruses, such as norovirus, infecting the digestive system.

263

- The influenza virus attacks mucous membranes, especially in the upper respiratory tract, causing sore throat, cough and fever.
- The spread of 'flu virus is not easy to control. It is inhaled in droplets from coughs and sneezes, in aerosol transmission. Mucus protects the virus. It also survives better when the air is dry and there is low ultra-violet light in the environment, which means it survives better in winter than in summer, giving rise to seasonal 'flu.
- There are several ways to reduce the risk of infection:
  - Regular hand washing.
  - Using and discarding tissues for coughs and sneezes.
  - Influenza vaccines can be effective, but are variable in their success. The surface antigens on the virus change and so a new vaccine is needed annually.
  - Quarantine.

Link There are more details of the immune response and vaccination on p274.

Key Terms

Antigenic type: Different individuals of the same pathogenic species with different surface proteins, generating different antibodies.

Epidemic: The rapid spread of infectious disease to a large number of people within a short period of time.

Antigenic types

There are many **antigenic types** of influenza viruses. Their differences have two main origins:

1. Antigenic drift

   There are no RNA proofreading enzymes, so following each round of replication, on average, every new virion has a new mutation. This produces a gradual change in the surface proteins, which is called antigenic drift. It explains why, each year, a new vaccine is needed.

2. Antigenic shift

   'Flu A has 16 different types of haemagglutinin, of which H1, H2 and H3 are the most common in humans. It has nine different types of neuraminidase, of which N1 and N2 are commonest in humans. If one cell is infected by viruses that have different combinations of H and N, e.g. H1N2 and H2N1, the separate strands of RNA can recombine, giving rise to new virus types, e.g. H1N1 or H2N2. This change is 'antigenic shift' and the new virus types can cause **epidemics**.

Exam Tip

Remember that antigenic drift produces gradual changes and the lack of adequate immune response means a new vaccine is needed each year. Antigenic shift is a recombination of H and N types and can cause epidemics.

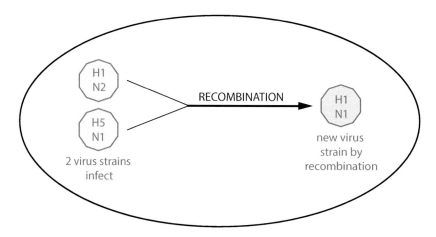

Antigenic shift: the formation of a new flu virus type by recombination

In some parts of the world, people live in close contact with animals. Chickens and pigs, in particular, are an **animal reservoir** for new virus infection in humans, i.e. a source of a new infection produced when an animal virus mutates and becomes able to infect humans. The influenza strains that infect people worldwide are continually monitored; H7N9 has been of particular concern since 2013. The table shows the H and N combinations in some recent influenza epidemics.

		Neuraminidase	
		1	2
Haemagglutinin	1	Spanish 'Flu 1918; Swine 'Flu 2009	Endemic in humans, pigs, birds
	2		Asian 'Flu 1957
	3		Hong Kong 'Flu 1968
	5	Bird 'Flu 2004	

Malaria

- Malaria is caused by a protoctistan, *Plasmodium*. There are five species which cause malaria, but *P. falciparum* causes the most deaths. *P. vivax* is also a major killer.

- *Plasmodium* is transmitted by over 100 species of *Anopheles* mosquitos, when they pierce the skin to take a blood meal. The females are **vectors** of malaria but the males are not, as they feed on plant nectar, not blood.

- Malaria occurs in habitats that support the *Anopheles* mosquito.

- Malaria:
  - Is endemic in some sub-tropical regions.
  - Can become epidemic during wet seasons.
  - Can also be regarded as pandemic. It affects millions of people worldwide and kills more people than any other infection, despite the years of research and drug development.

The transmission of malaria

- When a mosquito takes blood from an infected person, it takes in the sexually reproducing stage of *Plasmodium* called gametocytes. They produce zygotes, which develop into an infective stage, called sporozoites. Sporozoites migrate from the mosquito's gut to its salivary glands.

- When the mosquito takes another blood meal, *Plasmodium* sporozoites in the mosquito's saliva are injected into the human. They travel to the liver and reproduce asexually in the liver cells, producing merozoites.

- Merozoites are released into the blood and infect red blood cells, where they do more asexual reproduction.

- The red blood cells burst and release more merozoites, which infect more red blood cells. This cycle repeats every few days, and when the red blood cells burst, the fever recurs.

- Some merozoites become gametocytes.

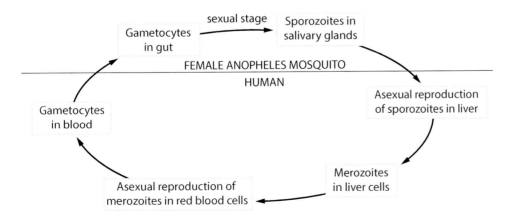

The life cycle of *Plasmodium*

> **Key Term**
>
> **Vector:** A person, animal or microbe that carries and transmits an infectious pathogen into another living organism.

> **Going further** ▶
>
> As global temperatures increase, malaria is already occurring at higher latitudes. It is feared that the mosquitos may survive in places that were previously unsuitable and that malaria will spread.

> *Exam tip*
>
> At A Level, you will not be tested on the names of the stages of the life cycle of *Plasmodium*.

Link The use of DNA technology to prevent and treat malaria is described on p238.

Treatment of malaria

- Drugs do not attack *Plasmodium* when it is inside cells, so their effectiveness is limited to when the *Plasmodium* is in the blood. Quinine was developed as a treatment for malaria in the 17th century, but it is less effective now than in the past. Artemisinin is a newer drug but resistance to that has now been seen. The current treatment of choice is artemisinin in combination with other drugs, because it is unlikely that *Plasmodium* would develop resistance to all the drugs at the same time.

- *P. falciparum* mutates frequently and it produces many antigenic types. It has therefore not yet been possible to produce a vaccine.

Prevention of malaria

There are several preventative measures that can be taken and these have seen the eradication of malaria in some parts of the world:

Preventative measure		Reason for effect
Responding to mosquito behaviour	Sleep under nets	Mosquitos feed between dusk and dawn
	Nets are treated with the pyrethroid insecticide	To kill mosquitos
	Spray indoor walls with insecticide	Kills mosquitos as they rest on walls after feeding
	Drain or cover stagnant water, e.g. water tanks, ponds	Removes mosquitos' access to egg-laying sites
	Film of oil on the water	Lowered surface tension prevents the larvae obtaining piercing the surface to obtain oxygen
Biological control	Fish introduced into water	Larvae are aquatic, so the fish eat them
	Infecting mosquitos with the bacterium *Wolbachia*	*Wolbachia* blocks *Plasmodium* development in the mosquito
	Male mosquitos are sterilised with X-rays	After they mate, no offspring are produced

Anopheles mosquito

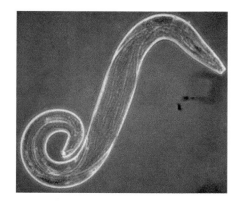

Plasmodium falciparum at the sporozoite stage is injected into the blood

44

Knowledge check

Match the diseases A–E with the types of pathogen 1–3 that cause them.

A. Cholera.
B. Tuberculosis.
C. Smallpox.
D. Influenza.
E. Malaria.

1. Virus.
2. Bacterium.
3. Protoctistan.

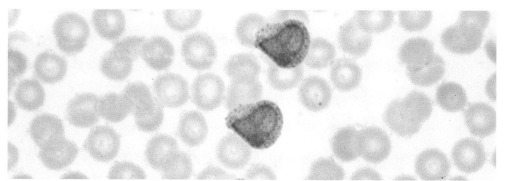

Blood film showing two red blood cells infected with *Plasmodium*

Each small pink dot in each cell is an individual *Plasmodium* at the merozoite stage

Virus pathogenicity and reproduction

Viruses have been described as 'the ultimate parasite' because outside living cells they are inert. They show none of the characteristics of life, except when inside a host cell, where they are replicated.

- In the **lytic** cycle, viruses immediately reproduce using the host's metabolism to copy their own nucleic acid and synthesise new coat protein.

  Their release may be by:

  - Lysis of the host cell, e.g. the common cold virus.
  - Budding, in which case, they acquire an envelope from the host's cell membrane, e.g. influenza virus.

- Some viruses are '**lysogenic**'. They integrate their nucleic acid into the host cell genome and may remain there for many cell generations with no clinical effect. They enter the lytic cycle at some later time, which is when they produce symptoms, e.g. herpes, HIV.

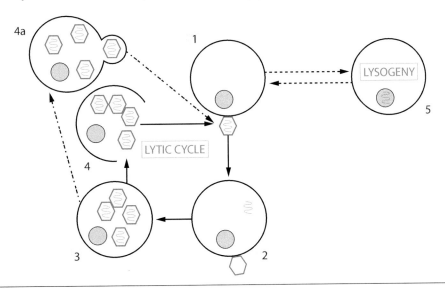

KEY
1. Virion attaches to cell
2. Viral nucleic acid injected into cell leaving protein coat outside
3. Nucleic acid and capsid protein are synthesised using the host's metabolism and they assemble to make mature virus particles
4. Cell lysis releases the viruses
or 4a. New virus particles bud from the cell surface
5. Viral nucleic acid integrates into a host cell chromosome

Viral life cycles

Viruses can be pathogenic in various ways:

- Cell lysis: when bacteria are infected with bacteriophage, the pressure of new virus particles inside causes the bacteria to burst. In contrast, in virus-infected animal cells, it is the inflammation caused by T-lymphocytes or antibodies that brings about lysis. An example is rhinovirus, one of the approximately 200 viruses that cause the common cold, which lyses cells in the upper respiratory tract.
- Toxins: many viral components and their by-products are toxic. The mechanisms are not well understood, but the following have been observed:
  - Measles virus can cause chromosome fusion.
  - Herpes virus can cause cell fusion.
  - Viral proteins can inhibit RNA, DNA and protein synthesis.

▼ **Study point**

Viruses do not have cell walls or metabolic pathways and so antibiotics do not affect them.

Link The genetic causes of some cancers are described on 214.

- Cell transformation: viral DNA can integrate into the host chromosome. If the DNA inserts into a pro-oncogene or tumour suppressor gene, it can result in the cell undergoing rapid, uncontrolled division, i.e. becoming cancerous. An example is HPV, the human papilloma virus, which can cause cervical cancer by inserting into the tumour suppressor gene, TP53.

- Immune suppression:
  - Suppression of the reactions that cause B and T lymphocytes to mature.
  - Reduction in antibody formation, e.g. HIV destroys a group of T helper cells, so B lymphocytes can no longer make antibodies. People with HIV infection are therefore immuno-compromised and highly susceptible to infection.
  - Reduction of phagocytic cells engulfing microbes.

YOU SHOULD KNOW ›››

››› Antibiotics can be bactericidal or bacteriostatic

››› The structure of the bacterial cell wall

››› The modes of action of penicillin and tetracycline

››› How antibiotics have given rise to antibiotic-resistant bacteria

Key Term

Antibiotic: A substance produced by a fungus, which diminishes the growth of bacteria.

Antibiotics

Types of antibiotic

Compounds that inhibit the growth of bacteria are called antimicrobials. They are:

- Antiseptics used on living tissue, e.g. Dettol.
- Disinfectants used on non-living surfaces, e.g. bleach.
- Antibiotics.

Antibiotics are produced by fungi and act on bacteria, but not on viruses and not on eukaryotic cells. They can therefore treat bacterial infection without harm to the patient. **Broad-spectrum antibiotics**, such as ampicillin and tetracycline, affect many different Gram-positive and Gram-negative species but **narrow-spectrum antibiotics** are much more selective, e.g. penicillin G kills Gram-positive bacteria only.

Different antibiotics affect different aspects of bacterial metabolism. Their use in medicine has allowed them to be classified:

- **Bactericidal** antibiotics kill bacteria, e.g. penicillin, which destroys bacterial cell walls.
- **Bacteriostatic** antibiotics prevent bacterial multiplication, but do not cause death, e.g. sulphonamides, which are competitive enzyme inhibitors and tetracycline, which inhibits protein synthesis. The bacteria resume their normal metabolism when the antibiotic is no longer present.

Bacterial cell walls

- Peptidoglycan, sometimes called murein, forms part of the bacterial cell wall. As the name suggests, it contains polysaccharide and short chains of amino acids. Transpeptidase enzymes cross-link the polysaccharide molecules by attaching them to the side chains of the amino acids. The cross-linking makes the cell wall strong, gives the cell its shape and allows it to resist bursting due to the osmotic uptake of water.

- The cell walls of Gram-positive bacteria are made of a thick layer of murein, which makes almost 90% of the cell wall. Pores in the murein close during the decolorisation stage of the Gram stain protocol, and so crystal violet is retained with in the cell, which stains violet. Safranin is used as a counter-stain and turns the violet Gram-positive cells purple. The murein is accessible to molecules outside the cell, making it susceptible to attack by lysozyme and penicillin.

- The cell walls of Gram-negative bacteria have a thin layer of murein, no more than 10% of the cell wall, surrounded by a layer of lipoprotein and lipopolysaccharide. These lipid-containing molecules are disrupted by the decolorisation stage of Gram staining and the crystal violet stain leaks out of the cell, leaving them unstained. Safranin, the counter-stain, turns the Gram-negative cells red. The lipid-containing layer protects the murein from antimicrobial agents including lysozyme and penicillin.

Link The structure of the bacterial cell wall is illustrated on p57.

Antibiotic mechanisms

Penicillin

The fungus *Penicillium* releases penicillin when its growth is inhibited and it is under stress. Penicillin was mass-produced from *P. notatum* initially, but now, high-yielding strains of *P. chrysogenum* are used. The fungus is grown aerobically in industrial fermenters and the penicillin is purified for use. The first available penicillin was penicillin G. This had to be injected, rather than ingested, because the stomach acid broke it down. Penicillin V and other derivatives, such as ampicillin, can be taken orally. They work by disrupting the bacterial cell wall:

- Penicillin readily diffuses through the cell wall of Gram-positive bacteria and it enters some Gram-negative bacteria through surface molecules called porins.

- Bacteria continually make and break down parts of their cell wall.

- The enzyme DD-transpeptidase catalyses condensation reactions that make cross-links between amino acid side chains joining peptidoglycan molecules. This enzyme is sometimes called PBP (penicillin binding protein) because penicillin binds to it, acting as an enzyme inhibitor.

- The breakdown by hydrolysis continues, so more cell wall is lost than gained.

- In addition, as no peptide cross-links are made, precursor molecules build up. These are hydrolysed also.

- The cell wall is so weakened that, as water enters the cell by osmosis, the wall is too weak to withstand the increased pressure potential and the cell lyses. Unlike some other antibiotics, such as vancomycin, the penicillin molecule is small enough to penetrate right through the murein, so the whole thickness of the cell wall is affected.

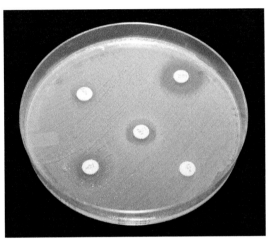

Bacteria growing on agar showing resistance to antibiotics diffusing from filter paper discs

Tetracycline

The fungus *Streptomyces* produces tetracycline, a broad-spectrum antibiotic. It acts against Gram-positive and Gram-negative bacteria and so it has a wide medical use against the bacteria that cause acne, against common infections, such as *Chlamydia*, but also against rare diseases, such as anthrax and plague. It even has some activity against some eukaryote parasites, including *Plasmodium*. Many bacteria, however, now show some resistance to tetracycline.

Tetracycline inhibits protein synthesis. It both diffuses and is pumped into bacteria cells. It binds to the small (30S) subunit of ribosomes and blocks tRNA attachment in the second position, the A site, so no new amino acids can be added to a polypeptide chain. Tetracycline binds reversibly, so its effect is bacteriostatic.

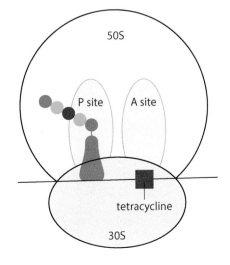

Tetracycline mechanism

Going further ▶

Penicillin binds irreversibly to a serine residue at the active site of DD-transpeptidase. It is not complementary to the active site shape and so it is a non-competitive inhibitor.

Exam tip

Make sure you can explain the difference between antibiotics that are: bacteriostatic or bactericidal; broad or narrow spectrum; injected or ingested.

◆Link▶ You learned about enzyme inhibitors in the first year of this course.

Going further ▶

Most amino acids in living organisms have the symmetry described as L-form. The alternative, D-amino acids, are rare in eukaryotes, but occur as a brain neurotransmitter and in some proteins. They are, however, abundant in the peptidoglycan wall of bacteria.

Going further ▶

Mammalian cells cannot pump tetracycline into their cells. A little may diffuse in but the effect on protein synthesis is negligible.

Key Term

Antibiotic resistance: Situation in which a micro-organism that has previously been susceptible to an antibiotic is no longer affected by it.

▼ **Study point**

Vancomycin is an antibiotic of last resort, used to treat highly resistant infections. But some resistance to vancomycin now occurs. A newer antibiotic, linezolid, is available but must be used sparingly.

Going further ▶

The M in MRSA is methicillin in the USA but its internationally approved name (IAN) is meticillin. It is no longer manufactured but the name has remained in use in MRSA.

Knowledge check

Fill in the gaps.

Antibiotics are produced by .... They affect .... but not viruses. Those that prevent bacteria reproducing are described as .... Their overuse has provided a selection pressure for bacteria that are .... to antibiotics, e.g. MRSA.

YOU SHOULD KNOW ›››

››› The natural barriers of the body to infection

››› The details of the humoral and cell-mediated response

››› The characteristics of the primary and secondary immune response

The overuse of antibiotics

Antibiotics have been used for thousands of years, and their presence is recognised in historical treatments of disease, such as the use of mouldy bread by the ancient Greeks to prevent wounds becoming infected. A major use has been to treat war wounds and they have saved huge numbers of lives, for example, in the Second World War. Since then, they have been used extensively to treat infection in people, but also to prevent infection in farmed animals. Freedom from infection allowed the animals' energy to go into growth, rather into fighting infection, so farms were more productive. In many farms now, antibiotics are not used in response to infection, but are used continually, to prevent infection.

Antibiotics in the environment kill susceptible individuals but any that have a mutation that makes them resistant will survive. They have a selective advantage in the presence of antibiotics. They reproduce, passing on the allele that confers **antibiotic resistance**, and build a resistant population.

There are two sources of antibiotic resistance alleles:

- Every time bacterial DNA replicates, a mutation conferring resistance may arise. Bacteria divide rapidly when conditions are suitable, so they have a high mutation rate.
- Bacteria may acquire plasmids that carry an allele conferring resistance from their environment. The plasmids replicate inside the bacterium and are passed on to the daughter cells when the bacterium replicates.

The mutated genes code for proteins that prevent antibiotics from working in ways that depend on the mechanism of the antibiotic:

Penicillin resistance	Tetracycline resistance
Secrete ß-lactamase, an enzyme that degrades penicillin	Pump tetracycline out of the cell
Altered PBP, so penicillin cannot bind	Dislodge bound tetracycline
Reduce penicillin entry with fewer or smaller porins	Prevent tetracycline attaching to a ribosome

If there were no antibiotics in the environment, using energy to synthesise these proteins would put the bacteria at a selective disadvantage and they would die. They only have a selective advantage while the antibiotic is present. So it is the continuous use of high levels of antibiotics that generates the resistance problem. Several clinically important bacteria show antibiotic resistance, including those causing leprosy, TB and gonorrhea. Some show resistance to several antibiotics, including:

- MRSA – meticillin-resistant *Staphylococcus aureus*. It is resistant to penicillin and all its derivatives, and is treated with vancomycin.
- *Clostridium difficile* – severe cases are treated with vancomycin.

Development of new antibiotics is a major priority. Without them, the control of infection following surgery will not be possible. Bacterial infection will become life threatening once again.

The immune response

The immune system enables the body to resist disease. There are physical barriers to protect against the entry of pathogens. If they fail, there are cellular and chemical responses. So, even though you will meet and be invaded by huge numbers of micro-organisms every day, most of the time, you remain healthy. The body must detect foreign, 'non-self' antigens and distinguish them from the 'self' antigens in its own tissues.

The innate immune system

The innate immune system is a group of natural barriers that resist infection in several ways:

- The skin covers the external surface of the body, except at openings such as the mouth and the eyes.
  - Keratin in epidermal cells makes the skin waterproof.
  - Collagen in the connective tissue of the dermis, maintained by vitamin C, makes the skin tough.
- The skin flora, or microbiota, comprises bacteria and fungi, which outcompete pathogenic strains. Unlike the pathogens, they are not easily washed off and so regular washing remains an important way of resisting infection.
- Inhaled air is a source of micro-organisms and their spores. Mucus traps them and the cilia of the epidermal cells lining the respiratory passages bring them up to the opening of the oesophagus, from where they are swallowed.
- When the skin barrier is breached:
  - If capillaries are broken, blood clots prevent the entry of microbes.
  - Inflammation sets in. Increased blood flow towards the site of infection brings large numbers of phagocytic cells. The broken capillaries heal and the raised temperature is unfavourable to microbes.
- If microbes enter the blood stream, phagocytic cells including macrophages and neutrophils, engulf and digest them.
- Tears, mucus and saliva contain lysozyme, an enzyme that hydrolyses peptidoglycan molecules in bacterial cell walls, and kills them.
- Stomach acid kills many of the microbes that are ingested in food and drink.

Going further ▶

Many organisms have an innate immune system: recognition of 'non-self' has been seen in all animal phyla and in plants. Phagocytic cells, e.g. macrophages and granulocytes, may have their evolutionary origin in organisms resembling *Amoeba*.

The adaptive immune system

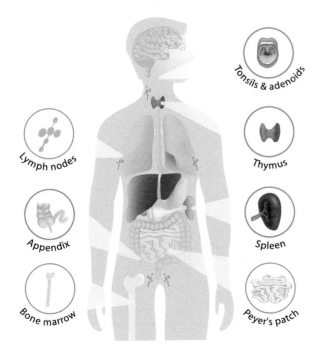

Important components of the adaptive immune system

Going further ▶

In auto-immune conditions, the adaptive immune system does not recognise the body's tissues as self, and treats them as non-self, so they are attacked by antibodies.

'Adaptive' means that the body produces a specific response to each antigen, so the body is adapting. Lymphocytes provide this response. They are derived from stem cells in the bone marrow and their precise role depends on their subsequent location.

**Key Term**

Antibody: An immunoglobulin produced by the body's immune system in response to antigens.

Exam Tip

There is a separate population of memory B lymphocytes and of memory T lymphocytes for each antigen.

The adaptive response

There are two components to this specific immune response.

1 The humoral response

The humoral response results in the production of **antibodies**.

B lymphocytes mature in the spleen and lymph nodes. The receptors on their cell membranes respond to a foreign protein in the blood stream and they divide, making:

- Plasma cells, which release antibodies.

- Memory cells, which remain dormant in the circulation and then divide to form more B lymphocytes if the same antigen is encountered in the future.

Antibodies are Y-shaped glycoprotein molecules, called immunoglobulins. They have quaternary structure as each molecule is made of four polypeptides, held together by disulphide bonds. The variable portions are specific to each antigen. An antigen molecule binds to an antibody and each antibody molecule can bind to two antigen molecules. Microbes with antigens on their surfaces are clumped together, i.e. they agglutinate.

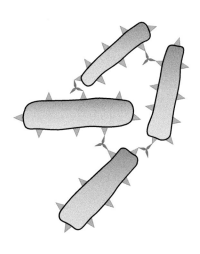

Microbes agglutinated by antibodies are engulfed more efficiently by phagocytic cells

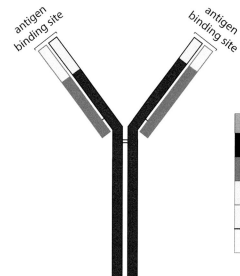

KEY	Polypeptide chain description	
■	Heavy	Constant
■	Light	Constant
□	Heavy	Variable
□	Light	Variable
—	Disulphide bond	

An antibody molecule

2 The cell-mediated response

The cell-mediated response refers to the activation of phagocytic cells, B lymphocytes and T lymphocytes.

T lymphocytes are activated in the thymus gland. The receptors on their cell membranes respond to antigens and they divide making:

- T memory cells, which remain dormant in the circulation and then divide to form more T lymphocytes if the same antigen is encountered in the future.

- T killer or cytotoxic T cells, which kill pathogenic cells with the antigens, by lysing them.

- T helper cells, which release chemicals including cytokines.

Cytokines stimulate:

- Phagocytic cells, including macrophages, monocytes and neutrophils, to engulf pathogens and digest them.

- B and T lymphocytes to undergo **clonal expansion**, i.e. to divide repeatedly into genetically identical cells, forming a large population of cells specific to a particular antigen. These cells differentiate into the various classes of lymphocyte.

- B lymphocytes to make antibodies.

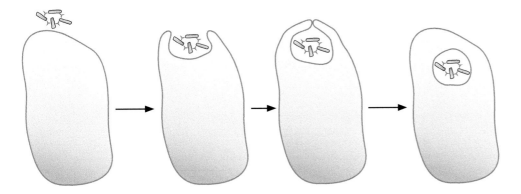

Microbes agglutinated by antibodies are engulfed by phagocytic cells

Fighting infection

The primary immune response

- On the first exposure to an antigen, there is a short latent period in which macrophages engulf the foreign antigen or the cell or virus to which it is attached, and incorporate the antigenic molecules into their own cell membranes. This is called **antigen presentation** and macrophages are therefore a type of antigen-presenting cell.

- T helper cells detect these antigens and respond by secreting cytokines.

- The B plasma cells secrete antibodies for about 3 weeks. The symptoms of infection subside.

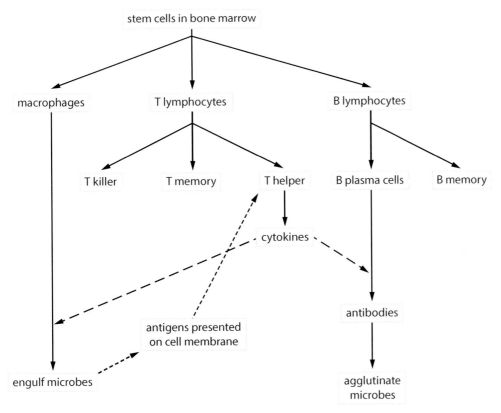

Summary of the primary response

The secondary immune response

The secondary response relies on memory cells and may protect against an identical antigen, even many decades after the first exposure.

- On re-exposure to even a small amount of an antigen, after a short latent period, memory cells undergo clonal expansion, but faster than in the primary response.

- Antibodies are made more quickly, and are up to 100 times more concentrated than in the primary response.

- They remain at high concentration in the circulation for longer and no symptoms develop.

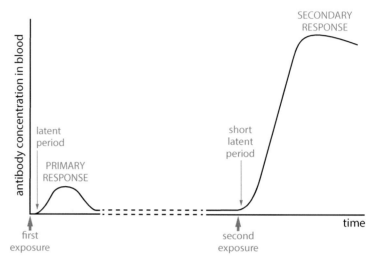

Graph showing immune response

 **Key Term**

Vaccine: A weakened or killed pathogen, or a toxin or antigen derived from it, which stimulates the immune system to produce an immune response against it without causing infection.

Active and passive immunity

Active immunity

The body makes its own antibodies in active immunity, stimulated by either infection or vaccination.

It is long lasting because the response produces memory cells.

Vaccination

When you are vaccinated, the cell-mediated and humoral responses are initiated, even though there is no harmful pathogen present. **Vaccines** can be:

- Antigens isolated from the pathogen, e.g. human papilloma virus (HPV) vaccine.

- Weakened or 'attenuated' strains of the pathogen, e.g. measles, mumps, rubella (MMR) combined vaccine.

- Inactive or killed pathogen, e.g. whooping cough vaccine.

- Inactivated toxin, e.g. tetanus vaccine.

The vaccine is recognised by the body as being non-self and the immune system behaves as it would if a pathogen were present. This is 'active' immunity as the body is making its own antibodies. Over time, the number of memory cells decreases if the body is not exposed to the antigen again. So in some cases, one or more boosters are given, e.g. a tetanus booster every 10 years is recommended. A booster is a further exposure to the vaccine and the response is faster, bigger and longer lasting than the first. More memory cells are made and so the body is protected for longer.

Passive immunity

The body receives antibodies produced by another individual in passive immunity. This happens naturally when they are transferred:

- From mother to foetus across the placenta.
- To the baby in breast milk.

Antibody injections, sometimes called Ig (immunoglobulin) replacement therapy, are given in various situations, including:

- When rapid resistance is needed, and there is no time for the active immune response to develop, e.g. when someone is bitten by an animal infected with rabies; a wound in which the risk of tetanus is very high; for an urgent visit to a part of the world where an infection is endemic, e.g. armed forces personnel who have to travel without warning.
- Cases of PIDD (primary immune deficiency disease) and cases of acquired immunity conditions, e.g. HIV-AIDS, in which patients do not make enough antibodies and so cannot provide enough protection for themselves against pathogens.

The protection given by injected antibodies is short lived because the body mounts an immune response against them, as they are recognised as foreign. In addition, the individual receiving the antibodies does not have any relevant memory cells.

WORKING SCIENTIFICALLY

With the understanding that antibodies bind selectively to specific molecules, uses have been devised which do not relate to their ability to attack pathogens, including:

- An advance in cancer treatment, which uses antibodies attached to chemotherapy drugs to deliver the drugs directly to tumour cells.
- A potential treatment for some types of migraine involves injecting antibodies against a characteristic protein.

Effectiveness of vaccination

Some vaccines are more effective than others. So to provide long-lasting immunity, vaccines against different diseases have different schedules. Some formulations of the HPV vaccine, for example, need only one dose; the vaccine against meningitis B is given in three doses; influenza vaccinations are given annually to older people.

For a vaccination to protect successfully against disease:

- The antigen should be highly immunogenic. This means that a single dose of the vaccine would cause a strong response from the immune system. It would rapidly make a large number of antibody molecules, specific for a particular antigen, e.g. *Vaccinia* virus, a close relative of *Variola major*, the smallpox virus, used to make smallpox vaccine.
- There should be only one antigenic type (serotype) of the pathogen, e.g. *Rubella*, the virus that causes one type of German measles. All *Rubella* viruses have the same surface antigens so only one vaccine is needed and the antibodies formed are able to attack all of them.

The 'flu A virus has many serotypes because it undergoes genetic recombination and frequent mutation. The memory cells produced in the first exposure may not be stimulated in a subsequent exposure if the virus has mutated. A different vaccine may, therefore, be needed for each antigenic type and the small changes in the antigens generated by mutation mean that a vaccine may not be totally effective.

▼ **Study point**

In active immunity the body makes its own antibodies. In passive immunity, the body receives antibodies made elsewhere.

Going further ▶

The word vaccination is derived from vacca, Latin for a cow, recalling the work of Edward Jenner, who vaccinated against smallpox using cowpox, a related but less dangerous pathogen.

46

Knowledge check

Match the terms 1–4 with their meanings, A–D.

1. Antibody.
2. Plasma cell.
3. Vaccine.
4. Passive immunity.

A. A differentiated B lymphocyte that secretes antibodies.

B. Short-term immunity due to the introduction of antibodies made by another person.

C. A Y-shaped protein produced in response to a specific antigen.

D. An inactive or weakened pathogen, or one of its proteins, used to produce active immunity against a disease.

▼ **Study point**

The most successful vaccines, e.g. the smallpox vaccine, are for pathogens with only one antigenic type, and are highly immunogenic.

Ethical considerations

Vaccination is a highly successful way of controlling infectious diseases and no other method is as effective. Indeed for some diseases, it is the only reliable means of protection. If enough people are vaccinated successfully against a contagious disease, there are fewer live pathogens in the population, which means that fewer people will become infected in the future. The spread of disease is controlled so even people who are not vaccinated have some protection. This is called 'community' or 'herd' immunity and requires a particular proportion of people to be vaccinated. If the proportion drops below this critical value, infection can spread. A moral argument can therefore be made for vaccination.

But not everyone is vaccinated. This could be because:

- Vaccination is not medically advised, including for those who are:
  - Immunocompromised, e.g. if the spleen is faulty and the immune system does not function normally
  - Taking chemotherapy
  - Living with HIV-AIDS
  - Very old
  - Very ill in hospital.
- Some people choose not to have vaccinations for themselves or their children for a variety of reasons not related to their medical condition. Arguments cited include:
  - Religious objections
  - A preference for 'natural' or 'alternative' medicine
  - Mistrust of pharmaceutical companies who manufacture the vaccines
  - Safety fears.

The cost effectiveness

When health authorities make decision about mass vaccination, they must consider if the financial cost outweighs the benefit provided. For dangerous conditions, such as measles, vaccination for all is advised, but for influenza, it is only recommended for target groups such as older people. In general, however, the UK government considers vaccination to be a cost-effective way of protecting the public.

Compulsory or voluntary vaccination

The policy in the UK is that the government cannot infringe on the right of the individual to choose what medication to take and so vaccination is not compulsory. It is hoped, however, that the benefits are so clear that parents will choose to have their children vaccinated. A complication arises with the issue of vaccination for health care workers, as they have a duty not to harm others, which may be contravened if they are not vaccinated.

Authorities around the world take different views. Where vaccination is compulsory, it is stressed that parents who choose against vaccination are not only withdrawing protection from their own children but also from all those with whom their children have contact. They argue that society has the right and duty to protect itself.

Side effects

Most medical drugs have side effects and many cite this as a reason not to vaccinate their children. Considering the complexity of living systems, it would be extraordinary if a molecule affected one aspect only of the body's biochemistry. Common side effects include soreness at the injection site, fever, fatigue and muscle or joint pain. Serious side effects have been documented, but because of the rigorous testing required for vaccines, these are extremely rare. The attention they get in the mass media makes them seem more common that they really are.

An example relates to the triple vaccine for measles, mumps and rubella, MMR. The percentage of children receiving the MMR vaccine fell from 92% to under 80% following a 1998 publication, in which a link between the vaccine and the onset of autism was suggested.

A case study

Measles is a highly contagious air-borne viral infection accompanied by a whole body rash and fever, sometimes in excess of 40°C. Complications occur in about 30% of cases and include diarrhoea, blindness, brain inflammation, pneumonia and even death. A vaccine was introduced in 1963. In the UK, about 97% of the population was vaccinated and since then, from being a common childhood disease, measles became extremely rare.

However, in 1998, misleading claims about the MMR (measles, mumps and rubella) combined vaccine, suggesting a link between the vaccine and the onset of autism, were published. They led to the percentage of the population vaccinated dropping. The claim was discredited but great damage was done. In Wales, the uptake of the MMR vaccine fell from 94% of two-year-olds in 1995 to 78% by 2003. In the Swansea area, uptake fell to 67.5%. Between November 2012 and July 2013, there were 1,455 measles notifications in the whole of Wales and 664 of them were in Swansea; 88 people were hospitalised for measles and a 25-year-old man died from complications. The cost associated with treating the sick and controlling that outbreak alone approached £500,000.

Going further ▶

Emerging diseases

Emerging diseases are infectious diseases in humans that have increased in frequency over the last 20 years or are likely to do so in the near future. Many such infections are correlated with increased global interconnectedness and close interactions between humans and animals.

They include:
- A new disease caused by an unknown microbe, e.g. HIV-AIDS
- A disease found in new areas, e.g. West Nile virus in the Western hemisphere, causing encephalitis.
- A microbe developing resistance to antimicrobial agents, e.g. meticillin-resistant *Staphylococcus aureus* (MRSA).
- A microbe from an animal that now infects humans, e.g. avian 'flu.
- Evolution producing a change in a microbe, e.g. most *E. coli* strains are relatively benign, but O157:H7 can cause severe illness.
- An organism that has been deliberately altered to cause intentional harm, e.g. *Bacillus anthracis*, which causes anthrax, was used to contaminate mail in the USA in 2001.

A recently emerged disease is Ebola hemorrhagic fever (EHF). It is a viral disease causing fever, sore throat, muscular pain and headaches, then, vomiting, diarrhoea and a rash, with decreased liver and kidney function. Some people bleed both internally and externally. It is fatal in 25–90% of cases, often because of low blood pressure from fluid loss. It was identified in tropical sub-Saharan Africa in 1976 in two simultaneous outbreaks and now occurs intermittently. The largest outbreak was the west African epidemic, from December 2013 to January 2016.

CH 15

Option B: Human musculoskeletal anatomy

Humans have an endoskeleton, an internal structure that provides support and protection and, being jointed, it allows movement. It is made of bone and cartilage, disorders of which may have devastating effects. Some are described here. Skeletal muscle is attached to bones by tendons, and bones are held together at joints by ligaments. The structure of muscle and the ultra-structure of its cells support the sliding filament theory, which describes how muscle fibres contract. The biochemistry of muscles explains the release of energy for contraction. Joints are highly specialised structures made from a variety of tissues. Faults leading to two types of arthritis are described.

Topic contents

By the end of this topic you will be able to:

- Describe the structure of three types of cartilage and two types of bone.
- Understand how the balance between osteoblasts and osteoclasts maintains bone structure.
- Describe Haversian systems.
- Understand the causes and treatments of rickets and osteomalacia.
- Explain the siding filament theory of muscle contraction in relation to muscle ultra-structure.
- Describe the differences between fast and slow twitch muscle fibres.
- Describe how muscles obtain energy, even when the oxygen availability is low.
- Describe the structure and functions of the human skeleton.
- Understand the types of fractures to which the skeleton is prone.
- Relate the structure of different vertebrae to their functions.
- Describe the causes and treatments of some postural deformities.
- Describe different types of joint.
- Understand the causes and treatment of osteoarthritis and rheumatoid arthritis.
- Describe joints as first, second and third order levers.
- Describe the structure of the synovial joint.
- Explain the concept of antagonistic muscles.

Skeletal tissues

The early embryos of most animals have cells arranged in germ layers. A germ layer contains the cells that eventually become the tissues of the body. In nine-day-old human embryos, cells form three germ layers, the ectoderm, mesoderm and endoderm. **Connective tissue**, including the skeletal tissues bone and cartilage, develops from the mesoderm.

Cartilage

Cartilage is a hard, flexible connective tissue. It is more compressible and flexible than bone, but less so than muscle.

- Being somewhat rigid it can hold open tubes, such as the trachea and bronchi, the Eustachian tube, between the middle ear and the throat, the nostrils and the pinna.
- Being flexible, it permits movement of the ribcage.
- It returns to its original shape after bending and has a role in load-bearing joints such as the hips and knees.

The cells in cartilage are **chondrocytes**. They secrete an extra-cellular matrix made of a transparent protein called chondroitin. It may contain fibres of collagen and of elastic material. Chondrocytes are in spaces in the matrix, called lacunae. Cartilage has no blood vessels so nutrient acquisition and gas exchange rely on diffusion. Because diffusion is slow, cartilage has a very slow turnover rate and healing is very slow. Cartilage has no nerves.

There are three types of cartilage, defined by their density and type of fibres they contain:

- **Hyaline cartilage** has a high proportion of collagen but is the weakest type of cartilage. It becomes bone in the foetus, i.e. it ossifies, but is retained in the adult as the articular cartilage at the ends of bones, joining the ribs to the sternum, in the nose, larynx, trachea and bronchi. If it is damaged, it is replaced by fibrocartilage scar tissue. A fibrous coat of connective tissue, called the perichondrium, surrounds hyaline cartilage.

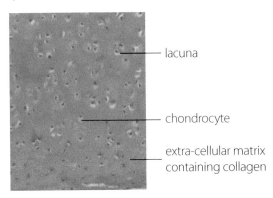

- lacuna
- chondrocyte
- extra-cellular matrix containing collagen

Hyaline cartilage in the light microscope

- **White fibrous cartilage** (fibrocartilage) is the strongest cartilage, and makes the intervertebral discs and ligaments. Its collagen is organised in dense fibres so it has greater tensile strength than other types of cartilage. The fibres are arranged in the direction of stress.

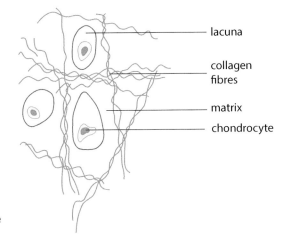

- lacuna
- collagen fibres
- matrix
- chondrocyte

Diagram to show structure of fibrocartilage

- **Yellow elastic cartilage** is intermediate in strength. In addition to collagen, its chondrocytes are surrounded by a network of fibres made of elastin. This makes it elastic but it maintains its shape, e.g. in the ear pinna and the epiglottis.

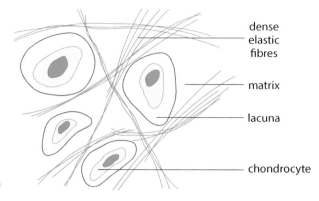

Yellow elastic cartilage

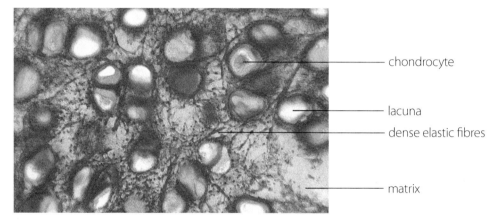

Yellow elastic cartilage in the light microscope

Bone

Bone is the most abundant skeletal material. Its functions include:

- Structural support in the skeleton
- Movement, because bones are attachment sites for muscles
- Physical protection of organs, e.g. brain
- Mineral regulation:
  - Storage, e.g. calcium and phosphorus
  - Trapping some minerals that cause harm elsewhere in the body, e.g. lead
  - Regulation of the calcium concentration in the blood.

There are two types of bone:

- Spongy or cancellous bone is found at the ends of the long bones and in the vertebrae. It has a network of spaces containing bone marrow, the flexible tissue in which blood cells are made.
- Compact bone represents 75% of the bone in the body. Haversian and Volkmann canals run through the bone, allowing blood vessels to penetrate. Compact bone surrounds most bones, and gives them their white, shiny appearance. It is strong and rigid. Cells called **osteoblasts** continually build it up and cells called **osteoclasts** continually degrade it. These cells are held in a matrix secreted by the osteoblasts, which is:
  - 30% organic, mainly collagen fibres, so it resists fracture.
  - 70% inorganic, mainly hydroxy-apatite, a mineral rich in calcium and phosphate ions, which is very hard and resists compression.

Haversian systems

Compact bone comprises units called **Haversian systems**, or osteons, each about 1 mm across and a few millimetres long, running longitudinally down a bone. Volkmann canals carry blood vessels from the bone surface through to the Haversian canal in the centre of each Haversian system. The Haversian canal contains an arteriole, a venule, lymph vessels and nerve fibres.

 **Key Term**

Haversian system: The structural and functional unit of compact bone; the osteon.

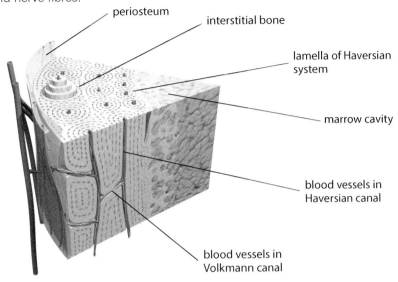

periosteum
interstitial bone
lamella of Haversian system
marrow cavity
blood vessels in Haversian canal
blood vessels in Volkmann canal

Bone structure

A Haversian system is built in concentric rings, or lamellae, around a Haversian canal. Osteoblasts are the bone cells that secrete the hydroxy-apatite that makes the lamellae. They sit in spaces called lacunae. Canaliculi are channels that radiate out of the lacunae into the bone lamellae. They contain processes from the osteoblasts, which are bathed in a fluid derived from the blood vessels in the Haversian and Volkmann canals. Osteoblasts get their nutrients and remove their waste in two ways:

- The matrix of the lamellae is slightly permeable, so food and oxygen brought by the blood can diffuse through the bone to the cells.

- The fluid and the processes from the osteoblasts in the canaliculi exchange materials. The canaliculi run through the bone matrix and those of neighbouring lacunae make contact in a three-dimensional network.

Haversian systems are separated by interstitial bone.

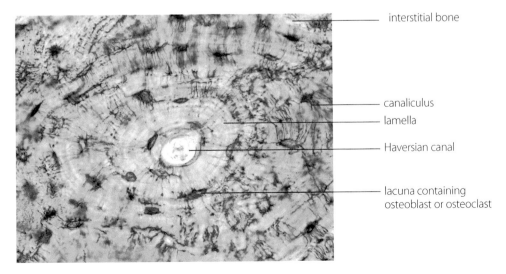

interstitial bone
canaliculus
lamella
Haversian canal
lacuna containing osteoblast or osteoclast

TS Haversian system

Bone formation

- Cartilage bones include the bones of the limbs, the vertebrae and the ribs. They are derived from hyaline cartilage in the embryo, in a process called ossification. Cartilage cells flatten and calcium salts are deposited around them. Osteoblasts secrete layers of bone matrix around the cartilage while osteoclasts break the cartilage down. Blood vessels invade.

In the long bones of the limbs, ossification begins at the caps at the ends (epiphyses) and in the middle (diaphysis). The cartilage remaining allows for growth, but eventually that is ossified too. Dense, fibrous connective tissue surrounding bone is called the periosteum. It develops from the perichondrium, the connective tissue surrounding the cartilage.

- Membrane bones, e.g. the collarbone (clavicle) and most of the skull (cranium) and facial bones form directly in embryonic connective tissue.

Bone is dynamic. That means it is constantly and simultaneously being built by osteoblasts and broken down by osteoclasts. The direction of stress on the bone when it is bearing weight, being bent or twisted, determines the direction in which the inorganic matrix is laid down. So bone is always built to give maximum support to the stress placed on it.

Bone disease

Rickets and osteomalacia

Rickets is a disease in which minerals are not adequately absorbed into children's bones, while the bones are still growing. The bones become soft and weak, leading to fractures and skeletal deformities. In adults, whose bones have stopped growing, the same condition is called **osteomalacia** and is milder. Signs and symptoms of rickets and osteomalacia include bone tenderness and fractures and skeletal deformities. Toddlers may have bow-legs and older children may have knock-knees and some have deformities in the skull, pelvis and spine. With insufficient calcium in the diet, the pelvic girdle grows to less than its normal size, and, as a result, childbirth may be difficult.

Rickets and osteomalacia usually result from a lack of vitamin D or calcium. Sunlight turns inactive vitamin D into active vitamin D, which incorporates calcium into bones.

- The increased use of sunblock has led to an increase in the incidence of rickets.
- Children whose skin is always covered or who spend a lot of time indoors may develop rickets.
- Breastfed babies do not receive adequate vitamin D if their mothers are vitamin D deficient from keeping their skin covered.
- People with darker skins need more exposure to sunlight to maintain vitamin D levels.

To prevent this condition, children and adults need exposure to sunlight and a diet with adequate calcium. Vitamin D is fat-soluble. Foods that contain it include butter, eggs, fish liver oil and oily fishes such as tuna, herring and salmon. Rickets and osteomalacia can sometimes be reversed with exposure to ultraviolet B light or cod liver oil, both of which provide vitamin D, but surgery may be needed to correct bone abnormalities.

Going further ▶

In the late 19th century, bustles were the height of women's fashion so their hips looked very wide. In an era with much poverty, bustles suggested, subliminally, that women wearing them were from rich families that could afford calcium-rich dairy products.

Brittle bone disease

Brittle bone disease is also called osteogenesis imperfecta (OI). It occurs in about 1 in 20 000 live births, producing susceptibility to fracture, poor muscle tone and loose joints. Testing collagen from a skin biopsy or a DNA analysis can confirm the diagnosis.

Collagen normally has a high proportion of the smallest amino acid, glycine. In OI, a mutation replaces glycine with bulkier amino acids so it cannot coil as tightly and so the hydrogen bonds holding the triple helix together are weaker. The distortion of the triple helix of collagen affects the way it operates: its interaction with hydroxy-apatite is altered and that makes the bones brittle.

There is no cure for OI. Treatment aims to increase bone strength to prevent fractures and maintain mobility:

- Drugs can increase bone mass, reduce bone pain and the tendency to fracture.
- Surgery, in severe cases, places metal rods inside the long bones, so that children can learn to walk.
- Physiotherapy strengthens muscles and improves mobility.

Osteoporosis

Bones grow and repair quickly in children and they reach their final length by the age of about 17. Their density increases up to the late 20s, but decreases above the age of about 35. **Osteoporosis** is the abnormal loss of density in spongy bone and compact bone, making the bones fragile and more likely to break. Fractures in the arm, wrist, hip and vertebrae are the most common. It is common, affecting around three million people in the UK. Diagnosis is generally by a bone scan to measure the mineral density in the hip. The lower the density, the more susceptible a patient is to a fracture. Osteoporosis is often only diagnosed if a fracture follows a minor fall, but it can be seen in older people whose osteoporotic spines decrease their height and prevent them from standing erect. It may cause chronic, debilitating pain.

Risk factors include:

- Age – post-menopausal women lose bone density rapidly, related to the drop in oestrogen; aging men are at increased risk, correlated with a drop in testosterone production
- Family history
- Inflammatory conditions, such as rheumatoid arthritis
- Medical conditions or long-term use of drugs that may affect hormone levels
- Alcohol use
- Smoking.

Treatment and prevention aim to prevent fractures to strengthen bones. Methods include:

- Regular load-bearing exercise to increases bone density
- Foods rich in calcium and vitamin D to increase bone density, especially for young children
- Drugs to enhance the calcium uptake into bones
- Giving up smoking
- Reducing alcohol consumption
- Learning to prevent falls.

Key Term

Brittle bone disease: An inherited disorder in the balance of the organic and inorganic components of bone, leading to an increased risk of fracture.

Going further ▶

Both dominant and recessive mutations can result in OI. They are generally inherited, but in about one-third of cases, the mutation is new.

Key Term

Osteoporosis: Abnormal loss of bone mass and density, resulting in increased risk of fracture.

Going further ▶

About 10% of bone is degraded by osteoclasts and rebuilt by osteoblasts at any one time. Losing bone mass could result from osteoclasts degrading too rapidly or osteoblasts rebuilding too slowly.

Going further ▶

Children should ingest a lot of calcium. They incorporate it in the centre of bones, making them strong. Older people lay calcium down at the bone edge, which is less beneficial.

48

Knowledge check

Fill in the gaps

If inadequate sunlight falls on the skin, insufficient vitamin D is synthesised and so bones lack .... In children, the bones deform in a condition called .... The equivalent in adults is .... In brittle bone disease, a mutation affects collagen and bones fracture easily. In people with .... , calcium is leached from bone, making it fragile and readily fractured.

Skeletal muscle

The defining property of muscle cells is that they contract and relax. Like bone and cartilage, muscles are derived from the embryonic mesoderm. There are three classes of muscle – skeletal, non-skeletal and cardiac. The role and mechanism of skeletal muscle will be described here.

Structure

Muscle tissue is made of cells called muscle fibres. A muscle, such as the biceps in the upper arm, can be thought of as a bundle of bundles:

- Muscle fibres are in a bundle called a fascicle, surrounded by connective tissue called perimysium.
- The fascicles are in a bundle, which is the muscle. They are surrounded by connective tissue called the epimysium.

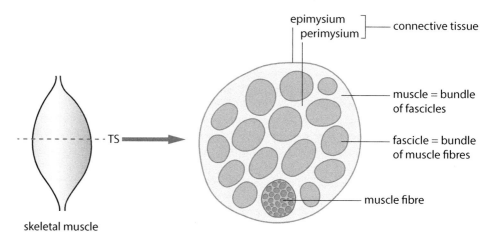

The structure of skeletal muscle

Ultra-structure of skeletal muscle

Muscle fibres

A muscle fibre is a long, thin cell, formed by fusion of several cells in the embryo, and so it has many nuclei. A cell with many nuclei is described as:

- Coenocytic, if the nuclei result from mitosis.
- A syncytium, if the nuclei result from fusion of many cells.

Muscle fibres are both. The muscle fibre grows to its final size after birth. The cell membrane of a muscle fibre membrane is the sarcolemma; its endoplasmic reticulum is the sarcoplasmic reticulum and its cytoplasm is the sarcoplasm.

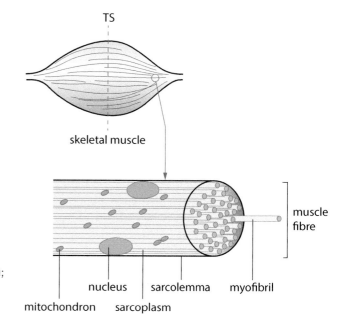

A muscle fibre

In the microscope, muscle fibres have transverse stripes, called striations. They are formed by structures called myofibrils, inside the muscle fibres.

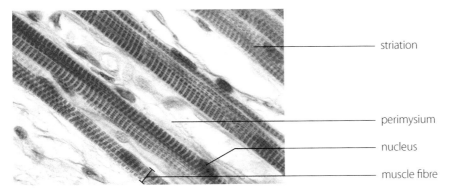

— striation

— perimysium

— nucleus

— muscle fibre

Striations in muscle fibres

Myofibrils

Inside a muscle fibre are longitudinal, cylindrical structures called **myofibrils**. They are packed together, surrounded by sarcoplasmic reticulum, with mitochondria lying between them. They are made of molecules of the proteins actin, troponin, tropomyosin and myosin, organised in long, thin structures called **myofilaments**. Thick myofilaments are largely myosin and thin myofilaments are largely actin. They have a geometrical arrangement in repeating units called sarcomeres, each about 3 μm long.

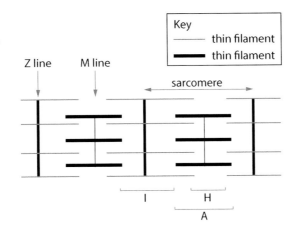

Portion of a myofibril showing its ultra-structure

The sarcomeres of adjacent myofibrils line up with each other, inside the muscle fibre, so that the thick myosin filaments of all the myofibrils are aligned, making a dark band, the A bands. The thin actin filaments of all the myofibrils are also aligned, making a pale band, the I band.

Z lines mark the ends of a sarcomere and are the attachment site for the actin filaments. The I band contains only actin. The myosin filaments overlap at their ends with actin in the A band. The portion of the myosin filaments with no actin overlap is the H zone. A line down the middle of the thick filament, the M line, where myosin molecules are attached, is sometimes seen.

Myofibrils in transverse section

If you cut a section through a myofibril, you will see the myofilaments inside. Their apparent arrangement depends on the position of the section:

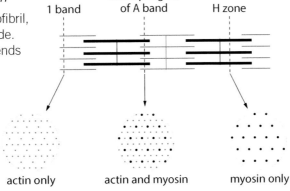

Section through a myofibril to show arrangement of myofilaments

---

Key Terms

Myofibril: Long, thin structure in a muscle fibre made largely of the proteins actin and myosin.

Myofilament: Thin filaments of mainly actin and thick filaments of mainly myosin, in myofibrils, that interact to produce muscle contraction.

Going further ▶

I stands for isotropic, meaning the band transmits light equally in all directions. A stands for anisotropic, which means that the light is not transmitted equally in all directions.

Exam tip

The structures of muscle in decreasing size order are:
1. Muscle
2. Fascicle
3. Muscle fibre
4. Myofibril
5. Myofilament.

WORKING SCIENTIFICALLY

In another example of scientific progress depending on available technology, in the early 1950s, Andrew Huxley used three types of microscope in developing his theories about muscle structure: the phase contrast microscope, the interference microscope and the electron microscope.

Knowledge check

Place the structures 1–5 in increasing size order:

1. Myofibril
2. Fascicle
3. Muscle
4. Myofilament
5. Muscle fibre

The T system

The cell membrane of a muscle fibre is called the sarcolemma. It invaginates making tubules across the myofibrils, called transverse tubules, or T tubules. As the T tubules cross the myofibrils, they transmit a nervous impulse through the muscle fibre very quickly, and all the myofibrils can contract at the same time. The sarcoplasmic reticulum makes a network around the myofibrils and dilates next to the T tubules to make a terminal cisterna. A T tubule and its two terminal cisternae constitute a triad.

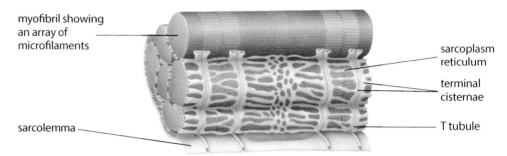

Structure of T system

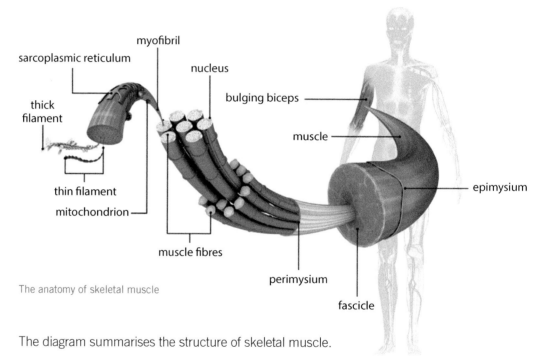

The anatomy of skeletal muscle

The diagram summarises the structure of skeletal muscle.

The sliding filament theory of muscle contraction

There are two types of filament in skeletal muscle:

- The thin filaments are largely actin but contain three different proteins:
  - Actin molecules are globular proteins, G-actin, joined in a long chain. Two chains twist around each other forming a fibrous strand, F-actin.
  - Tropomyosin is a protein that wraps around F-actin, lying in a groove between the two chains.
  - Troponin is a globular protein located at intervals along the thin filament.

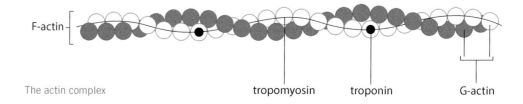

The actin complex

- The thick filaments are made of many myosin molecules. Each has a globular head, with ATPase activity, projecting from a fibrous tail. The globular heads of adjacent myosin molecules are 6 nm apart.

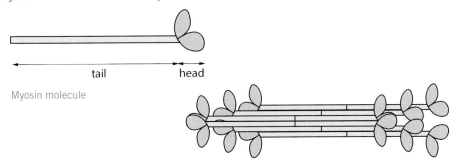

Myosin molecule

tail   head

Myosin molecules in a thick filament

Link In Unit 1, you learned that protein molecules can be globular or fibrous. Myosin is an intermediate protein as it has a globular head and a fibrous tail.

Muscles contract when the thin actin filaments slide between the thick myosin filaments. The lengths of actin and myosin molecules do not change but:

- The sarcomeres shorten so the myofibrils shorten and so the whole muscle fibre shortens.
- The A band remains the same length because it is defined by the length of the myosin filaments.
- The I band shortens.
- The H zone shortens and may get so small it cannot be seen.

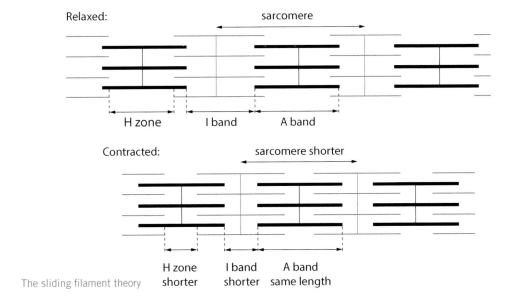

The sliding filament theory

WORKING SCIENTIFICALLY

Myosin, discovered in 1864, was found to be an ATPase in 1939. In 1942, ATP was shown to make myosin contract, but only if actin was present. The sliding filament theory was proposed in 1953.

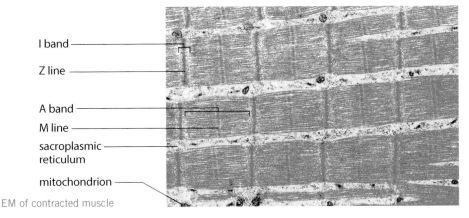

EM of contracted muscle

Sliding filament theory: The theory of muscle contraction in which thin, actin filaments slide between thick myosin filaments, in response to a nervous impulse mediated by the T system.

Nervous impulses cause muscles to contract, in a process called 'excitation-contraction coupling'. The molecular mechanism of muscle contraction is described by the **sliding filament theory**:

- The action potential of the nervous impulse (the excitation) crosses the neuro-muscular junction.
- The wave of depolarisation passes along the sarcolemma and T tubules, so that it penetrates all through the muscle fibre.
- Calcium channels in the sarcoplasmic reticulum open and Ca^{2+} ions diffuse into the myofibrils.
- Ca^{2+} ions bind to troponin molecules, which change shape.
- This moves tropomyosin molecules, exposing myosin binding sites on the F-actin.
- Myosin heads make cross-bridges to the actin by binding to these binding sites.
- ADP and a phosphate ion attached to the head are released changing the angle of the myosin head back to its relaxed shape and the myosin molecule rotates, pulling the actin past the myosin. This is the power stroke.
- Another ATP molecule binds to the myosin head and this breaks the cross-bridge to the actin.
- Hydrolysis of the ATP makes energy available and extends the myosin head again, ready to bind to actin.
- The sequence repeats until the Ca^{2+} ions, which exposed the myosin binding sites on the actin, are all pumped back into the sarcoplasmic reticulum.

Going further ▶

It has been calculated that a myosin head can make and break its cross-bridges 100 times a second.

Image position	Summary
1	myosin head extended
2	myosin binds to actin
3	ADP and Pi released
4	power stroke
5	ATP binds to myosin head
6	myosin unbinds from actin

Summary of the sliding filament theory

50

Knowledge check

Put these statements A–D, describing the sliding filament mechanism, in the correct order.

A. Myosin head pulls the actin past.
B. Ca^{2+} ions diffuse into the myofibrils.
C. The actin-myosin cross-bridge breaks.
D. Myosin heads bind to actin.

Fast and slow twitch muscle fibres

Striated muscles contain a mixture of fast and slow twitch muscle fibres. Their proportion and properties can be related to the role of the muscle:

	Slow twitch (Type I)	Fast twitch (Type II)
Use	Endurance, e.g. standing; long-distance running	Bursts of activity, e.g. jumping; eye movements; flight
Examples in humans	Back muscles and the soleus muscle in the calf have about 80% slow twitch fibres	Eye muscles have about 85% fast twitch fibres
Physiology	Contract for longer time	Contract then relax rapidly
	Contract slowly	Contract quickly
	Fatigue slowly	Fatigue rapidly
Myofibrils	Many mitochondria producing high ATP concentration	Few mitochondria
	Low density of myofibrils	High density of myofibrils
Respiration	Aerobic respiration	Anaerobic respiration
	High density of capillaries to deliver oxygen	Low density of capillaries as little oxygen needed
	Little glycogen, as blood brings glucose	High glycogen store to generate glucose for glycolysis
	High concentration of myoglobin, so oxygen available even at low oxygen partial pressure	Low concentration of myoglobin as no need for oxygen store
	Little lactate made because little anaerobic respiration; any made is removed slowly	Lactate removed quickly by oxidation to pyruvate or reversion to glucose
Colour	Muscle is dark in colour because of rich blood supply and high myoglobin concentration	Muscle is light in colour because little blood and low myoglobin concentration
Examples in chickens	Drumsticks contain the thigh muscles, used for walking	Breast meat contains muscles used for quick bursts of activity in flight

Anaerobic conditions

Energy source

When oxygen availability is limited, less ATP is made by oxidative phosphorylation in the mitochondria. The cell then relies on its creatine phosphate store to make ATP in the sarcoplasm. It transfers a phosphate group to ADP to maintain ATP levels in the cell: creatine phosphate + ADP $\rightleftharpoons$ creatine + ATP. This can continue until all the creatine phosphate in the cell has been used. It provides energy which can be used in intense bursts, for up to about 6 seconds. The hydrolysis of creatine phosphate is reversible and so creatine phosphate stores can be replenished when ATP is available.

When there is no high-energy phosphate store remaining, the cell uses lactate derived from glycolysis to reduce NAD and thus maintain the glycolytic pathway and a limited production of ATP. The cell's energy supply is limited.

Muscle fatigue

Muscle fatigue is the decline in ability of a muscle to generate force. Cardiac muscle, which must contract about every 0.8 seconds, from before birth for, in some cases, over a hundred years, never tires. But fatigue occurs in both striated and non-striated muscle. There are two recognised types:

- In neural fatigue, the nerves initiating contraction cannot generate a sustained signal.

- In metabolic fatigue there may be:
  - A shortage of substrates, e.g. glucose, glycogen, to provide energy.
  - An accumulation of metabolites, e.g. lactate. It may be that lactate lowers the pH in the muscle fibre, so contractile proteins are less sensitive to Ca^{2+} and their ability to contract is reduced. Other factors may be involved.

YOU SHOULD KNOW ›››

››› Properties of fast and slow twitch muscle fibres

››› The use of creatine phosphate

››› The causes of muscle fatigue and cramp

››› The sources of energy for muscle contraction

Exam tip

Be sure that you can relate these properties of the two muscle fibre types to their function and mode of respiration.

▼ **Study point**

Many aspects of muscles can contribute to athletic performance and be altered with training, e.g. fibre number and diameter; numbers of myofibrils and mitochondria; mass of myoglobin and glycogen stored.

Going further ▶

Many phosphorylated compounds release energy on hydrolysis. Some, e.g. creatine phosphate, release more energy than ATP. Others, e.g. glucose-6-phosphate, release less. ATP has a unique role because it is intermediate between two extremes.

Going further ▶

Creatine phosphate is a reservoir of phosphate in most vertebrates. Invertebrates use arginine phosphate for this purpose.

 Link Glycolysis is described on p44. Lactate metabolism is described on p48.

Cramp

A cramp is a severe, involuntary muscle contraction that can occur in striated or non-striated muscle. Skeletal muscles of the calves, thighs and arches of the foot are common sites of cramp. It happens during endurance events, such as a triathlon or marathon, or after exercise, even in elite athletes. But cramp may also occur when a muscle is relaxed or inactive. Its onset is usually sudden and it generally resolves over seconds, minutes or hours.

In part, muscle contraction is regulated by Cl^- ions, which inhibit and K^+ ions, which enhance it. Skeletal muscle cramps may be caused by fatigue as the lactate buildup can act as a trigger. Lactate inhibits the Cl^- effect but not the K^+ effect, so the muscles would tend to contract, leading to cramp. Lactate can also produce a burning sensation in active muscles, but the mechanism for that is unclear.

The biological value of cramp could be that it prevents lasting damage by preventing activity. When oxygen is available again, lactate is converted back to pyruvate and aerobic respiration can resume.

Energy sources during muscle contraction

Muscle glycogen provides most of the energy for contraction. It is only about 2% of the muscle mass, producing a much lower concentration than in liver but, unlike liver glycogen, when hydrolysed, the glucose it releases is only available for use within the muscle fibres, and does not enter the general circulation. The amount stored depends on:

- Physical training
- Basal metabolic rate
- Eating habits.

'Hitting the wall' occurs when long-distance athletes use up almost all their glycogen stores. They may experience extreme fatigue and even find it difficult to move. It can be avoided by:

- Eating foods with a high glycaemic index, i.e. that rapidly convert to blood glucose so that muscle glycogen is not used.
- Endurance training so that slow twitch (type I) muscle fibres become more efficient and use fats rather than glycogen for energy.
- Carbohydrate loading, i.e. increasing glycogen storage capacity by eating large quantities of carbohydrates after depleting glycogen stores.

Most dietary carbohydrates contain glucose and fructose. Fructose is metabolised into liver glycogen. It does not affect muscle glycogen so fruit and sweets, which have a high fructose content, are not suitable as energy sources. The classic carb-loading meal is pasta, but bread, rice, and potatoes are also suitable. This meal must also include protein, because muscles also use amino acids extensively in aerobic respiration.

One carb-loading system involves a normal diet and light training. The day before an event, the athlete does a short, high-intensity workout then eats 12 g carbohydrate / kg lean body mass over the next 24 hours. This increases the glycogen store by 90%. But carbohydrate loading must be carefully controlled so as not to limit performance because:

- If carbohydrate is eaten just before an event, the insulin secreted in response decreases blood glucose too much.
- If too much is eaten before a race, the digestive system does not have time to process the food and blood is diverted to the digestive system, away from the muscles.

Structure and function of the human skeleton

The axial skeleton comprises the skull, vertebral column, sternum and ribcage.

The appendicular skeleton comprises the pectoral girdle, the pelvic girdle, the forelimb and the hind limb.

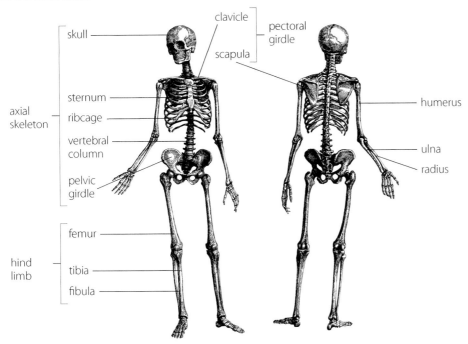

The human skeleton

Exam tip

Learn the names of the bones that are labelled on the diagram of the human skeleton.

Fractures

A fracture is a break in a bone. An X-ray, MRI or a CT scan can confirm the type of fracture a patient has.

Fractures are classified in different ways:

- Mechanism:
  - Traumatic fracture happens when the force on the bone is greater than its strength. It results from high impact or stress, e.g. a fall, a road traffic accident or a fight.
  - Pathologic fracture occurs when a medical condition makes the bone more fragile, e.g. OI, bone cancer and osteoporosis.
- The involvement of other tissues:
  - Closed or simple fracture: the overlying skin remains intact.

Type	Position
Greenstick	The bone is bent; common in children
Linear	Parallel to the bone's long axis
Transverse	At a right angle to the bone's axis
Oblique	At an angle to the bone's axis
Comminuted	The bone fragments into several pieces

  - Open or compound fracture: the bone may penetrate skin or be displaced inwards and damage other tissues. When the skin is broken, there is a high risk of infection.

- The position the bone fragments take:
  - Non-displaced – the bones are in their normal position.
  - Displaced – the bones are moved from their normal position.

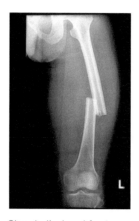

Closed, displaced fracture of a left femur

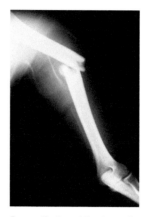

Open, displaced fracture of left humerus

Treatment

Pain management

It is very important to control pain and ibuprofen is recommended. There are no pain receptors in bone but fractures are extremely painful because:

- The connective tissue around the bone, the periosteum, has many pain receptors.
- Muscle spasms occur, which hold the bone fragments in place.

Immobilisation

Bone healing is a natural process. Immobilisation, for example with a cast or splint, immobilises above and below the fracture and holds the fractured pieces in their normal position while the bone heals. When the initial swelling goes down, a brace or surgical nails, screws, plates or wires can hold the fractured bone together more directly while it heals. An X-ray would be used to make sure the bones are aligned. Small bones in toes and fingers are hard to splint, so they are buddy-wrapped, which means one finger or toe is strapped to the adjacent one so its movement is very limited.

Surgery

Surgery is usually only used if immobilisation has failed. It is routine, though, for hip fractures because otherwise, the lengthy immobilisation could produce complications such as chest infections, pressure sores and deep vein thrombosis (DVT). Hip fractures are most commonly caused by osteoporosis and so, because these patients are usually older, the complications are more dangerous than the surgery.

A broken ankle immobilised with screws through the tibia to a metal plate on the fibula

Infection is especially dangerous in bones, because bones have few blood vessels, so few immune cells can be delivered. Treating open fractures, therefore, needs stringently sterile procedures and antibiotics to prevent further infection.

The stress of weight bearing encourages bone healing. It stimulates osteoblasts to produces new bone around the fracture, with osteoclasts removing any excess. Healing is hindered, however, by nicotine and by poor nutrition.

The vertebral column

The human vertebral column is a flexible support for the body, allowing bending and twisting, while protecting the nerve cord. It is composed of 33 vertebrae stacked on top of each other, held by ligaments and separated by intervertebral discs of cartilage. The vertebrae are named for the part of the body in which they occur:

- 7 cervical vertebrae in the neck, numbered C1–C7.

- 12 thoracic vertebrae in the thorax, T1–T12, with the ribs and sternum attached.

- 5 lumbar vertebrae in the lower back, L1–L5.

- 5 sacral vertebrae at the hips, fused into the sacrum.

- 4 coccyx vertebrae, the 'tailbone', with limited movement between them.

All vertebrae have:

- The vertebral body that bears weight and resists compression – it is made of spongy bone and is a site of blood cell formation

- The vertebral arch protecting the spinal cord

- Processes for muscle attachment

- Facets for articulation with adjacent vertebrae

- Thoracic vertebrae have facets for articulation with the ribs.

The structure of vertebrae in different parts of the vertebral column may be correlated with their roles:

- Cervical vertebrae have a longer spine and shorter transverse processes than thoracic and lumbar vertebrae. They each have a vertebrarterial canal, through which the vertebral artery and vein pass, serving the brain. The vertebral canal is proportionally wider than in lower vertebrae because the spinal nerve is widest in the cervical region. The top cervical vertebra (C1) is the atlas, which allows the head to nod. The second, C2, is the axis, which has a peg fitting into the atlas, allowing the head to shake.

- Thoracic vertebrae articulate with the ribs as well as with adjacent vertebrae, so they have several facets, or articulating surfaces. The facets of the vertebrae and the rib are shown in red in the diagrams:

Going further ▶

Mammals have the same number of vertebrae in the different regions as each other. Even giraffes only have 7 cervical vertebrae, although one thoracic vertebra has been co-opted into the neck.

vertebrarterial canal

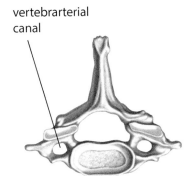

7th cervical vertebra

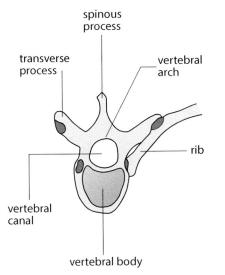

Thoracic vertebrae and rib (anterior view)

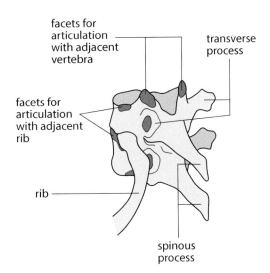

Two thoracic vertebrae and a rib (side view)

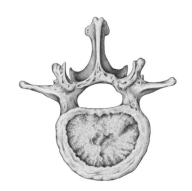

Second lumbar vertebra

- The lumbar vertebrae provide support to the body and they absorb the stress of lifting and carrying. They have more processes for muscle attachment than cervical and lumbar vertebrae. They are large with a large body, a thick vertebral arch and small vertebral canal.

Posture and postural deformities

The diagrams show the spine from the front and the side. The side view shows concave curves in the cervical and lumbar regions. The curves contribute to absorbing shock and to balance. They are maintained by muscles, ligaments and tendons. Good posture minimises strain, but excess body weight and weak muscles can distort the alignment of the spine, as can some genetic conditions.

Scoliosis

In scoliosis, in addition to the S-shape curve from front to back, there is also a sideways curve in the spine. It can be caused by muscle imbalance and the spine gets pulled in different directions. In about 30% of cases, there is a family history, which suggests there is some genetic component to scoliosis. At least one gene has been linked with the condition.

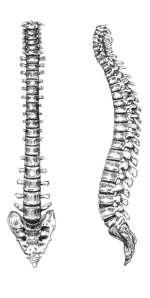

The human spine

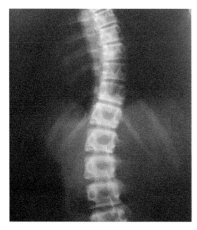

X-ray of spine showing scoliosis

Treatment of scoliosis depends on the maturity of the skeleton. For children and adolescents, physiotherapy exercises strengthen the spinal muscles. Bracing may be used or, depending on the severity of the curvature, surgery may be used to straighten the spine.

Flat foot

The arch of the foot, or instep, develops by the age of about 10, raising the inner side of the foot off the ground. With fallen arches, or flat foot, the foot rolls to the inner side and is flat on the ground. This is called 'overpronation'. The consequent strain on muscles and ligaments may cause pain in the foot, ankle, calf, knee, hip or back.

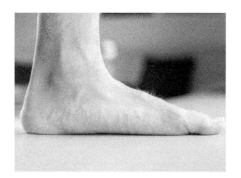

A flat foot

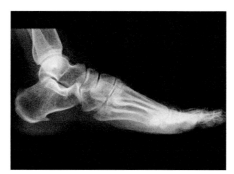

X-ray of normal foot

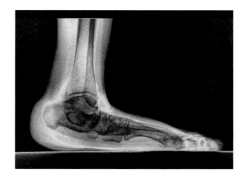

X-ray of flat foot

Causes:

- Genetic
- Congenital, e.g. abnormal joint or fused bones
- Arthritis
- Injury.

Treatment:

- Well-fitting shoes to ease foot pain
- A special shoe insole, made and fitted by a chiropodist, reduces pronation
- Exercises that stretch the calf and Achilles tendon may reduce pronation
- Surgery may be used for congenital defects, e.g. bones may be straightened or separated.

Knock-knees

A person has knock-knees if their feet do not meet when their knees touch. It is normal up to about 18 months of age but in older people it can result from:

- Rickets, caused by lack of vitamin D or calcium

- Bone infection

- Being overweight or obese

- Injured shinbone, but then, only the injured leg is knock-kneed.

In most cases, treatment is not necessary, and, unless it is caused by disease, it usually resolves naturally. Over the age of 7, some children use a brace at night but if the condition is severe and continues into adolescence, they may have surgery.

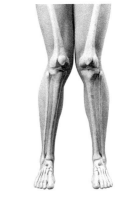

Knock-knees

Functions of the skeleton

- Support – the skeleton supports the body and determines its overall shape. Limbs are attached to girdles on an upright spine. The ribcage and sternum support the thorax.

- Muscle attachment – tendons join skeletal muscles to bone. In most cases, one end of the muscle, called the origin, is fixed. The other end, the insertion, moves when the muscle contracts. The insertion is usually further from the axis of the body. Projections, or processes, on bones are the muscle attachment points.

- Protection – the skeleton is a strong, rigid protection for many internal organs, preventing damage. The skull surrounds the brain and shields the eyes and the middle and inner ears. The vertebrae protect the spinal cord. The ribcage and sternum surround the heart, lungs and major blood vessels.

- Red blood cell production – bone marrow contains haemopoietic stem cells, the cells that can divide to form blood cells in a process called haemopoiesis. In children, the major sites are the humerus and femur, but in adults the pelvis and body of the vertebrae are most significant.

- Calcium store – 70% of the bone matrix is hydroxy-apatite, of which nearly 40% is calcium ions. The calcium in the bone is used to maintain serum calcium in a homeostatic process, in conjunction with the parathyroid gland and the kidneys. Calcium leaches out of bones in osteoporosis or if the phosphate concentration of the blood becomes too high, e.g. from drinking too many fizzy drinks. In rickets and osteomalacia, insufficient calcium is deposited in bone.

‹Link› In the first year of this course, you learned that carnivore skulls have prominent muscle attachment points, allowing their jaws to close more quickly and strongly than those of herbivores.

Joints

Joints are where bones meet. They can be classified:

- Immovable joints, or sutures, occur where bones grow together. The bones interlock and there is no movement between them, e.g. in the cranium.

Top of human skull showing sutures

YOU SHOULD KNOW ›››

››› The different types of joint

››› The causes and treatment of osteoarthritis and rheumatoid arthritis

Going further ▶

It has been said that elephants have four knees and camels have four elbows but all mammals have two of each. The positions of their wrists and ankles confuse people.

- Movable joints allow bones to move in relation to each other.
  - Gliding joints allow bones to glide over each other to give movement in many directions, e.g. between the vertebrae, at the ankle, eight small bones at the wrist.
  - Hinge joints allow movement in one plane, e.g. knees, pointing forwards, elbows pointing backwards.
  - Ball-and-socket joints allow movement in more than one plane, e.g. the hip and shoulder joints.

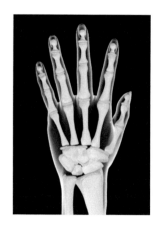

Gliding joint at wrist

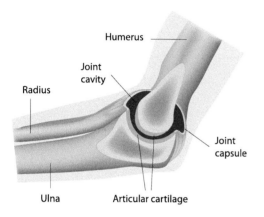

Hinge joint at the elbow

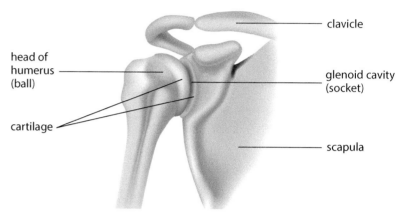

Ball and socket joint at the shoulder

▼ **Study point**

Any word containing 'arthr-' refers to joints. A word that ends in '-itis' describes an inflammation. 'Osteo-' refers to bones and 'rheumatoid' describes immune conditions affecting joints and connective tissue.

 Key Term

Osteoarthritis: Degenerative condition in which articular cartilage degrades and produces painful, inflamed joints.

Going further ▶

Regular, gentle exercise and maintaining a healthy weight can help prevent OA. Glucosamine, which many people take, is about as effective as a placebo. Cracking knuckles does not produce arthritis.

Arthritis

Arthritis is a group of conditions in which joints become inflamed.

Osteoarthritis

Osteoarthritis (OA) is the commonest joint disease. It is a degenerative condition. Glycoprotein and collagen in articular cartilage on ends of bones at joints are degraded faster than they are rebuilt. The breakdown products are released into the joint cavity and cells lining the joint remove them. But they cause inflamation and so the joint swells, becomes painful and stiffens. Ligaments may also thicken and spurs of bone may grow, and limit movement. The severity depends on the joint affected and on the individual.

Any joint can be affected but OA is commonest in the knees, hips and fingers. No genetic link has been established and OA is not caused by the body's own immune system attacking joint tissues, i.e. it is not an autoimmune condition.

Risk factors

- Age: OA is commonest in people over 45 years, although younger people can be affected.
- Being overweight, especially for knee and hip joints.
- Repeated flexing of a joint, such as with those who dance or play sport. When people knitted a lot, osteoarthritis was common in hands. Knitting fell in popularity and so did OA, but now OA is more common again: orthopaedic surgeons predict a massive increase in this and other hand and finger problems, associated with the use of mobile electronic devices.

Treatment

There is no cure for OA. The choice of treatments depends on its severity:

- Structured exercise plan with a physiotherapist.

- Pain control with NSAIDs, e.g. aspirin.

- Joint replacement, especially for hip and knee joints.
  - Advantages of replacement include pain relief, reduced drug use, restoration of mobility, movement, and enhanced quality of life.
  - Disadvantages include the risk of blood clots and infection, a long recovery period and a subsequent increased risk of hip dislocation. On current technology, such replacement joints last approximately 15–20 years, and a further replacement only increases these disadvantages.

Rheumatoid arthritis

Rheumatoid arthritis (RA) is an autoimmune disorder that attacks bone and cartilage at joints, especially in the wrists and hands. The joints become severely inflamed. Increased blood flow warms the joints and makes them swollen and painful, so movement is restricted. A diagnosis based on signs and symptoms can be confirmed with blood tests and imaging.

The causes appear to be both genetic and environmental. Smoking is a particular risk factor but cold, damp weather and a high intake of caffeine and red meat also increase the risk.

There is no cure for RA, but treatments can improve the symptoms and slow the progress of the disease.

Treatment aims to improve pain, decrease inflammation and improve a person's overall functioning:

- Physiotherapy, balancing rest and exercise to maintain muscles strength and overall physical function.

- Drugs – NSAIDs are injected into the joints to reduce inflammation and pain; disease-modifying anti-rheumatic drugs (DMARDs) slow the progression of disease, although side effects may make these unsuitable.

- Joint replacement surgery.

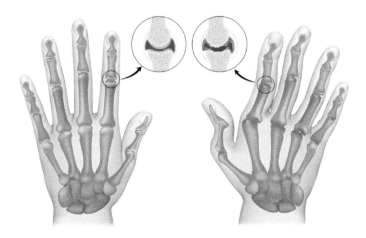

Normal   Rheumatoid arthritis

Diagrams to show a normal hand and a hand with rheumatoid arthritis

Joints as levers

A lever is a rigid, movable structure that pivots about a fixed position, the fulcrum. The effort is the force applied to the lever and the load is what the lever moves. In the body, a bone is a rigid, movable structure and a joint is a fulcrum. The contraction of muscles produces the effort and the part of the body that is moved is the load.

If a lever allows you to apply a small effort and get a larger force out, it is a force magnifier. An alternative is that you can apply an effort over a small distance and get the same force out, but it moves further than the effort you put in. It is a distance magnifier, which is a common type of lever in the body.

- First order levers have the fulcrum in the middle, like a seesaw. The body has a first order lever at the top of the spine. The neck muscles contract, providing the effort, which balances the load of the skull. The fulcrum is where the top vertebra, the atlas, meets the bone of the skull. When the neck muscles contract, the head tilts up.

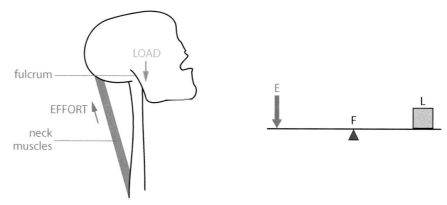

1st order lever

Imagine a small child balancing a large adult on a seesaw. If the child (the effort) is much further from the fulcrum, then the large adult (the load) will move. So a first order lever can be a force magnifier. But for the head, the load and effort are about the same distance as each other from the fulcrum, so, in this case, the lever acts to maintain posture, not to produce an increase in force.

- Second order levers have the load in the middle, like a wheelbarrow. They do not occur at any of the body's joints but they are used in certain situations, e.g. standing on tiptoes. The calf muscles provide the effort to lift the load of the body, with the toes as the fulcrum. A press-up uses a second order lever, with the feet as the fulcrum, the body weight, in the middle is the load and the effort is through the hands pressing on the ground.

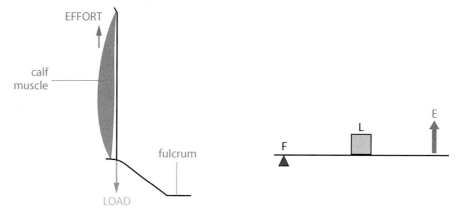

2nd order lever

- Third order levers have the effort in the middle, like using tweezers or chopsticks. An example in the body is the biceps. Contracting it provides the effort lifting the load of the hand, with the elbow as the fulcrum. The load moves in the same direction as the effort is applied. The load is further from the fulcrum than the effort and it so moves a greater distance than the effort. This is an example of a distance magnifier, i.e. the load moves a larger distance than the effort. Many of the body's levers are third order. They allow the muscle to be inserted near the joint, producing a large, rapid movement.

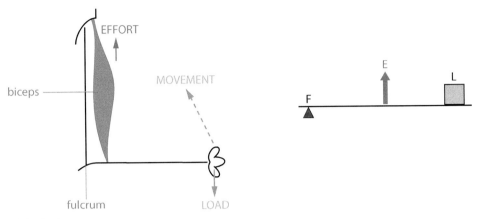

3rd order lever

Calculation

When a lever is at equilibrium, for example, when a seesaw balances or a bent arm is held steady:

$F_1 \times d_2 = F_2 \times d_2$, where F_1 and F_2 are the forces exerted by the load and the effort, measured in newtons; d_1 and d_2 are the distances of the load and the effort from the fulcrum, measured in metres.

A biological example: calculate the effort applied by your biceps when you are holding steadily a 1 kg bag of sugar in the palm of your hand. The distance from your elbow to your hand is 30 cm and from your elbow to the insertion of your biceps on the radius is 3 cm. Assume that a 1 kg mass gives a force of 9.8 newtons.

The paragraph tells us that $F_1 = 1 \times 9.8$ kg, $d_1 = 0.30$ m; $d_2 = 0.03$ m.

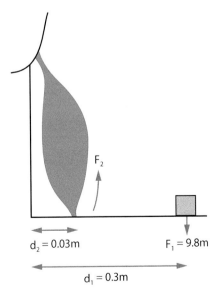

$$F_1 \times d_1 = F_2 \times d_2$$

$$\therefore 1 \times 9.8 \times 0.30 = F_2 \times 0.03$$

$$\therefore F_2 = \frac{9.8 \times 0.30}{0.03} = 98 \text{ newtons}$$

= effort provided by biceps.

$d_2 = 0.03$m $F_1 = 9.8$m

$d_1 = 0.3$m

Data for calculation

The biceps has to exert a large force but because the effort is close to the fulcrum, it can move the load quickly over a large distance. This rapid, large movement is what makes third order levers so useful in the body. The great speed and distance moved contribute to sporting excellence.

The synovial joint

Movable joints are also called **synovial joints**.

- The ends of the bones are covered with articular cartilage, which is hyaline cartilage, making a slippery coat.

- The synovial cavity is a small space between the ends of the bones, filled with synovial fluid, which is secreted by the synovial membrane that surrounds the space.

- The articular cartilage and synovial fluid lubricate the joint and act as shock absorbers.

- The synovial fluid also nourishes the chondrocytes. This is important because diffusion of nutrients from the blood vessels in the bones, though the articular cartilage, would be too slow to provide the chondrocytes with the nutrients they need.

- The synovial membrane is covered in a fibrous membrane that contains ligaments. The synovial membrane and ligaments make the joint capsule, which holds the joint together.

Link The circular and longitudinal muscles of the gut wall work antagonistically in peristalsis. Circular and longitudinal muscles in the iris work antagonistically controlling the pupil diameter.

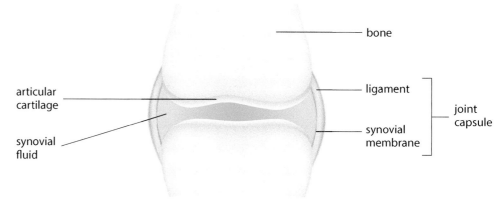

Synovial joint

Antagonistic muscles

When a muscle contracts it can pull on a bone and flex a joint. But when a muscle relaxes, it cannot push the bone. Separate muscles are needed to bend and extend joints. They work in a pair, a flexor contracting to bend a joint and a extensor contracting to straighten a joint. When one is contracted, the other is relaxed. The flexor and extensor constitute an **antagonistic pair**, which means they act in a co-ordinated, opposite fashion.

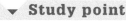
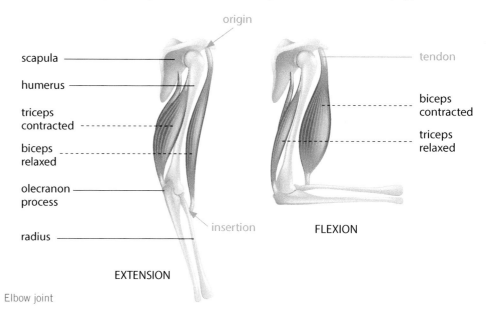

Elbow joint

The elbow has three flexor muscles, of which the biceps is the most important. Its origin is on the scapula, with two tendons. Its main insertion is on the radius.

The extensor muscle of the elbow is the triceps. It has origins on both the humerus and the scapula, just below the shoulder joint. Its insertion is on the bony projection at the back of the elbow, called the olecranon process.

Tendons

Tendons are 86% collagen, laid down in densely packed parallel fibres. They join muscles to bone and interweave at both ends to provide very strong connections. The connections are also inelastic, so the tendons remain the same length when the muscle contracts. The energy of contraction is, therefore, efficiently transmitted into movement at a joint, without losing energy in tendons changing length. Recent research provides evidence that tendons can store energy and act as springs, e.g. the use of the Achilles tendon in walking.

Like bone, tendons grow and remodel themselves as they are used, but they lose their properties if not used normally:

- Non-use – the diameter of the collagen bundles in the Achilles tendon of rats becomes smaller if the rats are inactive. This is likely to be true for other mammals, including humans.

- Micro-gravity – tendons lose their stiffness in micro-gravity, even with exercise, which has implications for treatment of bedridden patients and for the design of more effective exercises for astronauts.

52

Knowledge check

Fill in the gaps.

Articular .... and synovial fluid .... joints. Muscles operate joints in .... pairs, e.g. when the biceps is relaxed the triceps is .... and the elbow is ....

Going further ▶

Major functions of a skeleton include support and movement. Even bacteria contain filamentous proteins that perform these functions and, like other eukaryote cells, single-celled organisms, e.g. *Amoeba*, have a cytoskeleton of microfilaments and microtubules.

Many invertebrates have a skeleton, for example:

- Several sponges have an endoskeleton made of a modified collagen called spongin, and may have projecting spines of calcium carbonate or silicon dioxide. Some also have a calcium carbonate exoskeleton.

- Echinoderms, e.g. sea urchins and starfish, have external calcium carbonate spines, although these are not exoskeleton, but extensions of an endoskeleton just below the skin.

- Most Molluscs do not have either an endo- or exoskeleton, although their bodies may have hard parts, e.g. the beak of an octopus, made of chitin, and the cuttlebone of cuttlefish, made of aragonite, a form of calcium carbonate.

Vertebrates have an endoskeleton, the axial and appendicular skeletons, but the roof of the skull, teeth, scales and fin rays comprise, in addition, an exoskeleton. Little is known about the vertebrate skeleton's early evolution, because the organisms in which it first evolved are extinct. But skeletons are mineralised so they fossilise well, providing some evidence. A mineralised skeleton is thought to have appeared first in the exoskeleton of bony fish, e.g. in their scales and fin rays. In the ostracoderms, the armoured, jawless fishes of the Paleozoic era, which are now extinct, the exoskeleton completely surrounded the body. An exoskeleton, however, is very heavy and as vertebrates have evolved on land, it has been significantly reduced.

Option C: Neurobiology and behaviour

The structure and functioning of the brain have been examined with various imaging techniques. They have allowed specific areas of the brain to be correlated with particular responses and behaviours. Disorders, for example in speech, can be mapped to defined locations. The brain's neuroplastic properties allow it to change and adapt to new circumstances throughout life by modifying synapses in response to developmental stage and environmental triggers. Epigenetic changes explain how some conditions may be transmitted across generations. The study of behaviour and types of learning indicate how humans interact as individuals and in social groups. The organisation of some animal societies is described, including reference to territorial and courtship rituals.

Topic contents

By the end of this topic you will be able to:

- Describe the structure of the human brain.
- Describe the main functions of the cerebrum, hypothalamus, cerebellum and medulla oblongata.
- Describe the roles of the sympathetic and parasympathetic nervous systems.
- Understand how the hypothalamus links nervous and endocrine regulation.
- Describe the roles of the sensory and motor cortex.
- Understand the significance of the sensory and motor homunculi.
- Describe the roles of brain areas associated with language and speech.
- Describe brain imaging techniques.
- Explain the significance of critical periods in brain development.
- Describe the phenomenon of neuroplasticity.
- Describe how gene expression affects brain development and behaviour.
- Explain the differences between innate and learned behaviour.
- Describe social groupings and their advantages and disadvantages.
- Explain the evolutionary significance of territorial and courtship behaviours and how sexual selection has generated them.

The brain

The human brain is the most complex structure in the known universe. It is the co-ordinating centre of sensation and of intellectual activity. An adult's brain weighs about 1.5 kg and contains an estimated 8.6×10^{10} neurones, each making more than 1000 connections. It is thought that this high level of connectivity is significant in generating the properties of the brain. There are, in addition, about 10^{12} glial cells, supporting the functions of the neurones.

The evolutionary development of the human brain can be investigated by comparing it with the brains of other vertebrates. Animals that appear earlier in the fossil record have simpler brains, which may be correlated with their range of behaviours.

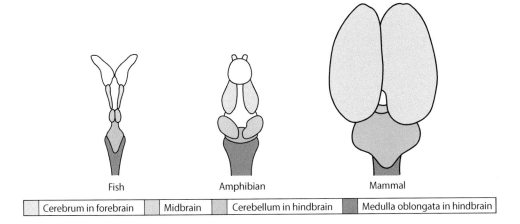

| Fish | Amphibian | Mammal |

| Cerebrum in forebrain | Midbrain | Cerebellum in hindbrain | Medulla oblongata in hindbrain |

Dorsal view of three vertebrate brains

Meninges

The brain is continuous with the spinal cord and, like the spinal cord, is surrounded by three membranes, or **meninges**: the delicate pia mater on the brain surface, the thick dura mater attached to the skull and the arachnoid mater between. An inflammation of the meninges is called meningitis. Viral meningitis is more common than bacterial meningitis and is often so mild that it is mistaken for 'flu. Bacterial meningitis can be fatal but a vaccination programme has reduced the number of cases of meningitis.

Brain structure

Ventricles

The brain is a swelling of the anterior end of the spinal cord. It has four spaces, called **ventricles**, which are continuous with the central canal of the spinal cord. Cells lining the ventricles produce cerebro-spinal fluid, which circulates between the ventricles and is passed into the spinal canal. It resembles plasma in its components, for example:

- It supplies nutrients, such as glucose.
- It supplies oxygen, carried in solution. Cerebro-spinal fluid does not contain red blood cells and is pale yellow in colour.
- It contains antibodies and white blood cells, giving it a role in resisting infection.

▼ Study point

When a word ends in '-itis' it refers to an inflammation.

Going further ▶

The Men B vaccine is available for babies. School students in Year 13 and first-time university students up to the age of 25 are offered the Men ACWY vaccine to protect against meningitis A, C, W and Y.

Key Terms

Meninges: Three membranes, the pia mater, arachnoid mater and dura mater that line the skull and vertebral canal, surrounding the brain and spinal cord.

Ventricles: Four connected cavities in the brain into which cerebro-spinal fluid is secreted.

Going further ▶

The ventricles of an adult contain about 80 cm³ of cerebro-spinal fluid, which is completely replaced about every six hours.

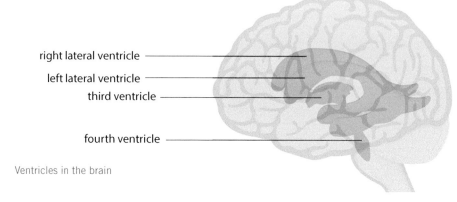

right lateral ventricle

left lateral ventricle

third ventricle

fourth ventricle

Ventricles in the brain

Key Terms

Cerebellum: Part of the hindbrain that co-ordinates the precision and timing in muscular activity, contributing to equilibrium and posture, and to learning motor skills.

Medulla oblongata: Part of the hindbrain that connects the brain to the spinal cord and controls involuntary, autonomic functions.

Going further ▶

There is evidence that the cerebellum has a role in some cognitive functions, including language.

Going further ▶

The hippocampus also has a role in spatial navigation and is one of the brain regions to have its function compromised early in the development of Alzheimer's disease.

◀Link▶ The hormones that control the menstrual cycle illustrate the link between the hypothalamus, the pituitary and endocrine glands as described on p170.

Key Term

Cerebrum: Two hemispheres responsible for integrating sensory functions and initiating voluntary motor functions. It is the source of intellectual function in humans, where it is more developed than in other animals.

Regions in the brain

The brain has three regions named for their positions: the forebrain, the midbrain and the hindbrain. Their evolutionary development in vertebrates is indicated in the diagram on p303. Evolution modifies what is already present, and species that appear later in the fossil record tend to have more sophisticated structures, not necessarily new ones.

- The hindbrain is the most primitive part of the human brain and is sometimes referred to as the reptilian brain. It sustains basic homeostatic functions:
  - The **cerebellum** has a convoluted surface, providing space for the cell bodies of a large number of neurones. It co-ordinates voluntary tasks requiring fine muscle control, e.g. writing and playing scales rapidly, and controls the muscles that maintain posture.
  - The **medulla oblongata** controls basic functions including ventilation, maintaining blood pressure, balance and the regulation of heartbeat.
- The midbrain contains nerve fibres linking the hindbrain and the forebrain, that relay information for vision and hearing.
- The forebrain
  - The **limbic system** is associated with emotion, learning, memory:
    - The **hippocampus** interacts with the cerebral cortex, contributing to learning, reasoning, personality and consolidating memory into a permanent store.
    - The **thalamus** is a relay centre, sending and receiving impulses to and from the cerebrum.
    - The **hypothalamus** controls general functions, e.g. body temperature, blood solute concentration, hunger, thirst, sleep. It is the main controlling region of the autonomic nervous system and links the brain to the endocrine system, through the pituitary gland.
  - The **cerebrum** controls voluntary behaviour, learning, reasoning, personality and memory. Most of its functioning is subconscious.

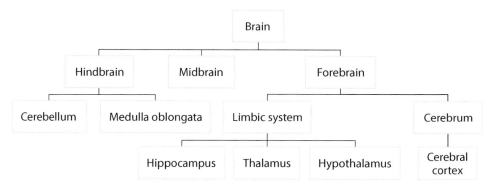

Organisation of the brain

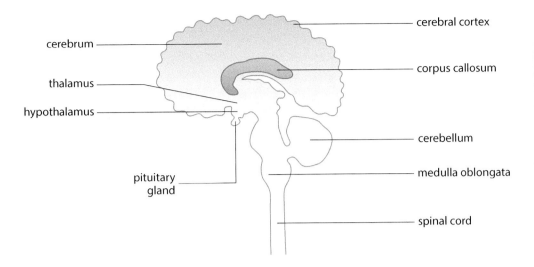

Sagittal section of brain. Note: the hippocampus is not visible in this section

53

Knowledge check

Match the structures 1–4 with their description A–D.

1. Meninges.
2. Ventricles.
3. Cerebellum.
4. Cerebrum.

A. Part of hindbrain associated with unconscious, fine muscle control.
B. Three membranes around brain and spinal cord.
C. Cavities in brain containing cerebro-spinal fluid.
D. Part of forebrain associated with thinking and decision making.

The autonomic nervous system (ANS)

The autonomic nervous system is a part of the peripheral nervous system and controls the continuous functioning of internal organs, without conscious intervention. The processes it controls are sometimes described as 'automatic' and include reflex actions, such as swallowing, coughing, vomiting and sneezing. The ANS is regulated in the hypothalamus. There are two antagonistic components and, as with other antagonistic systems, such as muscles, e.g. biceps and triceps, or hormones, e.g. insulin and glucagon, the dual aspect gives very precise control.

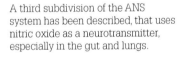

Key Term

Autonomic nervous system: The part of the peripheral nervous system that controls automatic functions of the body by the antagonistic activity of the sympathetic and parasympathetic nervous systems.

- The **sympathetic** nervous system uses the neurotransmitter noradrenalin. This acts in a similar fashion to adrenalin, and so the sympathetic system is, in general, excitatory. Its nervous impulses increase heart rate, blood pressure and ventilation rate. The cell bodies of sympathetic neurones are in the grey matter of the spinal cord and in ganglions lying outside the spinal cord.

- The **parasympathetic** nervous system uses acetylcholine as a neurotransmitter. It generally has an inhibitory effect, decreasing heart rate, blood pressure and ventilation rate. The cell bodies of parasympathetic neurones are in the brain and spinal cord and others lie close to target organs.

Going further ▶

A third subdivision of the ANS system has been described, that uses nitric oxide as a neurotransmitter, especially in the gut and lungs.

Exam tip

The sympathetic system has 'fight-or-flight' responses; the parasympathetic system has 'rest-and-digest' or 'feed-and-breed' responses.

The ANS and heart rate

A decrease in blood pH or, to a lesser extent, blood pressure increases heart rate. These responses are under autonomic control by the cardiovascular centre in the medulla oblongata.

During exercise:

- The cardio-acceleratory centre is stimulated.
- Nervous impulses travel along sympathetic nerve fibres to the sino-atrial node (SAN).
- The neurotransmitter noradrenalin is released.
- Noradrenalin binds to cell membrane receptors on SAN cells.
- The SAN's electrical discharge increases in frequency.
- The heart rate increases.

During sleep:

- The cardio-inhibitory centre is stimulated.
- Nervous impulses travel along parasympathetic nerve fibres to the SAN.
- The neurotransmitter acetylcholine is released.
- Acetylcholine binds to cell membrane receptors on SAN cells.
- The SAN's electrical discharge decreases in frequency.
- The heart rate decreases.

The cerebral cortex

Going further ▶

People with severe epilepsy may have the corpus callosum cut. This reduces the electric discharge from one hemisphere to the other, and seizures are reduced.

The brain is largely bilaterally symmetrical so the cerebrum has two hemispheres, although one, usually the left, is dominant. The cerebral hemispheres are connected by fibres, which run through the **corpus callosum**, the largest white matter structure in the brain. It is suggested that the special structure of the corpus callosum may explain why some areas of the brain are lateralised, i.e. appear to function somewhat differently in the two hemispheres, and may contribute to the understanding of the origins of human cognition.

The outer 2–3 mm constitutes the cerebral cortex. Its neurones have their axons deeper in the brain than their cell bodies. The axons are myelinated but cell bodies are not so, unlike in the spinal cord, in the cerebrum, grey matter surrounds white matter.

Going further ▶

Species that appear later in the fossil record, e.g. primates, have more convolutions in their neocortex than those that appeared earlier, e.g. fish, correlating with their more complex behaviour.

In reptiles and fish, the cerebral cortex is small and simple. In mammals, especially primates, in the recent evolutionary past, it expanded greatly in size and complexity, so it is also called the **neocortex**. Like the cerebellum, its surface is highly folded, producing a larger surface area, fitting in a larger number of neurones than if the surface were sooth.

The neocortex has of the order of 2×10^{10} neurones. Their high connectivity is thought to be responsible for the cerebrum executing higher cognitive functions including language, conscious actions, thought and processing sensory input. The cerebrum is strongly integrated with the thalamus, in the limbic system.

The lobes of the cerebral hemispheres

Each cerebral hemisphere has four distinct lobes. They were originally described based on their structure but it was found they have particular roles:

Lobe	Properties
Frontal	• Aspects of personality. • Site of reasoning, planning, emotions and problem solving. • The dominant hemisphere includes Broca's area, which controls the motor aspects of speech. • Contains motor cortex.
Temporal	• Includes the auditory cortex so generates the sense of sound. • Processes complex stimuli, e.g. face, scenes. • Has roles in learning and memory. • The left temporal lobe contains Wernicke's area, which is associated with written and spoken speech. • Temporal lobes are not lateralised, so, unlike other areas of the brain, neither is dominant.
Parietal	• Associated with the sense of taste. • Has a role in visuo-spatial processing. • Includes the somatosensory cortex.
Occipital	• Contains the primary visual cortex and is associated with vision.

Going further ▶

Temporal lobe epilepsy is correlated with profound and mystical feelings. Seizures in the left temporal lobes have been implicated in religious visions.

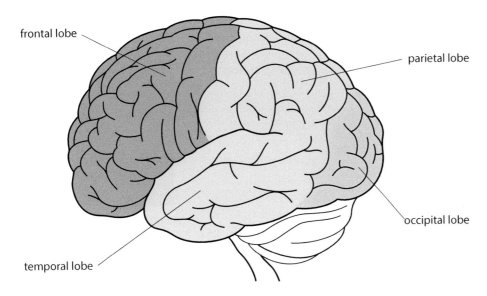

Lobes of the human brain

The areas of the cerebral hemispheres

There are three areas of the cerebral hemispheres, based on their functions:

- Sensory areas, also called the **sensory cortex**, or the somatosensory cortex, receive nervous impulses from the sense receptors of the body, via the thalamus. Nerve fibres from the two sides of the body cross over in the corpus callosum, so the sensory area in the cortex of one hemisphere processes information form the sense receptors of the other side of the body.

- **Motor areas,** also called the **motor cortex**, send nervous impulses through the corpus callosum, to effectors on the other side of the body.

- **Association areas** take up most of the cerebral cortex. They:
  - Receive nervous impulses from sensory areas
  - Initiate responses which are passed to motor areas
  - Associate new information with stored information, generating meaning
  - Interpret, process and store visual information in the visual association area
  - Interpret, process and store auditory information in the auditory association area.

▼ Study point

The cerebrum has four lobes, based on structure: the frontal, temporal, parietal and occipital lobes. It has three areas based on function: sensory, motor and association.

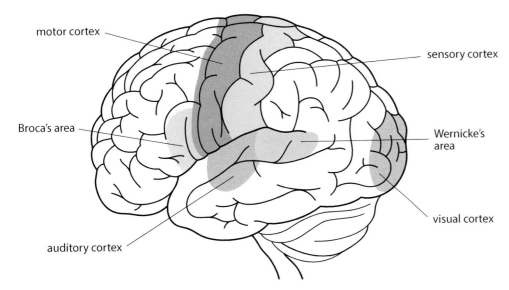

Functional areas of the human brain

Stroke

A stroke is an interruption to blood flow in the brain. A cerebral artery fails to deliver oxygen to the neurones of the brain and so they die. This is an infarction. There are two main causes:

- 87% of strokes are **ischaemic strokes**, where a blood vessel is blocked. It can be caused by:
  - a blood clot, or thrombus, forming at the site
  - an embolus, i.e. material travelling in the blood stream.

  If detected within about 4 hours, an ischaemic stroke may be treated successfully with 'clot-busting' medication.

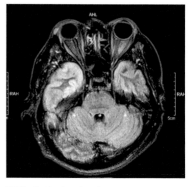

MRI of a brain showing stroke

- 13% of strokes are **haemorrhagic strokes**, caused by bleeding into the brain or the space around it, which can produce an intense headache.

When brain cells die, nervous impulses can no longer travel along their fibres from the brain's motor areas to effectors, causing paralysis. This may be on one side of the body only. Often, on that side, muscle tone is lost, because no muscle fibres are stimulated, and a stroke patient's face may droop. At the corpus callosum, nerve fibres cross over and so a stroke in one hemisphere of the cerebrum affects the body on the opposite side.

The main risk factor for stroke is high blood pressure but tobacco smoking, obesity and high blood cholesterol increase the risk. The risk increases with age, with about 70% of strokes occurring in people over 65. Strokes are diagnosed with a physical examination and imaging, e.g. a CT or MRI scan, while a blood test and electrocardiogram (ECG) can rule out other possible causes.

The cortical homunculi

The cortical **homunculus** is a drawing or model of the human body showing the positions in which parts of the body are represented in the cerebral cortex, and the area of the cortex devoted to them. It correlates the body's anatomy with a neurological map. There are two types of cortical homunculus:

1. The sensory homunculus

The sensory homunculus shows the relative sensitivity of different parts of the body, as represented in the sensory cortex. The most sensitive areas of the body are shown with the largest area in the drawing. The tongue, lips, fingers and genitals are the most sensitive areas and so they are largest in the sensory homunculus. They are the most sensitive because they have the highest density of sense receptors and sensory neurones coming from them. The areas of the brain that receives their nervous impulses are correspondingly larger.

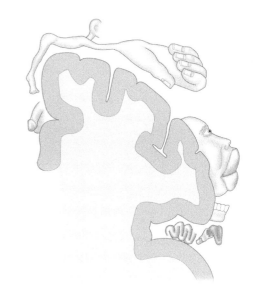

Sensory homunculus on cortex

2. The motor homunculus

The motor homunculus shows the motor control of different parts of the body, as represented in the motor cortex. It is very similar to the sensory homunculus but, as can be seen in the diagram, the areas of the cortex associated with some organs are different. The muscles of hands can make fine movements; it is estimated that up to 43 muscles control facial expressions, some of which are fleeting. A large number of motor fibres finely control the contraction and relaxation of the hand and face muscles and so much of the motor area in the cortex relates to them. Accordingly, the face and hands have a large representation in the motor homunculus.

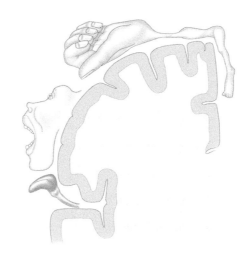

Motor homunculus on cortex

The images on the right show models of the body representing the degree of motor control (left) and sensitivity (right). They are also referred to as the motor and sensory homunculi, respectively, although they do not contain any representation of the cortex.

Models of the motor (left) and sensory (right) homunculi

Knowledge check

Complete the paragraph by filling in the gaps.

The cerebrum has three major areas: the sensory area, the motor area and the .... areas. In the sensory homunculus, the fingers are shown very large because they have a very .... density of sensory neurones. If blood supply to part of the brain is interrupted by a blood clot, cells die. This is an .... stroke.

Language and speech

Vocal animals, such as canaries, dolphins, chimpanzees and humans use only one cerebral hemisphere to control language and speech. In humans, it is usually the left hemisphere. The use of one hemisphere, rather than both, is called lateralisation and was discovered by physicians doing post mortem examinations of brains.

- Broca and Wernicke, in the 1880s, found damage in areas of the left hemisphere in the brains of patients with language difficulties. These areas are now named after them.

- Lateralisation was not confirmed for 70 years, when brain surgeons electrically stimulated the right hemisphere of a patient who was awake. When the patient was spoken to, they could reply. But if the left hemisphere was stimulated, the patient could not speak.

- These results were supported in the 1960s when similar experiments were performed using sodium amytal, which can anaesthetise one hemisphere, but not the other.

The current understanding of the location of the control of language is:

- The sensory areas are in the visual and auditory areas.

- The association area is Wernicke's area.

- The motor area is Broca's area.

Wernicke's area is an association area, which interprets written and spoken language. Damage may result in Wernicke's aphasia, in which patients do not understand the speech of other people. They speak fluently and grammatically but their sentences often use words that are irrelevant and unrelated, a phenomenon called 'word salad'.

WORKING SCIENTIFICALLY

Science progresses when new technology is applied. In the 1950s and 1960s, new anaesthetics and electrical techniques allowed the brains of patients who were awake to be investigated, confirming lateralisation.

▼ **Study point**

Broca's area has motor control over speech and deals with grammar. Wernicke's area controls the comprehension of written and spoken language.

Going further ▶

Neuroimaging suggests the some functions attributed to Wernicke's area may happen also in Broca's area.

Going further ▶

Most early studies were done on adult males. Women were later shown to be less lateralised. Children with left hemisphere damage may develop language in the right hemisphere instead.

Going further ▶

Golgi shared the Nobel Prize in Physiology or Medicine in 1906 for observations, descriptions and categorisations of neurones in the brain.

Going further ▶

Electrical stimulation of the brain has a long history. Scribonius Largus, the court physician for the first century Roman emperor, Claudius, used an electrical torpedo fish to treat headaches.

▼ **Study point**

Clinical assessment and brain imaging are non-invasive techniques, but electrical stimulation methods are invasive.

Broca's area was the first language area to be discovered. It is a motor area and its motor neurones innervate muscles of the mouth, larynx, intercostal muscles and diaphragm to produce the sounds we call speech. If the area is damaged, a patient has Broca's aphasia. They can understand speech but their own may be slow and slurred with incomplete, ungrammatical sentences.

Broca's and Wernicke's areas are in different lobes, but are quite close to each other. The **arcuate fisculus** is a bundle of nerve fibres generally thought to link them, although its exact location is still debated. If the arcuate fisculus is damaged, a patient can understand written and spoken words but cannot speak.

Neuroscience

Studying the brain

The effects of head injuries and the effects of various plant products on behaviour have been observed throughout history and, although ideas have changed, even in the ancient world, aspects of brain physiology were known. Ancient Egyptian battlefield surgeons observed aphasia and seizures following brain injury and, 1300 years later, the ancient Greeks described the cerebrum, cerebellum and ventricles. In the first century, Galen, a Greek physician who lived in the Roman Empire, deduced that the cerebellum controlled muscles, and the cerebrum processed senses. In the Renaissance period, Vesalius dissected brains and identified the corpus callosum, while Leonardo da Vinci made anatomically accurate drawings of brains and skulls. The 19th-century attempt to correlate bumps on the skull with aspects of personality, in the study of phrenology, is no longer credited. But battlefield observations of injury correlating with altered behaviour, especially during World War I, led to a better understanding of the role of different areas of the brain. As is so often the case, more detailed understanding relied on technological innovation.

Hieroglyph for the word 'brain' (c. 1700 BCE)

The brain is studied using three main approaches:

- Clinical or neuropsychological assessment, which is non-invasive. It tries to correlate impairment of function with a damaged area of the brain after injury or neurological illness. It assesses orientation, learning and memory, intelligence, language and visual perception. Its broad goals are diagnosis, understanding the nature of any brain injury and measuring any change over time.

- Stimulation methods, which are invasive. An electrical current is applied briefly to a specific area of the brain and the effect is observed. As patients are awake during these procedures, they can describe their subjective experience. The assumption is that the applied current is equivalent to a natural event, and the effect mimics what happens normally. In this way, the roles of speech and language areas of the brain have been identified, and areas associated with Parkinson's disease, dystonia and obsessive-compulsive disorder (OCD) have been studied. Deep brain stimulation (DBS) has been used to treat patients with these conditions and also those with chronic pain or depression.

- Brain imaging, which is non-invasive. Various imaging techniques indicate the relationship between brain structures and their function.

Electroencephalography (EEG)

EEG uses electrodes on the scalp to detect voltage fluctuations in the brain. It is used to provide information about normal electrical activity and the voltage fluctuations have been correlated with the functioning of the cerebral cortex and various types of behaviour. EEG contributes to the diagnosis of some brain disorders, including epilepsy, sleep disorders and brain death and it can assist in a prognosis in coma cases. EEGs also allow the monitoring of the depth of anaesthesia during surgery.

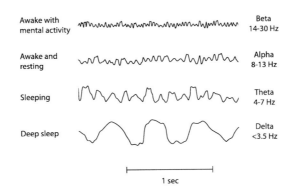

EEG traces of some normal brain waves

Going further ▶

The origins of some EEG waves have been identified, e.g. α- and β-waves originate in both sides of the cortex, but α-waves are posterior and β-waves are frontal.

The rhythms of an EEG trace are described by their frequency. An EEG of an adult is different from that of a child and the normal pattern varies throughout life. An EEG trace may have spikes and sharp waves. These happen normally but may also occur in a seizure.

Advantages and disadvantages

EEG is useful because it is non-invasive and the subject does not get claustrophobic or have to stay absolutely still. It is cheaper to perform than other scanning methods and there is a better understanding of what, neurologically, it is actually measuring. It shows changes in brain activity over the millisecond range, unlike CT or MRI scans. Unlike MRI, EEG is not noisy and does not use intense magnetic fields, which would be a problem if, for example, a patient has a pacemaker. But EEG can only detect activity in the cortex and, unlike PET, it cannot identify the synapses where drugs or neurotransmitters act.

Computerised tomography (CT)

CT scans were first used medically in the 1970s. Throughout their history, they have also been called X-ray computed tomography (X-ray CT) or computerised axial tomography (CAT) scans. Many X-ray images, taken from different angles, are combined using sophisticated software. This produces high-resolution cross-sections, or tomographic images called 'virtual slices', which make it possible to see inside the body without cutting.

A CT scan of the head can detect tumours, brain injuries such as skull fractures and can identify if a haemorrhage or blood clot has caused a stroke.

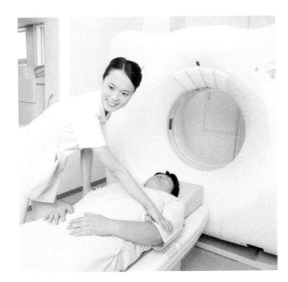

A patient about to enter a CT scanner

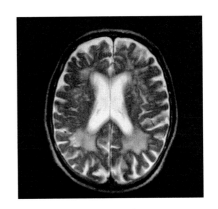

CT scan of the human brain

Going further ▶

CT scans are also used in non-medical environments such as in engineering, for non-destructive materials testing, and in archaeology, imaging the contents of sarcophagi.

Advantages and disadvantages

CT is useful because a whole-body scan can be done in seconds. Virtual slices show just the area of interest, without ghost images of other structures. The area can be viewed from different angles and the image has high contrast. But the radiation from a CT scan may lead to cancer and some people feel that the technique is greatly overused.

Magnetic resonance imaging (MRI)

MRI uses strong magnetic fields to align protons in water molecules. They are monitored as they move again following a pulse of radio waves. This produces detailed computer-generated images of soft tissues and organs and so it is used to investigate anatomy. Many images are taken milliseconds apart, so MRI can show how the brain responds to stimuli.

As MRI is more sensitive than CT, it shows some areas of the brain more effectively. It is therefore used to image brain cancers and may be used in their surgery. It gives good contrast between grey and white matter so it is useful for studying abnormal brain development, demyelinating diseases such as multiple sclerosis, and cerebrovascular disease, such as aneurysms, which may lead to stroke or dementia.

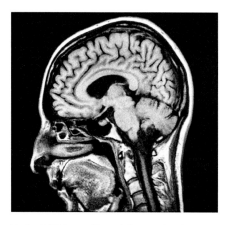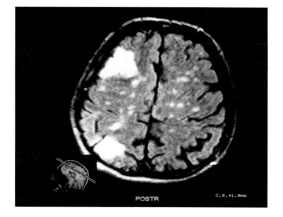

Vertical MRI scan through human head

MRI scan of patient with advanced Alzheimer's disease, showing lesions

Functional MRI (fMRI)

fMRI investigates the function rather than the structure of tissues because it uses a strong magnetic fields and a radio wave pulse to construct BOLD (blood oxygen level dependent) images. Oxyhaemoglobin and haemoglobin respond differently following the radio pulse so fMRI shows the flow of oxygenated blood. Statistical methods are used to construct a 3D map, in which areas of the cortex that are responding to particular stimuli produce a brighter area in the image. This is related to the greater demand for oxygen by active neurones and the greater proportion of oxyhaemoglobin in the blood in those areas. fMRI assesses brain activity on a second to second basis so it is the most commonly used type of brain scan. It has applications in psychology and is used in planning neurosurgery.

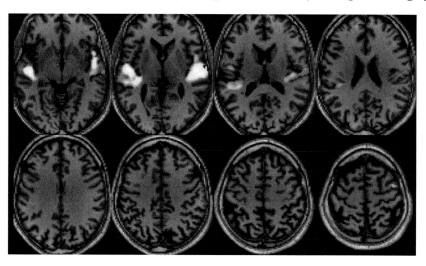

fMRI scan of subject listening to music

Advantages and disadvantages of MRI and fMRI

MRI and fMRI do not use ionising radiation and so are preferred to CT, but can be more expensive, time-consuming, and may be a problem for people with claustrophobia or who are sensitive to loud noises. The magnetic fields make it unsuitable for people who have metal in their bodies, e.g. cochlear implants or cardiac pacemakers. Demand for these techniques has risen and there are doubts about their cost effectiveness and the risk of overdiagnosis, i.e. recognition of an abnormality that would never actually become a problem.

Positron emission tomography (PET)

PET scans show the use of radioactive glucose in a 3D map of the brain. Active areas, using more glucose, release more radiation, which the computer translates into yellow and red. Less active areas are coloured green and blue. PET can detect biochemical changes in the brain before anatomical changes that accompany brain disease occur.

Mechanism

Fluorodeoxyglucose (FDG) is injected into a patient and it enters the respiratory pathway. Like glucose, it is phosphorylated by hexokinase, which is very active in rapidly growing tumours. ^{18}F in FDG replaces an oxygen atom in glucose that is needed for the next step in glucose metabolism, so there is no further reaction and FDG is trapped in any cell that takes it up. ^{18}F emits a positron as it decays. The emitted positron travels less than 1 mm, losing kinetic energy until it collides with an electron. The particles annihilate each other and produce γ-rays, which are detected in the scanner.

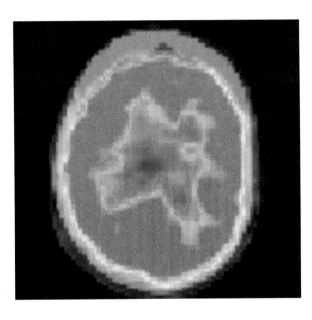

PET scan of a human brain

Uses of PET scans

- Diagnosis and monitoring of many cancers. Most PET scans are used to identify cancer metastasis, i.e. the development of secondary tumours.

- Neuropsychology:
  - PET scans confirm that during linguistic activities, areas in left hemisphere are more active than those in the right.
  - Conditions such as Alzheimer's disease decrease the rate of glucose metabolism. This means that PET using FDG can diagnose Alzheimer's disease early on and differentiate it from other types of dementia.

- Psychiatry: some tracers bind selectively to brain receptors for dopamine, serotonin or opioids. The brain activity of healthy controls can be compared with that of subjects with schizophrenia, substance abuse, mood disorders or other conditions.

- Radiosurgery: PET scans can be used to guide the surgical treatment of brain tumours.

Advantages and disadvantages

PET scanners have a high operating cost and the scan must take place within a short time of the FDG radiotracer being synthesised, before the isotope decays. The subject is exposed to gamma radiation and receives about double the ionising radiation dose of a CT-chest scan.

Neuroplasticity

In the past, the brain was considered a static organ. During most of the 20th century, people thought that its structure was relatively unchanging, after a critical period in early childhood. Now it appears that many aspects of the brain remain plastic, even into adulthood. Research indicates that experience can change both the brain's anatomy and its functional organisation.

The brain makes new neuronal connections throughout life, so it changes and adapts in response to sensory information, development and damage. It can compensate for injury by using other neurones for a particular task, and it adapts to a new environment. After stroke or brain injury, undamaged axons grow new dendrites to reconnect damaged neurones, forming new nerve pathways. This ability to make new connections is an aspect of **neuroplasticity**. Neuroplasticity occurs in healthy development, learning, memory, and recovery from brain damage.

Neuroplasticity refers to the changes in the neurones and their connections:

- Synaptic plasticity is the ability of a synapse to change in strength, for example by changing the amount of neurotransmitter released or changing the response in the postsynaptic neuron. These changes are a response to an alteration in the frequency of use of synaptic pathways, because of changes in behaviour, environment, neural processes, thinking and emotions. Synaptic plasticity has a significant role in learning and memory.

- Non-synaptic plasticity does not affect synapses but involves changing the excitability in the axon, dendrites or cell body of a neurone, e.g. through modification of voltage-gated channels. It may happen as a response to bodily injury.

In a young brain, neurones rapidly form synapses and during sensory processing, synapses weaken or strengthen. This developmental plasticity occurs mostly in the first few years. Childhood is a critical period – appropriate sensory inputs must be received for proper development. A baby is born with millions of neurones, each with about 2 500 synapses. By 2–3 years, this is increased to 15 000 per neurone. There is significant **synaptic pruning** in adolescence and by adulthood, each neurone has 1000–10 000 synapses.

The number of synapses has decreased and this fine-tuning of neurones continues throughout life, stimulated by interaction with the environment. The synapses that remain are efficient and strong. When they are repeatedly stimulated, that pattern of neural connections becomes 'hard-wired' in the brain. This makes an efficient, permanent pathway that allows signals to be transmitted quickly and accurately. Such changes can be confirmed by brain-imaging technology.

Brain development and language

Critical periods in brain development allow for the development of specific skills, such as language. The development of communication through language is an instinctive process and windows of opportunity for it occur throughout life. Most babies babble by 7 months, making speech-like sounds. Early experience is most significant in their language development:

- Deaf babies babble less than hearing babies and their language development is faulty, unless they are given another means of symbolic expression, such as signing.

- Repeated ear infections in the first few years compromise hearing and delay language development.

- A 'feral' girl with language deprivation up to the age of 13 years could only ever manage basic communication, even after intense training.

Key Terms

Neuroplasticity: The ability of the brain to modify its own structure and function following changes within the body or in the external environment.

Synaptic pruning: The elimination of synapses that happens between early childhood and maturity.

▼ Study point

Individual synapses within the brain are constantly being removed or recreated, depending on how they are used.

▼ Study point

There are periods in brain development when it is most receptive to acquiring specific skills. If that window of opportunity is lost, it is very hard to learn the skill subsequently.

Language starts to develop even before birth. At 29 weeks, a foetus responds to sound transmitted through bone conduction. It learns its mother's voice and the sound pattern of the language she speaks. There are reports of new-borns responding to the music played throughout the pregnancy.

Infants initially distinguish the sounds, or phonemes, of all languages. It is thought that, during brain development, experience selects certain synapses related to phoneme production and perception. As infants hear phonemes, different clusters of neurons in the auditory cortex of the brain respond to each sound. Absence of certain phonemes results in loss of the synapses representing those sounds, and correlates with decreased ability to perceive them. This may explain the lack of distinction between /l/ and /r/ in Japanese speakers of English. At 1–2 years of age, the brain organises synapses associated with language. Up to about 5 years, the brain learns syntax or grammar. It is much harder to learn these after about 5 years, although the brain is able to learn new words throughout life.

Parents in all cultures have a characteristic way of addressing babies and young children, with short, simple sentences, prolonged vowel sounds and high pitch. The speech patterns are recognisable, even in an unfamiliar language. This way of speaking is called 'parentese' or 'motherese'.

Genes, brain development and behaviour

At least a third of the approximately 20 500 genes that make up the human genome are expressed primarily in the brain. This is the highest proportion of genes expressed in any one part of the body.

As an organism grows and develops, chemical reactions, such as the methylation of DNA and the removal of acetyl groups from histone proteins, occur that activate and deactivate parts of the genome at particular times and in particular locations. Environmental factors may have a direct effect on brain development by influencing these epigenetic effects. They are most marked in the non-coding sequences that affect transcription. Environmental events therefore can induce long-term developmental changes.

Epigenetic changes can affect physiology, as shown in evidence from the Dutch famine of 1944. Children of women pregnant at the time had increased susceptibility to obesity and diabetes, which was ascribed to an epigenetic effect.

The epigenetic regulation of brain development seems to be unique to mammals, and appears to have an important role in brain development and in evolution. Epigenetic changes in brain cells are suspected in some cases of mental illness and addictive behaviour.

Human babies are undeveloped at birth compared with many animals. This means that most human brain development must occur after birth. The brain is therefore influenced by social and environmental factors, which may affect its function throughout life. Animals that have received good parental care have long-term brain changes, especially in the hippocampus, where genes that respond to stress are silenced. Presumably this applies also to humans, and these epigenetic effects are passed on to subsequent generations.

Gene expression and mental illness

For much of the 20th century, it was thought that conditions such as depression and schizophrenia were caused by poor parenting and damaged family relationships. Family and twin studies show that genetic factors have a role in the development of psychiatric disorders, e.g. people are more likely to suffer from depression and anxiety if their parents did. However, it has not been possible, in general, to link a specific gene with a specific disorder.

▼ Study point

Language needs extensive post-natal experience to produce and decode speech sounds; hearing and practice are crucial.

Going further ▶

There is evidence of a change in vocalisation when adult rats communicate with their young, analogous to parentese.

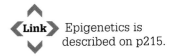 **Link** Epigenetics is described on p215.

Going further ▶

Autism, ADHD, bipolar disorder, major depressive disorder and schizophrenia were traditionally seen as distinct conditions. In 2013, mutations in particular genes were shown to link them.

▼ **Study point**

Cortisol production is controlled by negative feedback: the release of cortisol into the bloodstream results in the decline of cortisol production; this stops us from being constantly overstressed.

◀**Link**▶ The concentration of many hormones in the blood, e.g, ADH is controlled by negative feedback, as described on p128.

Going further ▶

Many other social factors are implicated in mental illness including poverty, bullying, social stress, traumatic events, employment problems, socioeconomic inequality, lack of social cohesion, migration and discrimination.

56

Knowledge check

Match the terms A–D with the characteristics 1–4.

1. Neuroplasticity.
2. Synaptic pruning.
3. Epigenetics.
4. Cortisol.

A. Reduction in the number of brain synapses between birth and maturity.
B. Modification of DNA and histones by environmental factors, altering gene expression.
C. Hormone secreted in response to stress.
D. Change in neural pathways in response to environmental and developmental triggers.

The genetic approach to understanding brain disorders has focused on genetic polymorphisms, but they do not explain the extent and variation in psychiatric illnesses that occurs. Many conditions are polygenic and the interaction of these many genes with each other and the environment contributes to the development of disease. An individual may inherit a susceptibility to a given condition but not actually develop it.

It is now realised that the causes of mental disorders are complex and varied, including genetics and epigenetics. The epigenetic causes have a variety of origins and their effects contribute to mental illness. Epigenetic changes can affect:

- The size of certain brain regions: bipolar disorder has been linked with increased amygdala volume.
- Neurotransmitter systems: abnormal neurotransmitters have been linked with schizophrenia, ADHD, OCD, phobias, post-traumatic stress disorder and generalised anxiety disorder.
- Correlations with drug use: alcohol can damage white matter in the brain in areas affecting thinking and memory and alcoholism is linked to depression; amphetamines and LSD can produce paranoia and anxiety; cannabis can worsen depression.
- Childhood experiences: when abused children beome adults, they are more likely than the general population to suffer from serious depression and their recovery is poorer. They have a significantly higher risk of other psychiatric conditions including schizophrenia, eating disorders, personality disorders, bipolar disease and general anxiety. They are more likely to abuse drugs or alcohol. It is possilbe that appalling experiences in childhood made epigentic changes to the brain, affecting gene expression at crucial times during development, which made them more susceptible to mental illness as an adult.

Mental illness and cortisol

Increased stress increases the release of cortisol from the adrenal glands into the blood. Cortisol production is controlled by the hippocampus through the HPA (hypothalamus – pituitary – adrenal) axis. In response to stress, the hippocampus sends nervous impulses to the hypothalamus, which releases two hormones, corticotrophin-releasing hormone and arginine vasopressin (ADH). These two hormones stimulate the anterior lobe of the pituitary gland to release adrenocorticotrophic hormone (ACTH), or adrenocorticotrophin, into the blood. When the cells of the adrenal glands take up ACTH, they release cortisol. As the cortisol circulates in the bloodstream, it binds to glucocorticoid receptors on the hippocampus, which responds by sending inhibitory nervous impulses to the hypothalamus, reducing cortisol secretion.

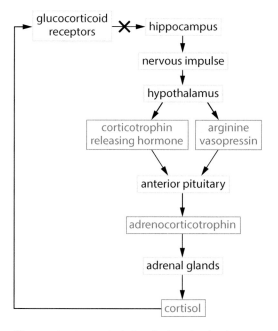

Diagram showing control of cortisol production by negative feedback in the HPA axis

Adults who suffered traumatic childhoods remain constantly over-stressed. They produce too much cortisol all of the time and the feedback loop is over-ridden. Either the hippocampus or hypothalamus or both do not respond appropriately to high cortisol levels, resulting in continued production of corticotrophin-releasing hormone. This demonstrates a higher background stress level, which may be part of the reason why they are more vulnerable to mental illness.

Behaviour

Behaviour describes the range of actions made by individuals, in relation to themselves or their environment.

Innate behaviour

Innate behaviour is the inborn complex patterns of behaviour that exist in most members of a species. They are instinctive. Innate behaviour is more significant in animals with less complex neural systems as they have less ability to learn and to modify their own behaviour.

Innate behaviours include:

- **Reflex**: a rapid, automatic, protective response to a given stimulus that improves the organism's chances of survival, e.g. withdrawing a hand from a hot object.

- **Kinesis**: the movement of a whole organism, but not in a particular direction in relation to the stimulus. Kineses are non-directional responses. They are more complex than reflex actions and include responses to light intensity, humidity and temperature. In response, the individual moves faster or slower and changes direction with a different frequency. Individuals of the same species have the same response, so the stimulus brings individuals together, although not because they are attracted to each other. In the type of kinesis called orthokinesis, the speed of movement depends on the intensity of the stimulus, e.g. with decreased humidity, woodlice move more.

- **Taxis**: the whole organism moves towards a stimulus, in a positive taxis, or away from the stimulus in a negative taxis. Organisms may sample the environment all over the body to determine the direction of a stimulus, or may use bilateral sense organs to do so. A taxis is named for the stimulus, e.g. in chemotaxis, the bacterium *E. coli* shows positive **chemotaxis** when it swims up a concentration gradient of glucose. In moving away from light, cockroaches show negative phototaxis but *Euglena viridis*, the photosynthetic protoctistan shows positive **phototaxis**.

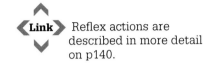

Link Reflex actions are described in more detail on p140.

Going further ▶

Woodlice that move more when the air is dry have a selective advantage as they are more likely than those that move less to encounter humid air, which prevents their dehydration.

Going further ▶

Chemotaxis is described in prokaryotes and eukaryotes, but the receptors, intracellular signalling and effectors are significantly different.

Learned behaviour

Learning can be defined as a relatively permanent change in behaviour or skills due to past experience. Learning is a conscious or unconscious process that builds on and modifies existing knowledge. Humans, animals and some machines can learn. Children learn through play: they experiment, learn rules and learn how to interact with others. As in other young mammals, play is crucial for their intellectual development.

- **Habituation** involves learning to ignore stimuli because they are not followed by either reward or punishment. An example in humans is not being startled by a repeated loud noise, and this occurs even in a foetus. Nearly 100 pregnant women were repeatedly given a very low sound that makes a vibration. Using ultrasound, 30–34 week old foetuses were seen to be startled for the first 14 stimuli only. This indicates that the brain is already primed for learning and memory at 30–34 weeks.

- **Imprinting** is a type of learning that occurs rapidly during a critical, very early stage in the life of birds and some mammals. Konrad Lorenz studied instinctive behaviour in animals, especially in greylag geese, which have young that leave the nest early. They respond to the first larger, moving object they see, smell, touch or hear, which is usually the mother, and attach to this object. The association is reinforced as they receive warmth and food. Imprinting on a member of their own species is important when they seek to identify a suitable mate when sexually mature. Young birds can imprint on many different types of moving object, especially if they make sounds. Lorenz became famous for being followed by a line of goslings that had imprinted upon him.

Young geese imprinted on their parent

- **Associative behaviours** produce conditioning, in which animals learn to associate a particular stimulus with a particular response.
  - **Classical conditioning** involves the association of a natural and an artificial stimulus bringing about the same response. A famous example is that of Pavlov's dogs:

    In the 1890s, the Russian physiologist Ivan Pavlov noticed that his dogs salivated when he came into the room, even if he was not bringing them food. He found that any object or event that the dogs learned to associate with food would trigger salivation. Pavlov used a bell as a 'neutral stimulus'. When he fed his dogs, he rang the bell. After doing this a number of times, he used the bell alone and they salivated, even in the absence of food. The dogs had learned an association between the bell and the food and a new behaviour, salivation, was produced. This response was learned, or conditioned, and so it is called a conditioned response. The food is the 'unconditioned stimulus', because it always provoked salivation. The initially neutral stimulus, the bell, became a 'conditioned stimulus' when the dogs learned it heralded the food.

  - **Operant conditioning** involves the association between a particular behaviour and a reward or punishment. BF Skinner's original experiments involved rodents learning that if they pressed a lever, food would appear. The food was a positive reinforcer for the learning. An example using a negative reinforcer was playing a loud noise until they performed a task, e.g. pressing a lever, which would make it stop. Punishment, such as a loud noise or an electric shock, following a certain behaviour decreases the behaviour. Operant conditioning is the basis of behaviourism, which seeks to alter behaviour with rewards and punishments. Experiments with people seek to explore the differences between people who choose a small reward immediately or a larger reward at a later time.

A rat in a Skinner box presses a lever for food

- **Latent (exploratory) learning** is not done to satisfy a need or obtain a reward. Many animals explore and learn about their new surroundings as they do so. This may provide information that, at a later stage, may mean the difference between life and death. In a classic study by Tolman, three groups of rats in mazes were observed each day:

Rat group	1	2	3
Food availability	always at the end of the maze	none	none for 10 days, but present on 11th, at the end of the maze
Behaviour	quickly learned to rush to the end of the maze	wandered but did not seek the end of the maze	on Day 11, ran to the end of the maze as quickly as Group 1 rats

This showed that the Group 3 rats had learned about the organisation of the maze for 10 days, but without the reinforcement of food, and used what they had learned on Day 11.

- **Insight learning** is a complex behaviour based on information previously learned. Köhler's experiments with chimpanzees that defined insight learning were in the 1920s. Fruit was hung just out of reach. The chimpanzees were given sticks and boxes and, after displaying anger and frustration, found ways to reach the fruit, by climbing and knocking it down. Köhler called this learning through personal interactions with the environment 'insight learning'. Brain imaging studies of insight learning in humans shows activity in in the right temporal cortex coinciding with the solution occurring, the 'aha moment'.

- **Imitation** is an advanced behaviour. It is a form of social learning that allows learned behaviour patterns to spread rapidly between individuals and to be passed down from generation to generation without a need for genetic inheritance. It involves copying the behaviour of another individual, usually a member of the same species. Japanese monkeys, however, have been seen to wash potatoes after seeing humans washing them. Differences can arise between populations as a result of imitation of different behaviour patterns in different areas, e.g. different populations of chimpanzees use stones or sticks or branches to crack nuts. Learned behavioural differences between populations are described by some as 'culture'. Such behaviours may be adaptive, and put groups at a selective advantage in comparison with others.

Experiments with apes and birds, especially the Japanese quail, show imitation. Evidence from studies of whale song and dolphin behaviour suggests that cetaceans are also capable of imitation. There is little evidence, however, to show imitation in other mammals or in other animal groups.

57

Knowledge check

State the type of learning involved in these situations.

1. If you tap a snail it retreats into its shell, but after several taps, it no longer retreats.
2. An octopus opens a box to eat a crab, having watched another, trained octopus do it first.
3. The males of birds called Japanese quail choose sexual partners who are similar to their mother.

Living in social groups

Social behaviours involve interactions between individuals of the same species and many species form highly structured social groups, or societies. Within these groups the behaviour of one individual can influence the behaviour of others within the group.

Social behaviour relies on the ability of animals to be able to communicate with one another, although communication between members of the same species is not necessarily a social behaviour. Animals communicate in many different ways but they always involve one individual producing a signal, the **sign stimulus** that can be detected by another. The sign stimulus may trigger an innate response in the second individual. An example is the begging behaviour of the chick of the kelp gull, *Larus dominicanus*. It pecks at the red spot on the mother's beak, which causes her to regurgitate food for her chick.

Kelp gull or Dominican gull, *Larus dominicanus*

Study point

Aggression, altruism, scapegoating and shyness are examples of social behaviours in humans.

Male 3-spined sticklebacks demonstrate social behaviour in the spring. They change colour, establish a territory and build a nest. Using very basic models, Tinbergen showed that the sign stimulus provoking the male to attack is anything that resembles the red abdomen of another male. The sign stimulus of a model resembling the swollen abdomen of the female provokes the male to encourage it into the nest.

Male 3-spined stickleback, *Gasterosteus aculeatus*

Key Term

Fixed action pattern: An innate behaviour which is a complex sequence of events that runs to completion, once it has been initiated by a sign stimulus.

These behaviours are often referred to as stereotyped behaviours or **fixed action patterns** (FAPs). FAPs produce well-defined, co-ordinated movements. The sign stimulus activates nervous pathways which bring about a predictable sequence of ready-made motor patterns, causing movements that do not need any decision making. The cerebellum is heavily involved with FAPs because it co-ordinates rapid detailed movements without conscious thought.

Going further ▶

Tinbergen said that four questions must be asked of any animal behaviour: its cause, its development, its function and its evolution.

Humans and animals have innate FAPs, e.g. a newborn's motions for sucking, crying and yawning, and learned FAPs, e.g. those to produce leg motions for walking and for riding a bicycle and finger motions for playing a violin. They are more complex than reflexes, although some are highly automatic, but others are activated by choice. FAPs can be changed: they can be learned, remembered, and perfected.

The response of the individual depends on its motivational state. If, for example, a cheetah is hungry (motivational state) it will initiate stalking behaviour when it sees a prey animal (sign stimulus); if the cheetah is not hungry the sight of prey does not induce stalking behaviours.

Insect social structures

There are various ways of arranging social groupings. The method called eusociality, seen in ants, bees, wasps and termites, has the highest level of organisation. **Eusocial colonies** have:

- Brood care, including caring for offspring of other individuals.

- Overlapping generations in a colony.

- Division of labour creating specialised groups, called castes. Individuals cannot perform the tasks of individuals in other castes. The success of insect societies is largely due to the division of labour, which increases the overall efficiency of the group.

Going further ▶

Eusocial groupings also occur in Crustaceans, e.g. the shrimp *Synalpheus regalis* and mammals, e.g. the naked mole-rat.

Honeybee colonies

A colony of the honeybee, *Apis mellifera*, has one queen and a few hundred reproductive males called drones. These are the only colony members that reproduce. There are tens of thousands of sterile female workers which:

- Build new wax combs

- Find food

- Make the environment suitable by cleaning the hive, altering the temperature and humidity

- Defend the colony

- Care for young.

Queen bee surrounded by workers

Honeybee genetics

▼ **Study point**

50% of a worker's genes are the same as all other workers as they come from the father; 25% of her genes are the same as they come from the mother, making workers' genes 75% identical.

Caring for the young of others has been explained by noting that female workers and the brood they raise share 75% of their genes, rather than the 50% shared by the diploid siblings of diploid parents. Workers are diploid. 50% of their genes come from the male, which, being haploid gives all his genes to each of the offspring, so these are the same in every worker. The other 50% of the workers' genes are from the queen. She is diploid and gives 50% of hers to each. But each worker has a different 50% from the queen so, on average, each sister will have the same 25% of the queen's genes as other workers. Identical genes do not provide a complete explanation for eusociality because not all eusocial animals have this trait, and some species that do are not eusocial.

Honeybee communication

Karl von Frisch won the Nobel Prize in 1973 for his work on honeybee communication. Individuals in a colony communicate by touch, pheromones and by displays called dances. Worker bees called scouts, find nectar and pollen. They return to the hive and communicate the distance, direction and odour of the source to other workers, called foragers, in a dance performed on the walls and floor of the hive.

- The 'round dance' tells the foragers that the source of food is less than about 70 m from the hive, but gives no indication of direction.
- The 'waggle dance' tells the foragers that the food is more than 70 m away. The dance communicates the distance of the food source from the hive and its direction relative to the hive and the position of the sun.

Vertebrate social structure

Vertebrate social interactions range from being solitary to living in large, complex groups. Social groups have evolved independently many times. Birds, mammals, and to a lesser extent, fish are the most social vertebrates with large, stable groups. Amphibians have complex acoustic communication systems involved in breeding and, in some, parental care. Social animals exhibit one of more of these behaviours:

- Co-operative rearing of young.
- Overlapping generations living together permanently.
- Co-operative foraging or hunting.
- Co-operative defence against predators.
- Social learning, e.g. a young chimpanzee copying the use of a twig to acquire termites.

Three types of social group are recognised:

- Egalitarian: all individuals have equal rank, e.g. herring.
- Despotic: one member is dominant and all others are equally submissive, e.g. meerkats, wolves.
- Linear: **dominance hierarchies** exist in which higher-ranking individuals are dominant over lower-ranking individuals. This is the commonest type of organisation in vertebrates and is sometimes called a 'pecking order', as it was observed first in hens in a hen-house. Dominance hierarchies can only exist where animals can recognise each other and can learn.

Dominance hierarchies

Advantages of a dominance hierarchy

Being dominant may have advantages for the dominant individuals:

- In baboons, higher-ranking males have greater access to females and high-ranking females produce more surviving offspring.
- In vervet monkeys, high-ranking individuals have greater access to water.
- In monogamous bird species, the dominant pairs have the best territories, so both offspring and adults have better health.
- In oyster catchers, marine birds, high-ranking individuals have greater access to food.
- In dunnocks, woodland birds, the highest-ranking male in a group has all the matings.

Dominance hierarchy decreases the aggression associated with feeding, mate selection and breeding-site selection. The hierarchy is relatively stable and is normally maintained by aggression. Fighting is a last resort and, as an alternative, there is a series of ritualised actions where each action acts as a sign stimulus for the next action of the other animal in the encounter. Mature red deer stay in single-sex groups for most of the year. During the mating season, called the rut, mature stags compete and challenge each other by roaring and walking parallel with each other. They can assess each other's antlers, body size and fighting prowess. Their roar has low-frequency sounds and the volume of these is an indication of the stag's strength and size. The weaker stag generally backs down, avoiding a fight. But if neither does, they clash antlers and sometimes receive serious injuries.

Going further ▶

Another type of dance, with rapid side-to-side vibrations, tells other workers that she needs dust and pollen to be cleaned off her body.

 Key Term

Dominance hierarchy: Ranking system in an animal society, in which each animal is submissive to animals in higher ranks but dominant over those in lower ranks.

WORKING SCIENTIFICALLY

Observations of modern humans in hunter-gatherer groups or in settled chiefdoms without agriculture, show that they practise egalitarianism, usually very successfully. This suggests our distant ancestors did so too.

Going further ▶

The blue-footed booby, a marine bird, shows 'brood hierarchy'. The first egg is laid 4 days before the second. The older chick is dominant and kills the younger if food is scarce.

A red deer roaring

Going further ▶

Some squirrels claim 10 hectares of territory. Golden eagles have territories of 9 000 hectares. The sanderling, a shore bird, has a linear territory, up to 120 metres along the shore.

Ring-tailed lemur using visual display

Going further ▶

Wolves advertise territories by scent marking and howling. When howling together, they harmonise and create the illusion of there being more wolves than there actually are.

A possible disadvantage of dominance

It had traditionally been thought that higher ranking individuals had less stress, but observations of the eusocial naked mole-rat suggests that may not always be so. The naked mole-rat queen suppresses testosterone and luteinising hormone production in males and suppresses the entire ovarian cycle in females. If she is removed, they become fertile. The queen's dominance may be related to a high concentration of glucocorticoids in her blood. These are stress hormones and it may be that animals with subordinate ranks have less stress, contrary to past beliefs. High corticosteroid concentration in the blood is correlated with more disease, e.g. osteoporosis and immunosuppression. This may, in part, explain why subordinates rarely challenge the dominant for her position.

Territorial behaviour

A **territory** is the area that an animal consistently defends against other members of the species.

Territoriality is only shown by a minority of species. More commonly, an individual or a group of animals will use an area, called the home range, that it does not necessarily defend. The territory enhances an animal's ability to reproduce because it provides food, nesting sites or mating areas.

Many animals create signposts to advertise their territory. They communicate information by smell, sight and sound:

- That the territory is occupied
- Their sex
- Their reproductive status
- The dominance status of the territory-holder.
  - Scent marking is by urine, faeces or from specialised scent glands. The scent may contain pheromones. It can indicate the presence of the territory-holder to prey species, e.g. leopards and jaguars mark by rubbing themselves against vegetation. Ring-tailed lemurs scent-mark surfaces with their anogenital scent glands. To mark vertical surfaces, they do a handstand, grasping the highest point with their feet while applying the scent.
  - Visual signposts include the colouration of the animal or long-term visual signals, such as faecal deposits.
  - Many animals use vocalisations to advertise their territory, including birds, frogs and members of the wolf family. Male European robins are aggressively territorial. The red breast is highly visible when it sings at the edge of its territory. They attack other males that enter their territory and other small birds, without apparent provocation. This accounts for up to 10% of adult robin deaths in some areas.

These methods are sometimes combined, e.g. the ring-tailed lemur holds its tail high up when urinating, using a visual signal, while it advertises its territory with urine scent marks.

Animals use ritualised aggression to defend their territory without the need for fighting. If an intruder enters the territory, both animals begin ritualised aggression, with postures, vocalisations and displays such as spreading the wings or gill covers, showing claws, head bobbing, tail and body beating. Domestic cats are very territorial and defend with ritualised body posturing, stalking, staring, spitting and yowling. This often ends when one of the animals, usually the intruder, flees but if it does not, fighting may occur as a last resort.

Courtship behaviour

Courtship is the behaviour by which animals select a mate for reproduction. Courtship allows individuals to recognise others of the same species, of the opposite sex and whether they are sexually mature and receptive. It can stimulate sexual behaviour and synchronise the activity of the potential partners. Courtship routines are innate and they ensure intraspecific mating, which is likely to produce fertile offspring. Animal courtship involves dances, touching, vocalisations, displays or fighting.

When animals choose a mate there are two opposing processes operating over the generations:

- Sexual selection will make a characteristic more conspicuous and, therefore, more attractive to the opposite sex. Alleles for a more exaggerated form of the characteristic are selected and transmitted to the next generation.

- Natural selection will make the characteristic less conspicuous as those with an exaggerated form are more likely to be predated upon and have fewer offspring to carry the relevant alleles.

There are two main mechanisms driving sexual selection:

- **Intra-sexual selection** involves male-male competition. In some species, males compete for sexual access to many females. This can be by combat, e.g. among African lions and southern elephant seals. Sexual selection then favours the evolution of larger, more aggressive males. In lek mating systems, common in birds but also seen in some amphibians and mammals, males gather together and perform their courtship rituals before the assembled females. In some species, e.g. black grouse, they fight to display their dominance.

- **Inter-sexual selection**. In most species, the male starts the courtship and the female chooses to either mate or reject him. There are two main models for this:
  - The physical attractiveness model, in which the more attractive a male is, the more likely he is to be chosen by a female, e.g. the 3-spined stickleback: in spring, males and females move to shallow water and each male defends a territory, where he digs a pit, which he then fills with algae, sand and debris. He makes tunnel through the nest and courts females with a zigzag dance. He first swims short distances left and right, and then swims back to the nest in the same way. If she approaches, he may swim through the tunnel. She follows him through and deposits her eggs. Males have a red patch on the throat that is a **sign stimulus**, releasing aggression in other males and attracting females. Females are more attracted to males with brighter colouration.

The male bowerbird's courtship ritual involves building a U-shaped platform with twigs and grass, adding blue objects, such as berries and flowers and even bottle caps and string. A visiting female crouches in the bower and the male performs the 'buzz-wing-flip', in which he fluffs up his feathers, produces buzzing vocalisations and runs backwards and forwards. The dance is performed four times. The female does not make her choice until she has visited bowers and watched the performance of several males.

Bowerbirds

- The male handicap model: sexual dimorphism occurs in some species in which males are chosen by females. This means males and females have different ornamentation, colouration or behaviour, e.g. male peacocks have elaborate, brightly coloured tail feathers and females are dull brown; male lions have a mane but females do not. The theory suggests that reliable signals, such as the peacock's tail, are costly to the male. An inferior male could not afford such a wastefully extravagant signal so the female knows that the signal indicates quality.

Going further ▶

Complex brains, as in vertebrates, are not necessary for courtship behaviour. Male *Drosophila* perform a mating dance for the female and if it is not 'correct', she does not mate.

Going further ▶

The stickleback's red pigment is from dietary carotenoids and it is brighter in males with fewer parasites. The female can therefore assess the male's health and his ability to find food.

▼ **Study point**

There may be more than one explanation for a characteristic, e.g. the peacock's tail may be explained by the male handicap model and by the physical attractiveness model in a sexually dimorphic species.

58

Knowledge check

Match the description with an example of the type of behaviour.

1. Territorial behaviour
2. Inter-sexual selection
3. Dominance hierarchy

A. Higher-ranking baboons have access to more mates and reproduce more successfully.
B. Rhinoceroses squirt urine on to trees and bushes every few minutes and scatter their dung throughout a region.
C. The tail of the male widowbird is about twice the length of his body. Female widowbirds select males with longer tails.

1. The male reproductive system makes mature spermatozoa in the testes, under the influence of the male hormone, testosterone.

    (a)   The diagram shows a vertical section through part of the human male body from the left side.

          Label the structures indicated. (2)

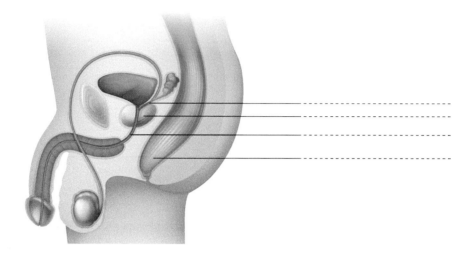

Diagram of the male reproductive system

    (b)   Spermatozoa develop in the seminiferous tubules of the testes, supported by Sertoli cells. Predict physical characteristics of the Sertoli cells that you might expect, on the basis of their function. (3)

    (c)   There has been pressure over many years to develop a hormonal contraceptive to be used by men. Current attempts are based on the use of testosterone and its derivatives. Results of early trials show that men treated with a testosterone derivative, testosterone enanthate (TE), fall into one of three groups. Data are shown in the table:

Description of response	Sperm count/cm³ before treatment	Sperm count/cm³ semen after 16 weeks of treatment	Sperm count/cm³ semen 16 weeks after treatment stopped
Azoospermic	2×10^7	< 200	2×10^7
Oligospermic	2×10^7	$<3 \times 10^6$	2×10^7
Non-responders	2×10^7	10^6	2×10^7

          (i)   How might maintaining a high blood concentration of testosterone reduce sperm count? (5)

          (ii)   Why is important to assess the sperm count of the volunteers in these experiments before TE injections are given? (1)

          (iii) Suggest a side effect you might expect to see in a man using TE. (1)

    (d)   Data from three experiments on the effectiveness of TE as a contraceptive gave the following results:

Number of azoospermic men	447
Number of oligospermic men	249
Total number of pregnancies	12

          (i)   What is the percentage of pregnancies occurring in this cohort of volunteers? (2)

          (ii)   Untreated controls have a pregnancy rate of 85%. Considering all the information given above, discuss the issues in deciding whether TE is suitable for use as a contraceptive for men. (3)

                                                                                              (Total 17 marks)

2. There are nine species of *Adansonia*, the baobab tree, of which six are native to Madagascar. The photograph shows a group of Madagascan baobabs. The tree has spectacular, fragrant flowers which open around dusk. They release all their pollen on their first night.

Baobab (*Adansonia*)

(a) What is meant by the term pollination? (2)

(b) Baobab flowers are pollinated by lemurs and by bats, which are both in the class Mammalia.

(i) Name another group of organisms that are pollinators of flowering plants. (1)

(ii) Based on the method of pollination of baobab flowers, describe two features of their flowers that you might expect to find based on a visual inspection. (2)

(iii) Bats and lemurs are described as biotic pollinators. Name an abiotic pollinator. (1)

(c) (i) The electron micrographs below show pollen grains from two species of flowering plant. Suggest the main method of pollination for each and give a reason for your answers. (2)

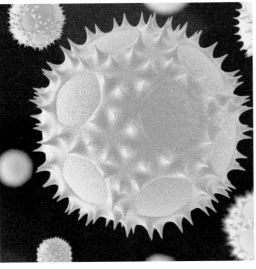

Species A

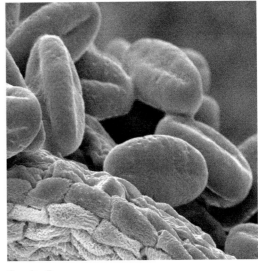

Species B

(ii) Wood sorrel flowers are described as cleistogamous, which means the flowers do not open. How might they be pollinated? (1)

(d) A farmer who grew courgettes noticed that some plants in his field produced a higher yield than others and he wondered if the difference was related to how the flowers were pollinated. He knew that courgette pollen is 0.12 mm in diameter.

The farmer set up three test beds of plants in which the flowers in each bed were treated in one of three different ways:

- Bed 1: Flowers covered with a net with holes 0.20 mm across.
- Bed 2: Pollen was transferred from the flower of one to another, using a fine paintbrush. The pollinated flowers were then covered with a net with holes 0.20 mm across.
- Bed 3: Flowers were untouched and left uncovered.

He counted the number of flowers that produced courgettes. The table shows the treatment in each bed of plants and the number of flowers that produced courgettes.

Bed	Treatment	Number of flowers treated	Number of flowers that produced courgettes	Percentage of flowers that produced courgettes
1	Flowers covered with a net with holes 0.20 mm across.	391	0	0
2	Pollen was transferred from the flower of one to another, using a fine paintbrush. The pollinated flowers were then covered with a net with holes 0.20 mm across.	367	356	
3	Flowers were uncovered and had no treatment.	425	378	88.9

(i) Calculate the percentage of flowers that produced courgettes in Bed 2, to one decimal place. (1)

(ii) What word describes the plants in Bed 3? (1)

(iii) Explain how the percentage of flowers that produced courgettes provides evidence for deducing the methods by which the flowers in each of the three conditions may have been pollinated. (6)

(Total 17 marks)

3. Capuchin monkeys have colour vision. Like all primates, they have an autosomal gene for a retinal pigment called S opsin, which codes for a protein that allows them to see short wavelengths, e.g. blue. The X chromosome carries a second gene coding for different opsins. This gene has two co-dominant alleles, X^M, which codes for a protein that responds to medium wavelengths, e.g. green and X^L, which allows sensitivity to long wavelengths, e.g. red.

Capuchin monkey

Sex determination in monkeys has the same mechanism as other primates, including humans.

(a) (i) Describe the mechanism by which sex is determined in capuchin monkeys. (1)

    (ii) State the colours that male capuchins can distinguish and explain why. (3)

(b) (i) Explain the term 'co-dominant'. (1)

    (ii) Explain why different female capuchins may see different colours from each other. (4)

(c) (i) Construct a genetic diagram showing the inheritance of red and green colour vision in a cross between a male capuchin that has red vision and a female that sees red and green. (4)

    (ii) By observing many crosses between males with red vision and females with red and green vision, data were gathered to test if this genetic explanation could be supported. The following phenotypes were observed:

A null hypothesis was constructed: red and green vision is controlled by a single gene on the X chromosome.

Colours distinguished	Male	Female
Red	37	30
Green	24	0
Red and green	0	21

A chi^2 test was performed and a calculation showed $\chi^2 = 5.4$

Use the table below to determine if the null hypothesis can be accepted. (4)

Level of significance	0.90	0.10	0.05	0.01
χ^2	1.61	9.24	11.07	15.09

(iii) How might your conclusion have differed if the calculation had given $\chi^2 = 15.4$? (2)

(iv) What might be a biological reason for the results that generated a χ^2 of 15.4? (1)

(Total 20 marks)

4. The mouse has been a useful organism for studying the genetics and evolution of coat colour. At least five genes control the coat colour of mice. They vary from white to black but in the wild, they range from light to dark brown. Their fur, although short, shows continuous variation in its length.

   The common house mouse, *Mus musculus*, weighs only about 20 g. Females begin breeding at about 7 weeks of age. They have a gestation period of 20 days and give birth to litter of up to 12 young, called pups. They can produce a litter up to 10 times each year.

   Mice live about a year in the wild, but in a protected environment may live three times as long. They are prey species and are eaten by wild cats, such as lynx and even by jaguars and tigers, if their other food sources are scarce; by birds, such as owls and hawks; by snakes and, in some cultures, by humans.

Light and dark brown mice

   (a) Suggest why mice have been such useful organisms for studying genetic and evolutionary changes. (2)

   (b) (i) Explain the meaning of the term selection pressure. (1)

   (ii) Contrast the features of continuous variation with discontinuous variation. (3)

   (iii) Predict and explain the consequences for the hair length of wild mice as the average global temperature increases. (5)

   (c) Consider a single gene that determines whether a mouse is white or coloured, where the allele producing a coloured mouse is dominant. If there are 240 white mice in a population of 1800, calculate the frequency of the heterozygous genotype. (4)

   (Total 15 marks)

5. The chain termination method of DNA sequencing uses fragments about 500–800 base pairs long, cut by restriction endonucleases. These are bacterial enzymes and they can be used to isolate fragments of DNA that contain genes. They cut the DNA at specific base sequences and, if the gene of interest occurs between the cuts, the gene can be isolated.

   (a) The human genome contains 3×10^{12} base pairs. What is the minimum number of DNA fragments that would be needed to sequence the human genome if they were all the maximum size of 800 base pairs. (2)

   (b) A well-known restriction endonuclease, *Hind*III, is isolated from the bacterium *Haemophilus influenzae*. The enzyme makes a staggered cut in the DNA between two adenine nucleotides in the sequence:

   5' ....AAGCTT.... 3'

   3' ....TTCGAA.... 5'

   leaving sticky ends:

   5' ....A AGCTT....3'

   3' ....TTCGA A....5'

   (i) Use clearly drawn arrows to indicate where *Hind*III might cut this sequence of double-stranded DNA:

   5' G G G C C A A A G C T T G T A C G T C A T T T A A A G C T T A A

   3' C C C G G T T T C G A A C A T G C A G T A A A T T T C G A A T T (2)

   (ii) Write the base sequence for the double-stranded DNA fragment that this restriction enzyme could isolate from the DNA fragment that includes the above sequence. (1)

   (iii) Explain the significance of the base sequences of the sticky ends. (1)

(c) (i) If this fragment contained a gene that could increase muscle mass, why would it be controversial to insert it into an oocyte prior to fertilisation, to create an embryo? (2)

(ii) If such an embryo were allowed to develop, it would contain stem cells. Why would it be more useful to use these embryonic stem cells to grow muscle tissue in culture, rather than adult stem cells, once the embryo tissues have differentiated? (2)

(d) (i) If these stem cells were used to grow new tissue, a scaffold would have to be used. Explain the function of a scaffold in tissue engineering. (2)

(ii) Many different materials have been used for scaffolds, including polylactic acid and carbon nanotubes. Explain why polylactic acid is likely to be more useful as a scaffold in the body that carbon nanotubes. (1)

(Total 13 marks)

6. **A made-up story**

Emyr and Rhys began nursery school on 3 September. On 4 September Rhys was kept at home, with a high temperature and a rash, feeling very unwell. His older sister, Betsi, had picked up an infection from her school and passed it on to him. This particular infection was endemic to their part of Wales. Emyr was distraught and, to cheer him up, his mother took him to visit his new friend. Of course, Emyr developed a temperature and a rash a short time later, but within a few days, both were back at school and the incident was forgotten.

60 years later, Emyr and Rhys, now old friends, were members of the same U3A group. (The University of the Third Age is an educational institution for retired people. Life-long learning is very important.) Gruffydd, the group leader, left during the meeting and was later reported to have a very high temperature and a rash. Emyr and Rhys visited him, bringing delicious Welsh cakes to make him feel better, but they did not catch his infection and their health was unaffected. Gruffydd recovered within a few days and they all lived happily ever after.

(a) Define the term 'endemic'. (1)

(b) (i) Describe the role of Emyr's macrophages in initiating an immune response to the infection. (2)

(ii) Emyr's B and T lymphocytes were involved in the primary immune response. Describe the role of his T helper cells in this. (5)

(iii) Emyr and Rhys were not infected by Gruffydd, as their secondary immune response was effective. Describe the main differences between the primary and secondary immune responses. (4)

TB is a notifiable disease, which means that the law requires cases of it to be reported. The table below shows that number of notifications at 10-year intervals since 1920. It also shows the number of people who died in those years, where TB was a contributory factor in their deaths.

Streptomycin, the first antibiotic effective against *Mycobacterium tuberculosis* has been used against TB since 1944, and other antibiotics have been introduced since then. The BCG vaccine became widely available in 1953, and 13-year-olds were routinely vaccinated until the late 1990s.

Year	All TB notifications in England and Wales	Deaths in which TB was a contributory factor
1920	73 332	33 000
1930	67 401	29 500
1940	46 572	23 900
1950	49 358	11 100
1960	23 605	2 450
1970	11 901	1 088
1980	9 142	605
1990	5 204	390
2000	6 575	373
2010	7 797	299

Data from Centre for Infectious Disease Surveillance and Control, Public Health England

(c) (i) Calculate the percentage decrease in deaths in which TB was a contributory factor between 1920 and 2010. Give your answer to 1 decimal place. (2)

(ii) The percentage decrease in the number of TB notifications in England and Wales fell by 89.4% between 1920 and 2010. Account for the difference in the proportion of cases and the proportion of deaths between the years 1920 and 2010. (2)

(d) Describe a laboratory test that might distinguish a suitable antibiotic with which to treat a bacterial infection. (4)

(Total 20 marks)

7. (a) (i) The image shows the bones in a human leg that is swinging from the knee. The large, front, thigh muscle is attached to the tibia, just below the knee. The knee joint acts as a lever. Indicate and label on the leg the position of the fulcrum and the positions and directions of the load and effort for this lever. (3)

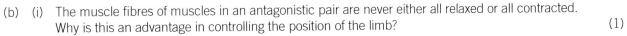

(ii) State which order of lever the knee represents and give another example of this order of lever in the human body. (1)

(iii) The quadriceps and the hamstrings are antagonistic groups of muscles in the front and back of the thigh, respectively. Explain their action in moving the knee joint. (2)

(b) (i) The muscle fibres of muscles in an antagonistic pair are never either all relaxed or all contracted. Why is this an advantage in controlling the position of the limb? (1)

(ii) Muscle contraction is explained by the sliding filament theory. Describe the differences in appearance of a contracted and a relaxed sarcomere. (4)

(iii) If a relaxed sarcomere is 2.40 µm long and it contracts by 38% of its initial length, what is its new length? Give your answer in suitable units, to 2 decimal places. (3)

(c) (i) Some athletes experience osteoarthritis as they age. Describe the changes to the joints that define this condition. (3)

(ii) Sometimes the condition is so serious that a joint replacement is required. What properties are required in the materials used to construct a new joint? (3)

(Total 20 marks)

8. The brains of mammals have similar structures to the brains of other vertebrates but a major difference is in the size of the forebrain, relative to the size of the body. The diagram below shows a sagittal section through the brain of the rat, which is a mammal.

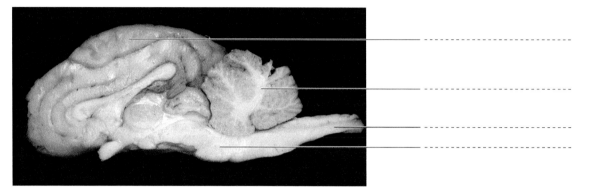

Sagittal section through the brain of a rat

(a) (i) Name the structures labelled on the diagram of the brain. (2)

(ii) The outer surface of the forebrain is highly folded, though less so than in the human brain. Explain the significance of this observation. (2)

(iii) The brain is an expansion of the spinal cord. Other than folding, how might the surface of the cerebrum differ from that of the spinal cord? (1)

(iv) Describe an additional difference that might be seen in the brain of a fish, in comparison with that of a mammal. (1)

(b) Read the following passage and answer the questions below:

Research into mothering skills in rats provides useful models for human behaviour. If pups are licked and groomed often by their mothers, they develop into relaxed individuals with relatively low concentrations of adrenocorticotrophin and cortisol and if subjected to mild stress, they remain calm. Neglected pups that were licked and groomed little by their mothers have relatively high concentrations of adrenocorticotrophin and cortisol and if subjected to mild stress, they react strongly.

Analysis showed that relaxed pups had more glucocorticoid receptors in the hippocampus than the anxious rats. Their hippocampus would therefore be very efficient at detecting low concentrations of cortisol in the blood, leading to inactivation of the hypothalamus.

Marmosets and tamarins are types of monkey that live in family groups. Observations show that if a marmoset or tamarin mother is isolated from the older, experienced females of her family, she may not know how to feed her young, may hold them upside down and does not tolerate holding and carrying them.

(i) Describe the different behaviours shown by rat mothers as opposed to tamarin and marmoset mothers in relation to raising their young and explain how each mode of behaviour is useful. (4)

(ii) Describe an experiment that would test whether the response of the rat pups described above was genetic or an effect of their environment. (5)

(iii) Ethical treatment of animals is essential. How would you ensure that the treatment of these rats was ethical? (1)

(iv) Deduce how licking and grooming may result in pups having more glucocorticoid receptors in the hippocampus than the anxious rats. (4)

(Total 20 marks)

Knowledge check answers

Chapter	KC	Answer
1	1	A – 1; B – 4; C – 3; D – 2
2	2	photosystems; antenna complex; chlorophyll a; carriers
	3	ribulose bisphosphate; NADP; light dependent; triose phosphate
	4	1 – D; 2 – C; 3 – A; 4 – B
3	5	glycolysis; cytoplasm; pyruvate; mitochondria; Krebs; electron transport chain
	6	1 – B; 2 – A; 3 – B; 4 – A; 5 – A+B
4	7	1 – B; 2 – A; 3 – C
	8	2; 3; 4
5	9	1 – D; 2 – A; 3 – C; 4 – B
	10	1 – C; 2 – B; 3 – D; 4 – A
	11	1 – A; 2 – D; 3 – C; 4 – B
	12	1 – D; 2 – B; 3 – C; 4 – A
	13	algal bloom; light; photosynthesis; saprobiontic; oxygen; nitrite
6	14	1 – D; 2 – C; 3 – A; 4 – B
	15	1 – C; 2 – D; 3 – A; 4 – B
7	16	1 – B; 2 – C; 3 – A; 4 – D
	17	hypothalamus; posterior pituitary; ADH; collecting duct; medulla
	18	diet; pressure; counter-current; blood/HLA
	19	1 – B; 2 – C; 3 – A
8	20	brain; spinal cord; sensory; motor
	21	sensory; motor; cell body; axon; myelin sheath; electrical insulator
	22	4; 3; 1; 7; 6; 2; 5
9	23	1 – C; 2 – D; 3 – B; 4 – A
	24	seminiferous tubules; spermatogenesis; primary spermatocytes; spermatids; Sertoli
	25	oogenesis; mitosis; meiosis; Graafian; ovulation
	26	1 – D; 2 – A; 3 – C; 4 – B
	27	low; Graafian follicle; ovulation; corpus luteum; progesterone
10	28	1 – B; 2 – A; 3 – D; 4 – C
	29	1 – B; 2 – C; 3 – D; 4 – A
	30	1 – C; 2 – A; 3 – B; 4 – E; 5 – D
	31	oxygen; enzymes; radicle; endosperm/cotyledon; plumule; photosynthesis
11	32	1 – C; 2 – B; 3 – A; 4 – D
	33	homozygous; heterozygous; recessive; co-dominant
	34	1 – B; 2 – A; 3 – D; 4 – C
12	35	species; discontinuous; environment; polygenic
	36	1 – D; 2 – A; 3 – B; 4 – C
	37	1 – D; 2 – C; 3 – B; 4 – A
	38	A – 2; B – 3; C – 1

Chapter	KC	Answer
13	39	1 – B; 2 – C; 3 – A
	40	1 – A; 2 – C; 3 – B; 4 – D
	41	plasmid; deletions; reading; antisense
	42	A – 3; B – 4; C – 1; D – 2
	43	differentiate; marrow; tissue; drug
14	44	A – 2; B – 2; C – 1; D – 1; E – 3
	45	fungi; bacteria; bacteriostatic; resistant
	46	1 – C; 2 – A; 3 – D; 4 – B 3
15	47	1 – C; 2 – B; 3 – D; 4 – A
	48	calcium; rickets; osteomalacia; osteoporosis
	49	4; 1; 5; 2; 3
	50	B; D; A; C
	51	A – 3; B – 1; C – 2
	52	cartilage; lubricate; antagonistic; contracted; extended
16	53	1 – B; 2 – C; 3 – A; 4 – D
	54	association; high; ischaemic
	55	1 – D; 2 – A; 3 – E; 4 – C; 5 – B
	56	1 – D; 2 – A; 3 – B; 4 – C
	57	1 – habituation; 2 – imitation; 3 – imprinting
	58	1 – B; 2 – C; 3 – A

Exam practice answers

Unit 3

1. (a) (i) Chlorotic

   (ii) Mineral / iron / magnesium deficiency

   (iii) Chemical testing of soil / Add minerals / fertiliser / iron salts / magnesium salts to the soil and determine if further leaves were chlorotic

   (b) Yield decreases from 2005–2015; global warming so crops grow less well / soil progressively depleted of minerals; Y always less than Ya; because Ya incorporates values for previous 5 years, which have higher yields

   (c) (i) H^+ ions from the photolysis of water in thylakoid space; proton pump in thylakoid membrane; removal of H^+ in stroma by reduction of NADP

   (ii) X = carbon dioxide and Y = glycerate-3-phosphate / glycerate phosphate

   (iii) Converting triose phosphate to RuBP

   (iv) Regenerates RuBP; so cycle can continue

2. (a) (i) catabolic = the breaking down of a molecule

   pathway = a series of enzyme-controlled reactions in which the product of one is a substrate for the next

   (ii) Series of carriers adjacent to each other so electrons passed directly from one to the next

   (b) (i) As the DNP concentration increases, the ATP production decreases / negative correlation;

   Approximately linear relationship

   (ii) No proton gradient so no protons flow through ATP synthetase;

   So no ATP made

   (iii) Respiration rate increases because there is no negative feedback by ATP;

   Stored carbohydrate and fats are oxidised

   (iv) The only ATP production is by substrate level phosphorylation in glycolysis;

   So insufficient ATP can be produced;

   Energy from electron transport chain not incorporated into ATP;

   Excessive heat loss denatures the proteins of the cells, which die

3. (a) (i) It contains components the contents of which are not known; yeast extract, egg extract and peptone

   (ii) As an energy source and as a carbon source

   (b) Growth medium contains penicillin;

   Susceptible bacteria killed by penicillin and those with resistance allele not killed

   (c) (i) Crystal violet;

   Lugol's iodine;

   Acetone – alcohol;

   Safranin;

   Those staining purple are Gram-positive;

   They are penicillin resistant;

   Count a sample and score for staining red (Gram-negative) and purple (Gram-positive);

   Percentage liable to be susceptible to penicillin

   $= \dfrac{\text{number of Gram–positive}}{\text{total}} \times 100\%$

   (ii) (Number at 40 h, before penicillin added $= 8.2 \times 10^{10} / cm^3$)

   From prediction, after penicillin added, expect 87% of number before penicillin added.

   After penicillin added, expect $(100 – 87) = 13\%$ to remain;

   $= \dfrac{13}{100} \times 8.2 \times 10^{10} = 13 \times 8.2 \times 10^8 = 1.1 \times 10^{10}\ cm^{-3}$

   remain / $\dfrac{(8.2 - 0.71) \times 10^{10}}{8.2 \times 10^{10}} \times 100 = 91.3\%$ killed;

   Answer to 1 dp;

   So the prediction was valid because the number decreased approximately as predicted;

   But validity limited as actual killing rate was exceeded

4. (a) Other plants present;

   Bare ground present

   (b) Trees were coppiced in 2000 after data collection

   (c) (i) All plants increased in % cover until 2010 then decreased;

   Willowherb, soft rush and grasses all decreased to 0;

   Leaf litter increased

   (ii) Increased humus; increased water availability; from 2010 decreased light intensity

   (d) $\dfrac{7.2 - 2.1}{2.1} \times 100 / \dfrac{5.1}{2.1} \times 100$;

   $= 242.9\%$;

   + / increase; (1 dp);

   (e) Secondary succession

   (f) Development of plant community much slower

5. (a) High light / energy input so long food chains / complex food webs;

   Warm and moist so conducive to plant growth;

   Many plant types so many animal types;

   (b) Legislation to protect the habitat;

   Collect seed for a seed banks;

   Breed in botanic gardens and reintroduce;

   Educate people who use the area

(c) (Exposed topsoil eroded and) what remains has lower mineral content;

(Slow water evaporation so) soil becomes wetter;

Soil is colder;

Growth of denitrifying bacteria, and so the soil loses its nitrates

(d) (i) Subsistence farming increased about three-fold;

Intensive farming approximately doubled;

Ranching and pasture diminished to about one fortieth

(ii) Extract fatty acids;

React with methanol / alcohol;

Synthesise biodiesel

(e) Limits between which global systems must operate to prevent abrupt and irreversible environmental change

6. (a) Because the water volume taken in would be variable

(b) Many individuals of each species and calculate a mean;

Individuals of a species to be genetically similar;

Individuals to be the same age;

Individuals to be the same sex;

Individuals to be in good health

(c) (i) Rat = 322 and human = 311

(ii) Plasma – values similar / all between 300 and 360 mOsmol dm^{-3};

Urine concentrations all higher;

Urine concentrations more variable

(d) (i) More water in the habitat is correlated with more dilute urine

(ii) Ascending limb – ions pumped out;

Medulla tissue fluid water potential decreases;

Descending limb – water out by osmosis;

Water from medulla reabsorbed into vasa recta;

Collecting duct passes through low water potential region of medulla;

Water leaves collecting duct by osmosis

(iii) The rat's ascending limb may be more efficient at pumping ions into the medulla / may have a lower water potential in its medulla / the rat may have a higher proportion of juxtamedullary nephrons

7. (a) A = Schwann cell + forms myelin sheath / electrically insulates axon

C = Nissl granules + protein synthesis

(b) (i) Thickness = $\dfrac{\text{thickness on image}}{\text{magnification}} = \dfrac{15}{50\,000} \times 1000$;

= 0.3 μm (1dp)

(ii) A wraps its membrane around the axon many times; and withdraws its cytoplasm (leaving B)

(iii) Each winding of B around the axon leaves two thicknesses of cell membrane;

(Each membrane is a phospholipid bilayer) each winding is 4 phospholipid molecules thick;

Assume the number of dark bands in B = number of windings of the Schwann cell around the axon;

∴ Thickness of each dark band = $\dfrac{\text{thickness of B}}{\text{number of windings}}$;

∴ Length of 1 phospholipid molecule

= $\dfrac{\text{thickness of each dark band}}{4}$

(c) (i) Wider diameter gives faster transmission;

The squid giant axon has a faster conduction speed than a motor neurone;

Wider diameter allows ions to flow through the bulk of the axon so there is less physical resistance from contact with the membrane;

Myelination increases transmission speed;

Human touch sensor fibre has faster impulses than temperature sensor fibre;

Action potential has saltatory conduction

(ii) Temperature increases conduction speed;

The earthworm giant fibre transmits faster than the squid's giant fibre;

Ions diffuse faster at higher temperatures

Unit 4

1. (a) ejaculatory duct; prostate gland; urethra; rectum

(b) Large nucleus; many mitochondria; much ER; many Golgi bodies

(c) (i) Negative feedback; high testosterone concentration inhibits LH production; so interstitial cells targeted less; so less testosterone produced; so mature sperm not produced

(ii) To calculate the degree to which it inhibits sperm formation

(iii) Increased muscle mass / increased body hair

(d) (i) % pregnancies = $\dfrac{12}{696} \times 100$

= 2% (0 dp) / 1.7% (1 dp)

(ii) Reduction in pregnancy rate ∴ suitable;

Return to normal sperm production within 16 weeks ∴ suitable;

Side effects ∴ unsuitable

2. (a) Transfer of pollen from the anther of one flower to the ripe receptive stigma of a flower of the same species

(b) (i) Insects / bees / moths

(ii) Carpels/stamens held within flower; large petals

(iii) Wind / water

(c) (i) Species A: Insects + spines attach to insect body / named body part

Species B: Wind + smooth / aerodynamic shape

(ii) Self-pollination

(d) (i) Percentage of flowers that produced courgettes

= $\dfrac{356}{367} \times 100 = 97.0\%$

(ii) Control

(iii) Courgettes are the fruit of the plant and are only produced following pollination;

1 – wind and self-pollination could occur;

1 – no fruit has formed so it may be deduced that courgettes are not wind- or self-pollinated;

2 – hand-pollination mimics insect pollination + self-pollination could occur;

2 – the very high yield suggests that the plants are insect-pollinated in the field;

3 – wind-, insect- and self-pollination could occur;

3 – the yield is high but lower than bed 2, as chance dictates which flowers visited by insects

3. (a) (i) Male XY and female XX

(ii) All see blue because they have the S opsin gene;

If they have the X^M allele, they also see green;

If they have the X^L allele they see red (instead of green)

(b) (i) Both alleles are expressed in the heterozygote so phenotype differs from either homozygote.

(ii) Females have two X chromosomes / sex-linked alleles for colour vision;

If homozygous / X^MX^M, they see green (in addition to blue);

If homozygous X^LX^L, they see red (in addition to blue);

If heterozygous, X^MX^L, they see red and green (in addition to blue);

(c) (i) Parents X^L Y X^M X^L

Gametes (X^L)(Y)   (X^M)(X^L)

F1

Gametes	(X^M)	(X^L)
(X^L)	X^MX^L female: green and red	X^LX^L female: red
(Y)	X^MY male: green	X^LY male: red

(ii) Level of significance = 0.05;

χ^2 = 5.4 < critical value / 11.07;

Null hypothesis accepted;

Any deviation from Mendelian ratio due to chance

(iii) χ^2 = 15.4 > critical value / 11.07;

Null hypothesis rejected

(iv) Other gene(s) involved / epigenetic effects / sample too small

4 (a) Small;

Rapid reproduction / large, frequent litters

(b) (i) Feature in the environment that gives some phenotypes an advantage so they reproduce more successfully than others

(ii) Continuous has many values within a range and discontinuous has a small number of values;

Continuous is polygenic and discontinuous is monogenic;

Continuous has genetic and environmental influence and discontinuous has only genetic influence

(iii) Mice with shorter hair will remain cooler;

So they are at a selective advantage;

They will have more offspring than mice with longer hair;

They will contribute the alleles for shorter hair to the next generation in greater proportion;

Over many generations, the proportion of alleles for shorter hair will increase in the population

(c) If the allele for white mice is a, f(aa) = $\frac{240}{1800}$ = 0.13;

q^2 = 0.13 ∴ q = √0.13 = 0.36 (2 dp);

p + q = 1 ∴ p = 1 – 0.36 = 0.64;

If the allele producing coloured mice is A,

f(Aa) = 2pq = 2 × 0.64 × 0.36 = 0.46 (2 dp)

5. (a) $\frac{3\,000\,000\,000\,000}{800}$ = 3.75 × 10⁹; (2dp)

(b) (i) 5′ GGGCCAAAGCTTGTACGTCATTTAAAGCTTAA

3′ CCCGGTTTCGAACATGCAGTAAATTTCGAATT

(ii) AGCTTGTACGTCATTTAA

ACATGCAGTAAATTTCGA

(iii) They are complementary so two DNA sequences with the same sticky ends can readily base pair with each other

(c) (i) It changes germ line / affects future generations;

It might interact with other genes in an unpredictable way

(ii) More stem cells in an embryo/easier to isolate from embryo;

Adult stem cells are not totipotent so may not differentiate as readily

(d) (i) Allows cells to attach;

Allows diffusion of nutrients / waste products

(ii) Carbon nanotubes are not biodegradable and would remain in the engineered structure but polylactic acid can be metabolised to lactic acid and removed

6. (a) A disease is endemic when it occurs frequently, in a specific location or population

(b) (i) Engulf pathogen;

present antigens on membrane

(ii) Secrete cytokines;

Causing clonal expansion of lymphocytes;

B lymphocytes differentiate into plasma cells;

Which secrete antibodies;

More macrophages engulf pathogens

(iii) Shorter latent period;

Faster antibody production / more antibodies produced in shorter time;

Fewer antigens needed to stimulate response;

More antibodies produced /antibody concentration in blood remains higher for longer

(c) (i) Actual decrease = 33000 – 299 = 32701;

% decrease = $\frac{32701}{33000}$ × 100 = 99.1 (1 dp)

(ii) Vaccine available now but not in 1920;

So fewer cases occur;

Antibiotics available now but not in 1920;

So a smaller proportion are fatal

(d) Grow lawn of bacteria;
Place filter paper discs impregnated with antibiotics on lawn;
Culture at 25°C;
Widest halo given by most effective antibiotic

7. (a) (i)

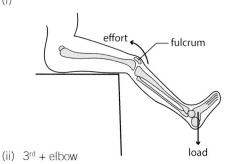

(ii) 3rd + elbow

(iii) Quadriceps contracts and hamstrings relax to flex knee;
Quadriceps relaxes and hamstrings contract to extend knee

(b) (i) Finer control of limb position

(ii) Relaxed: thick / myosin filaments surrounded by thin / actin filaments; with some overlap
Contracted: thin / actin filaments slide between thick / myosin filaments
I / H bands shorter;
A band the same length

(iii) Contracts by $\dfrac{38}{100} \times 2.40 = 0.91$ μm;
∴ new length = 2.40 − 0.91;
= 1.49 μm

(c) (i) Collagen and proteoglycan / glycoprotein degrade faster than they are synthesised;
Depletion of articular cartilage;
Swelling;
Thickened ligaments / spurs of bone develop

(ii) Strong; light; chemically inert

8. (a) (i)

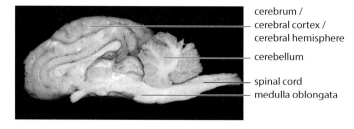

cerebrum /
cerebral cortex /
cerebral hemisphere
cerebellum
spinal cord
medulla oblongata

(ii) Folded therefore more area for cell bodies so more neurones
Humans have greater intellectual ability than rats

(iii) Cerebrum has grey matter outside and spinal cord has white matter outside

(iv) Cerebrum smaller in relation to rest of brain / body in fish

(b) (i) Rats show innate / instinctive behaviour;
Immediate response;
Monkeys show imitation learning;
Information passes from one generation to the next

(ii) Independent variable = presence or absence of licking and grooming;
Dependent variable = level of anxiety of rat when adult;
Controlled variables – two of stress factor / age / sex / siblings;
Many rats per group

(iii) Mild stress only

(iv) (Licking and grooming produces) epigenetic changes;
For example, decreased DNA methylation / de-acetylation of histones;
To the glucocorticoid receptor gene;
Associated with high levels of gene expression

Glossary

Abiotic A part of the environment of an organism that is non-living e.g. air temperature, oxygen availability.

Absolute refractory period Period during which no new action potential may be initiated.

Absorption spectrum A graph showing how much light is absorbed at different wavelengths.

Abundance The number of individuals in a species in a given area or volume.

Acrosome reaction Acrosome enzymes digest the corona radiata and the zona pellucida allowing the sperm and oocyte cell membranes to fuse.

Action potential The rapid rise and fall of the electrical potential across a nerve cell membrane as a nervous impulse passes.

Action spectrum A graph showing the rate of photosynthesis at different wavelengths.

Aerobic respiration The release of large amounts of energy, made available as ATP, from the breakdown of molecules, with oxygen as the terminal electron acceptor.

Allele A variant nucleotide sequence for a particular gene at a given locus, which codes for an altered phenotype.

Allele frequency The frequency of an allele is its proportion, fraction or percentage of all the alleles of that gene in a gene pool.

Allopatric speciation The evolution of new species from demes isolated in different geographical locations.

Anaerobic respiration The breakdown of molecules in the absence of oxygen, releasing relatively little energy, making a small amount of ATP by substrate-level phosphorylation.

Antagonistic pair When one muscle of an antagonistic pair is contracted the other is relaxed. The contraction and relaxation are co-ordinated.

Antenna complex An array of protein and pigment molecules in the thylakoid membranes of the grana that transfer energy from light of a range of wavelengths to chlorophyll a, at the reaction centre.

Antibiotic A substance produced by a fungus, which diminishes the growth of bacteria.

Antibiotic resistance Situation in which a micro-organism that has previously been susceptible to an antibiotic is no longer affected by it.

Antibody An immunoglobulin produced by the body's immune system in response to antigens.

Antidiuretic hormone Hormone produced in the hypothalamus and secreted by the posterior pituitary; it increases permeability of the cells of the distal convoluted tubule and collecting duct walls to water, increasing water reabsorption.

Antigen A molecule that causes the immune system to produce antibodies against it. Antigens include individual molecules and those on viruses, bacteria, spores or pollen grains. They may be formed inside the body, e.g. bacterial toxins.

Antigenic type Different individuals of the same pathogenic species with different surface proteins, generating different antibodies.

Aseptic technique Laboratory practice that maintains sterility in apparatus and prevents contamination of the equipment and the environment.

Autonomic nervous system The part of the peripheral nervous system that controls automatic functions of the body by the antagonistic activity of the sympathetic and parasympathetic nervous systems.

Biofuel A fuel made by a biological process such as anaerobic digestion, rather than by geological processes, such as long-term heat and compression, that formed fossil fuels.

Biomass The mass of biological material in living, or recently living, organisms.

Biotic A part of the environment of an organism that is living, e.g. pathogens, predators.

Birth rate The reproductive capacity of a population; the number of new individuals derived from reproduction per unit time.

Brittle bone disease An inherited disorder in the balance of the organic and inorganic components of bone, leading to an increased risk of fracture.

Capacitation Changes in the sperm cell membrane that increase its fluidity and allow the acrosome reaction to occur.

Carbon footprint The equivalent amount of carbon dioxide generated by an individual, a product or a service in a year.

Carrier (Pathology) An infected person, or other organism, showing no symptoms but able to infect others.

Carrier (Genetics) A phenotypically normal female with one normal, dominant allele and one mutant, recessive allele.

Carrying capacity The maximum number around which a population fluctuates in a given environment.

Cartilage Hard, flexible connective tissue in the respiratory airways, at the ends of bones, at the anterior end of the ribs, at the nose and the outer ear.

Cerebellum Part of the hindbrain that co-ordinates the precision and timing in muscular activity, contributing to equilibrium and posture and to learning motor skills.

Cerebrum Two hemispheres responsible for integrating sensory functions and initiating voluntary motor functions. It is the source of intellectual functions in humans where it is more developed than in other animals.

Chemiosmosis The flow of protons down an electrochemical gradient, through ATP synthetase, coupled with the synthesis of ATP from ADP and a phosphate ion.

Climax community A stable, self-perpetuating community that has reached equilibrium with its environment, and no further change occurs.

Clone A population of genetically identical cells or organisms formed from a single cell or parent, respectively.

Coenzyme A molecule required by an enzyme in order to function.

Colony (of bacteria or fungus) A cluster of cells, or clone, which arises from a single bacterium or fungal spore by asexual reproduction.

Commensalism An interaction between organisms of two species from which one benefits but the other is not affected.

Community Interacting populations of two or more species in the same habitat at the time.

Connective tissue Tissue that connects and supports organs or other tissues. It has cells embedded in a matrix with fibres of collagen and elastic tissue.

Conservation The protection, preservation, management and restoration of natural habitats and their ecological communities, to enhance biodiversity while allowing for suitable human activity.

Coppicing Cutting down of trees close to the ground and leaving them for several years to re-grow.

Core boundary Crossing this planetary boundary would drive the earth into a new and unpredictable state with severe consequences for the biosphere.

Cortical reaction Fusion of cortical granule membranes with the oocyte cell membrane, releasing their contents, which convert the zona pellucida into a fertilisation membrane.

Cyclic photophosphorylation ATP can be produced by electrons that take a cyclical pathway and are recycled back into the chlorophyll a in PSI.

Deamination The removal of an amine group from a molecule. Excess amino acids are deaminated in the liver, and the amine group is converted to urea.

Decarboxylation The removal of a carboxyl group from a molecule, releasing carbon dioxide.

Dehiscence The opening of the anther, releasing pollen grains.

Dehydrogenation The removal of one or more hydrogen atoms from a molecule.

Depolarisation A temporary reversal of potential across the membrane of a neurone such that the inside becomes less negative than the outside as an action potential is transmitted.

Desalination The removal of minerals from saline water.

Disease reservoir The long-term host of a pathogen, with few or no symptoms, always a potential source of disease outbreak.

Distribution The area or volume in which the individuals of a species are found.

DNA ligase Enzyme that joins together portions of DNA by catalysing the formation of phosphodiester bonds between their sugar-phosphate backbones.

Dominance hierarchy Ranking system in an animal society, in which each animal is submissive to animals in higher ranks but dominant over those in lower ranks.

Dormant Describes a seed when its active growth is suspended. Germination will only occur when specific conditions are met.

Ecosystem A characteristic community of interdependent species interacting with the abiotic components of their habitat.

Ecotourism Responsible travel to natural areas that conserves the environment and improves the well-being of local people.

Electrophoresis A lab technique that separates molecules on the basis of size, by their rate of migration under an applied voltage.

Endemic Disease occurring frequently, at a predictable rate, in a specific location or population.

Environmental resistance Refers to environmental factors that slow down population growth.

Epidemic The rapid spread of infectious disease to a large number of people within a short period of time.

Epigenetics The control of gene expression by modifying DNA or histones, but not by affecting the DNA nucleotide sequence.

Equilibrium species Species that control their population by competition rather than by reproduction and dispersal.

Eutrophication The artificial enrichment of aquatic habitats by excess nutrients, often caused by run-off of fertilisers.

Evolution A change in the average phenotype of a population.

Excretion The removal of metabolic waste made by the body.

Fertilisation The fusion of a female and male gamete, producing a zygote.

Fixed action pattern An innate behaviour which is a complex sequence of events that runs to completion, once it has been initiated by a sign stimulus.

Founder effect The loss of genetic variation in a new population established by a very small number of individuals from a larger population.

Fresh water Fresh water has a low concentration of dissolved salts i.e. <0.05% (w/v). It is sometimes called sweet water, to contrast it with seawater, which is salty.

Fruit A structure developing from the ovary wall, containing one or more seeds.

Gene A sequence of DNA that codes for a polypeptide and which occupies a specific locus on a chromosome.

Gene pool All the alleles present in a population at a given time.

Genetic drift Chance variations in allele frequencies in a population.

Germination The biochemical and physiological processes through which a seed becomes a photosynthesising plant.

Global warming The increase of average global temperature in excess of the greenhouse effect caused by the atmosphere's historical concentration of carbon dioxide.

Gram stain A method of staining the cell walls of bacteria as an aid to their identification.

Gross primary productivity The rate of production of chemical energy in organic molecules by photosynthesis in a given area, in a given time, measured in $kJ\ m^{-2}\ y^{-1}$.

Habitat The place in which an organism lives.

Haversian system The structural and functional unit of compact bone; the osteon.

Homunculus Drawing of the relationship between the complexity of innervation of different parts of the body and the areas and positions in the cerebral cortex that represent them.

Hybrid The offspring of a cross between members of different species.

Immigration The movement of individuals into a population of the same species.

Implantation The sinking of the blastocyst into the endometrium.

Infection A transmissible disease often acquired by inhalation, ingestion or physical contact.

Light compensation point Sometimes called just compensation point: The light intensity at which a plant has no net gas exchange as the volume of gases used and produced in respiration and photosynthesis are equal.

Limiting factor A factor that limits the rate of a physical process by being is short supply. An increase in a limiting factor increases the rate of the process.

Linked Description of genes that are on the same chromosome and therefore do not segregate independently at meiosis.

Liposome Hollow phospholipid sphere used as a vehicle to carry molecules into a cell.

Medulla oblongata Part of the hindbrain that connects the brain to the spinal cord and controls involuntary, autonomic functions.

Meninges Three membranes, the pia mater, arachnoid mater and dura mater that line the skull and vertebral canal, surrounding the brain and spinal cord.

Metabolic water Water produced from the oxidation of food reserves.

Monoculture The growth of large numbers of genetically identical crop plants in a defined area.

Mutualism An interaction between organisms of two species from which both derive benefit.

Myofibril Long, thin structure in a muscle fibre made largely of the proteins actin and myosin.

Myofilament Thin filaments of mainly actin and thick filaments of mainly myosin, in myofibrils, that interact to produce muscle contraction.

Natural selection The increased chance of survival and reproduction of organisms with phenotypes suited to their environment, enhancing the transfer of favourable alleles from one generation to the next.

Negative feedback A change in a system produces a second change, which reverses the first change.

Net primary productivity Energy in the plant's biomass which is available to primary consumers, measured in $kJ\ m^{-2}\ y^{-1}$.

Neuroplasticity The ability of the brain to modify its own structure and function following changes within the body or in the external environment.

Neurotransmitter A chemical secreted in response to an action potential, which carries a chemical signal across a synapse, from one neurone to the next, where a new action potential is initiated.

Niche The role and position a species has in its environment, including all interactions with the biotic and abiotic factors of its environment.

Nitrification The addition of nitrogen to the soil, most commonly as nitrite (NO_2^-) and nitrate (NO_3^-) ions.

Nitrogen fixation The reduction of nitrogen atoms in nitrogen molecules to ammonium ions, by prokaryotic organisms.

Non-cyclic photophosphorylation ATP can be produced by electrons that take a linear pathway from water, through PSII and PSI to NADP, which they reduce.

Non-disjunction A faulty cell division in meiosis following which one of the daughter cells receives two copies of a chromosome and the other receives none.

Oncogene A proto-oncogene with a mutation that results in cancer.

Osmoregulation The control of the water potential of the body's fluids by the regulation of the water content of the body.

Osteoarthritis Degenerative condition in which articular cartilage degrades and produces painful, inflamed joints.

Osteomalacia A disease in which calcium is not absorbed into bones of adults. The bones become softer, weaker and in some cases deformed.

Osteoporosis Abnormal loss of bone mass and density, resulting in increased risk of fracture.

Overfishing The rate at which fish are harvested exceeds the rate at which they reproduce.

Pandemic An epidemic over a very wide area, crossing international boundaries, affecting a very large number of people.

Pathogen An organism that causes disease in its host.

Photolysis The splitting of water molecules by light, producing hydrogen ions, electrons and oxygen.

Photophosphorylation An endergonic reaction bonding a phosphate ion to a molecule of ADP using energy from light, making ATP.

Pioneer species The first species to colonise a new area in an ecological succession, e.g. algae, lichens and mosses in a xerosere.

Planetary boundary Limits between which global systems must operate to prevent abrupt and irreversible environmental change.

Plasmid Small circular loop of self-replicating double-stranded DNA in bacteria.

Pollination The transfer of pollen grains from the anther to the mature stigma of a plant of the same species.

Polyploidy Having more than two complete sets of chromosomes.

Population An interbreeding group of organisms of the same species and occupying a particular habitat.

Primary productivity The rate at which energy is converted by producers into biomass.

Primary succession The change in structure and species composition of a community over time in an area that has not previously been colonised.

Primer A strand of DNA about 10 nucleotides long that base pairs with the end of another, longer strand, making a double-stranded section, to which DNA polymerase may attach, prior to replication.

Probe Short piece of DNA that is labelled with a fluorescent or radioactive marker, used to detect the presence of a specific base sequence in another piece of DNA, by complementary base pairing.

Protandry The stamens of a flower ripen before the stigmas.

Recombinant DNA DNA produced by combining DNA from two different species.

Reproductive isolation The prevention of reproduction and, therefore, gene flow between breeding groups within a species.

Resting potential The potential difference across the membrane of a cell when no nervous impulse is being conducted.

Restriction enzyme Bacterial enzyme that cuts the sugar-phosphate backbone of DNA molecules at a specific nucleotide sequence.

Reverse transcriptase Enzyme, derived from a retrovirus, that catalyses the synthesis of cDNA from an RNA template.

Rheumatoid arthritis Autoimmune condition in which bone and cartilage at joints is attacked, producing pain, swelling and stiffness.

Rickets A childhood disease in which calcium is not absorbed into bones, making them soft, weak and in some cases deformed.

Saltatory conduction Transmission of a nervous impulse along a myelinated axon, in which the action potential jumps from one node of Ranvier to the adjacent node.

Saprobiont A micro-organism that obtains its food from the dead or decaying remains of other organisms.

Secondary active transport The coupling of diffusion, e.g. of sodium ions, down an electrochemical gradient providing energy from the transport, e.g. of glucose, up its concentration gradient.

Secondary productivity The rate at which consumers convert the chemical energy of their food into biomass.

Secondary succession The changes in a community following the disturbance or damage to a colonised habit.

Seed Structure developed from a fertilised ovule, containing an embryo and food store enclosed within a testa.

Selection pressure An environmental factor that can alter the frequency of alleles in a population, when it is limiting.

Selective cutting Felling only some trees, leaving the others in place.

Selective reabsorption The uptake of specific molecules and ions from the glomerular filtrate in the nephron back into the bloodstream.

Sex linkage A gene is carried by a sex chromosome so that a characteristic it encodes it seen predominately in one sex.

Sliding filament theory The theory of muscle contraction in which thin, actin filaments slide between thick myosin filaments, in response to a nervous impulse mediated by the T system.

Soil erosion The removal of topsoil, which contains valuable nutrients.

Speciation The formation of a new species.

Species A group of phenotypically similar organisms that can interbreed to produce fertile offspring.

Stem cell An undifferentiated cell capable of dividing to give rise to daughter cells, which can develop into different types of specialised cell or remain as undifferentiated stem cells.

Sticky end A sequence of unpaired bases on a double-stranded DNA molecule that readily base pairs with a complementary strand.

Succession The change in structure and species composition of a community over time.

Sympatric speciation The evolution of new species from demes sharing a geographical location.

Synaptic pruning The elimination of synapses that happens between early childhood and maturity.

Synovial joint A joint at which bones' movement is lubricated by articular cartilages and synovial fluid, secreted by a synovial membrane. The joint is held in a ligamentous joint capsule.

Test cross, back cross Cross between an individual with the phenotype of the dominant characteristic, but unknown genotype, with an individual that is homozygous recessive for the gene in question.

Toxin A small molecule, e.g. a peptide made in cells or organisms that causes disease, following contact or absorption. Toxins often affect macromolecules, e.g. enzymes, cell surface receptors.

Transamination An enzyme-catalysed reaction that transfers an amino group to an α-keto acid, making an amino acid.

Transgenic An organism that has been genetically modified by the addition of a gene or genes from another species.

Trophic level Feeding level; the number of times that energy has been transferred between the Sun and successive organisms along a food chain.

Trophoblast Cells forming the outer layer of the blastocyst.

Ultrafiltration Filtration under high pressure.

Ultra-structure The detailed structure of a cell as seen in the electron microscope. It is also called fine structure.

Vaccine A weakened or killed pathogen, or a toxin or antigen derived from it, which stimulates the immune system to produce an immune response against it without causing infection.

Variation The differences between organisms of the same species.

Vector (Pathology) A person, animal or microbe that carries and transmits an infectious pathogen into another living organism.

Vector (Genetics) A virus or plasmid used as a vehicle for carrying foreign genetic material into a cell.

Ventricles Four connected cavities in the brain, into which cerebro-spinal fluid is secreted.

Z scheme The pathway taken by electrons in non-cyclic photophosphorylation.

Index